DRUGS ON THE PAGE

DRUGS on the PAGE

PHARMACOPOEIAS
and
HEALING KNOWLEDGE
in the
EARLY MODERN ATLANTIC WORLD

Edited by Matthew James Crawford & Joseph M. Gabriel

UNIVERSITY OF PITTSBURGH PRESS

Published by the University of Pittsburgh Press, Pittsburgh, Pa., 15260
Copyright © 2019, University of Pittsburgh Press
All rights reserved
Manufactured in the United States of America
Printed on acid-free paper
10 9 8 7 6 5 4 3 2 1

Cataloging-in-Publication data is available from the Library of Congress

ISBN 13: 978-0-8229-4562-8
ISBN 10: 0-8229-4562-2

COVER ART: Theodor De Bry, *How the Indians Treat Their Sick*. 1591. In *Brevis Narratio Eorum Quae in Florida Americae Provi[n]cia Gallis Acciderunt : Secunda in Illam Navigatione, Duce Renato De Laudo[n]niere Classis Praefecto, Anno MDLXIIII . . .*, by Jacques Le Moyne De Morgues. Francoforti Ad Moenum: Typis Ioan[n]is Wecheli, Sumtibus Vero Theodori De Bry, Venales Reperiu[n]tur in Officina S. Feirabe[n]dii, 1591
COVER DESIGN: Joel W. Coggins

CONTENTS

Acknowledgments
ix

Introduction. Thinking with Pharmacopoeias
MATTHEW JAMES CRAWFORD & JOSEPH M. GABRIEL
3

PART I. Pharmacopoeias and Textual Traditions

1. Pharmacopoeias and the Textual Tradition in Galenic Pharmacy
PAULA DE VOS
19

2. Authority, Authorship, and Copying: The *Ricettario Fiorentino* and Manuscript Recipe Culture in Sixteenth-Century Florence
EMILY BECK
45

3. An Imperial Pharmacopoeia? The *Pharmacopoeia Matritensis* and *Materia Medica* in the Eighteenth-Century Spanish Atlantic World
MATTHEW JAMES CRAWFORD
63

PART II. Pharmacopoeias and the Codification of Knowledge

4. Beyond the Pharmacopoeia? Secret Remedies, Exclusive Privileges, and Trademarks in Early Modern France
JUSTIN RIVEST
81

5. Crown Authorities, Colonial Physicians, and the Exigencies of Empire:
The Codification of Indigenous Therapeutic Knowledge in India
and Brazil during the Enlightenment Era
TIMOTHY D. WALKER
101

6. Imperfect Knowledge: Medicine, Slavery, and Silence in Hans Sloane's
Philosophical Transactions and the 1721 *London Pharmacopoeia*
WILLIAM J. RYAN
121

PART III. Pharmacopoeias and the Construction of New Worlds

7. The Flip Side of the Pharmacopoeia:
Sub-Saharan African Medicines and Poisons in the Atlantic World
BENJAMIN BREEN
143

8. Consuming Canada: *Capillaire du Canada* in the French Atlantic World
CHRISTOPHER PARSONS
160

9. Rethinking Pharmacopoeic Forms:
Samson Occom and Mohegan Medicine
KELLY WISECUP
177

PART IV. Pharmacopoeias and the Emergence of the Nation

10. National Identities, Medical Politics, and Local Traditions:
The Origins of the *London, Edinburgh,* and *Dublin* Pharmacopoeias, 1618–1807
STUART ANDERSON
199

11. The Codex Nationalized: Naming People and
Things in the Wake of a Revolution
ANTOINE LENTACKER
222

12. Indian Secrets, Indian Cures, and the
Pharmacopoeia of the United States of America
JOSEPH M. GABRIEL
240

Afterword. The Power of Unknowing: Early Modern Pharmacopoeias and the Imagination of the Atlantic
PABLO F. GÓMEZ
263

Notes
269

Bibliography
323

Contributors
361

Index
365

ACKNOWLEDGMENTS

THIS VOLUME has benefited from the support and encouragement of many people. Financial support for the project was provided by the American Institute of the History of Pharmacy and the George Urdang Chair in the History of Pharmacy endowment fund at the University of Wisconsin–Madison. We would like to think Greg Higby at the AIHP and David Mott, Michelle Chui, Kevin Look, Olayinka Shiyanbola, Betty Chewning, and Kristen Huset for their support, and for making the Division of Social and Administrative Sciences at the University of Wisconsin–Madison School of Pharmacy a hospitable and intellectually stimulating place to work. Financial support was also provided by the Department of Behavioral Sciences and Social Medicine at Florida State University, and the Department of History and the University Research Council at Kent State University. We would also like to thank Abby Collier at the University of Pittsburgh Press for her enthusiasm and editorial guidance on this project, as well as the two anonymous reviewers for their thoughtful feedback and suggestions. In addition to the contributors to the volume, Florence Hsia, Luce Giard, Greg Higby, Margaret Flood, Hugh Cagle, and Florencia Pierri participated in the initial workshop for the volume and provided important contributions to the conceptualization of the project. The Center for the Humanities at the University of Wisconsin–Madison provided space and support for the initial workshop, for which we are thankful. Joseph Gabriel would like to thank Susan Lederer for her support as well as members of both the Department of History and the Department of Behavioral Sciences and Social Medicine at Florida State University. He would also like to thank Claudia Sperber and Jacob Sperber for their love and support. Matthew Crawford would like to acknowledge Jacob House for research and editorial assistance; Brian Hayashi, chair of the Department of History at Kent State University, for his support of this project; the staff of the Lloyd Library, especially Betsy Kruthoffer; and the Science History Institute for providing a stimulating working environment while the final touches were put on this volume. Finally, we would both like to thank the contributors to this volume for their dedication and exceptional scholarship. It has been an honor working with all of you.

DRUGS ON THE PAGE

INTRODUCTION

THINKING WITH PHARMACOPOEIAS

MATTHEW JAMES CRAWFORD & JOSEPH M. GABRIEL

ONE YEAR after his death, the Danish physician, naturalist, and collector Ole Worm (1588–1654) published a descriptive catalogue of the various natural objects and artifacts that he had collected over the course of his life.[1] An engraving of the interior of Worm's museum accompanied the catalogue (figure I.1). The miscellany of the objects depicted in the image signals the sumptuousness of Worm's collection and the diversity of his interests; as noted in the subtitle, the catalogue promised readers a "history of rare things," including objects that were "natural as well as artificial, domestic as well as exotic." Curiosity and a desire for rarities motivated Worm to build this collection. He was particularly interested in objects that presented a challenge to the natural philosophy of his time and, as a result, Worm's museum is often cited as a key example of the phenomenon of the *Wunderkammer*, or cabinets of curiosities, that were prevalent among scholars and elites in early modern Europe.[2]

One often overlooked feature of this engraving is the series of labeled trays and boxes on the shelves and floor in the right side of the image. They seem positively mundane amid the many other striking objects, such as the stuffed polar bear suspended from the ceiling. Yet these trays and boxes offer further clues

FIGURE I.1. Engraving of Ole Worm's museum from Ole Worm, *Museum Wormianum* (1655). Courtesy of the Science History Institute.

to the nature and purpose of Worm's collection. The topmost set of boxes bear the labels *terrae* (earths), *salia* (salts), *sulphura* (sulphurs), and two trays labeled *lapides* (stones). The set of boxes on the next shelf down bear the labels *radices* (roots), *herbae* (herbs), *cortices* (barks), *ligna* (woods), *semina* (seeds), *fructus* (fruits), and *succi* (juices). And finally, the trays at the back of the image are labeled *metallica* (metallic objects), *metalla* (metals), *mineralia* (minerals), *turbinata* (turbinates), *conchillata* (conches), *animalium partes* (parts of animals), and *marina* (marine objects).

Visitors to Worm's museum in the seventeenth century would have been familiar with this way of grouping natural specimens. In addition to being used in works of natural history, this system of classification appeared in contemporary pharmacopoeias—a genre of texts that listed medicinal substances and provided the recipes to prepare them for therapeutic use. Several pharmacopoeias of the time used virtually the same categories in organizing the master list of *materia medica* (medicinal substances) used in their prepared medicaments. For example, the first edition of the *Pharmacopoeia Londinensis* (1618) grouped its materia medica under the following headings: *radices* (roots), *cortices* (barks), *ligna* (woods), *folia* (leaves), *flores* (flowers), *fructus et germina* (fruits and buds), *semina sive grana* (seeds or grains), *lachrimae* (saps), *succi* (juices), *plantarum excrementa* (excretions of plants), *animalia* (animals), *animalia partes* (parts of animals), *marina* (marine objects), *sales* (salts), and *metallica* (metallic objects).[3] While these categories do not map precisely onto those of Worm's museum (notably several types of earths and the shells are missing), they are close enough to illustrate how early modern pharmacopoeias reflected and were integrated into the broader intellectual culture of their time.

The organization of natural specimens under categories that appeared in pharmacopoeias would have been especially important for Worm, who made his collection an integral part of his teaching at the University of Copenhagen, where he was a professor of medicine. In his capacity as a physician and professor of medicine, Worm would have been required to inspect apothecary shops to assess the identity and quality of their materia medica.[4] It was yet another reason why Worm would have found it useful and convenient to have a collection of these medicinal substances at hand for reference. Not surprisingly, in his "Preface to the Reader," Worm mentioned the use of certain materials for the "preservation and restoration of the health of the human body," as well as the importance of having knowledge of such materials in order to detect medicaments that have been adulterated.[5] Worm's series of labeled trays and boxes were, in a sense, a physical embodiment of the same impulse to name and classify materia medica that appeared in numerous manuscripts, printed books, and other types of pharmacopoeias that circulated in the early modern Atlantic World.

Collections of curiosities and pharmacopoeias served, in part, as tools for making sense of the encounters with a diversity of peoples, places, and things provoked by the commercial and colonial expansion of early modern Europe. In this way, these places and texts bear the traces of the worlds beyond Western Europe. Many, but not all, of the specimens and phenomena from West Africa and the Americas were new to early modern Europeans, and so, museums and pharmacopoeias served the function of attempting to assimilate these previously unknown phenomena to the existing understandings of the natural world and of healing goods. For example, Worm's catalogue of his museum (and presumably the museum itself) included dozens of entries and descriptions of mineral, botanical, and animal specimens from the Americas. According to book two of the catalogue, which focused on plants, Worm's collection included almost fifty botanical specimens unique to the Americas. Most of these items were identified as originating from New Spain or Brazil, but the catalogue also included specimens identified as originating from Peru, Florida, Cuba, Jamaica, Virginia, and Canada.[6] Given that Worm's collection had an explicit focus on "rare things," it is perhaps no surprise that American plants are so prominent.

For information about these "rare" American plants, Worm relied on a variety of early printed natural histories of different regions of the Americas. He also benefited from information from one of earliest pharmacopoeias printed in the Americas: *Quatro libros. De la naturaleza y virtudes de las plantas y animales que estan recividos en el uso de Medicina en la Nueva España*, a work by Francisco Ximénez (ca. 1560–1620), a Dominican who served as a nurse at the convent of San Domingo in Mexico. Ximénez developed and adapted his text from the earlier work of Francisco Hernández (1514–1587), a physician who was appointed by the Spanish Crown to lead a natural history expedition to New Spain in the sixteenth century.[7] The networks of the Atlantic World provided the foundation for all these activities. When Ximénez traveled from Spain to New Spain as a missionary, he relied on these networks. And when Worm acquired botanical specimens from the Americas and gained access to information from Ximénez's book, which was printed in Mexico City in 1615, he relied on the networks of the Atlantic World as well.

As much as early modern pharmacopoeias were the products of long-distance social and cultural networks as well as local ones, these texts had a complex relationship with the contexts in which they were produced and consumed. Take, for example, two books published by the English astrologer and apothecary Nicholas Culpeper (1616–1654). As Michael Flannery notes, in 1649 Culpeper challenged the authority of the College of Physicians of London by translating the *Pharmacopoeia Londinensis* into English, thereby undermining their effort to monopolize medical knowledge through the use of Latin.[8] Three years later, Culpeper published the first edition of his medical text *The English Physician*. Culpeper presented descriptions of 277 plants that either grew wildly or were cultivated in England, such as barley, garlic, mint, quince, wormwood, and valerian. Such plants, he noted, were "the Works of God" and "common and cheap, and easie to be found." Culpeper gave detailed information on how to prepare and use the plants, the conditions they could be used to treat, and astrological information about their relationship to various planetary bodies. Notably, Culpeper excluded remedies from *The English Physician* that needed to be imported, such as cinnamon, under the belief that the natural bounty of England provided all that was needed for the care of his fellow countrymen. Culpeper also included explicit references in his text to the social, political, and economic context in which his work was produced, including comments noting his disdain for the College of Physicians, who, "being as Ignorant in the Knowledge of Herbs as a Child of four years old," hide their ignorance behind their learned airs and harm the common man with their complex and expensive remedies. "He that hath but half an eye may see their pride without a pair of Spectacles."[9]

Culpeper's book proved immensely popular and was republished countless times over the next century, often with significant variations and sometimes in forms that seem to have had little to do with the original work.[10] Indeed, the complexity of the publishing history of texts printed under the name *The English Physician* suggests the difficulty of stabilizing the relationship between names and things—a problem that is well known among historians of pharmaceuticals. In 1708, for example, a printer in Boston named Nicholas Boone published what was purportedly a reprinted version of Culpeper's book.[11] The exact relationship between Culpeper's original text and Boone's text is not at all clear, but at a minimum there were a variety of significant differences between the two: Boone's version was arranged according to medical problem instead of plant name, for example, and included formulas made from both imported goods (such as cinnamon) and from a variety of other substances that Culpeper had ignored—including a cure for colic made from powdered human skull mixed in a syrup of violets.[12] In 1720, Boone also reprinted Culpeper's translation of the *London Pharmacopoeia*, this time apparently with little modification. Boone's edition was probably the first formal pharmacopoeia published in the North American colonies, and while its influence on the daily practice of both physicians and apothecaries is difficult to ascertain, it undoubtedly served at least some practical use. Yet its utility was also clearly limited by the fact that the materia medica available in the colonies differed significantly from that available in Europe, in part because of the diversity of indigenous plants in North America and in part because of the influence of Indian and enslaved people's knowledge and practice on colonial medicine. Sassafras, for example, was used medicinally by Indigenous peoples in North America, and by the late seventeenth century it had been incorporated into the medical practices of white colonialists. Yet Boone's edition of the *London Pharmacopoeia*, not surprisingly, contained no mention of this useful plant.

Worm's and Culpeper's work thus illustrate how European efforts to organize knowledge about materia medica into useful taxonomies was deeply intertwined with both the larger scientific effort to understand the natural world and the specific social, political, and economic contexts in which these efforts took place. At the same time, they also point to the fact that European efforts at classification took place within emerging Atlantic and global contexts in which goods, peoples, and texts circulated and often resisted European ways of knowing and classifying the world. Taken as a whole, the essays in this volume thus demonstrate that examples from Europe represent but one portion of a diverse and complex set of activities centered on the knowing and using of different substances for medicinal purposes in the early modern Atlantic World. The motives behind such enterprises were varied, including the desire to understand natural phenomena; the need to keep a record of effective remedies; the interest in controlling commercially valuable resources; and the aspiration to promote civic, national, and imperial goals. At the same time, these essays show that the impulse to produce knowledge about healing goods involved a diverse cast of characters and took a variety of textual forms, from ephemeral lists drawn up on scraps of paper, to official texts printed under the auspices of municipal or royal governments. Some of these enterprises represented the idiosyncratic visions of a single author, while others reflected the views of the collective efforts of imperial agents, local healers, or a committee of learned physicians. Some of the texts produced from such efforts aimed at formally standardizing medical practice, while others served as informal representations of individual intellectual authority or of the imagined expanse of an empire's natural wealth.

REFRAMING PHARMACOPOEIAS AND THEIR HISTORIES

In light of this diversity, this volume takes a broad view of what counts as a "pharmacopoeia." Such an approach embraces the ambiguity of the term itself. On the one hand, the term *pharmacopoeia* can refer specifically to a genre of medical writing that lists simple and compound medicaments as well as the techniques for preparing and administering these medicaments according to a specific medical tradition. As noted in this volume by Paula De Vos (chapter 1), some of the earliest exemplars of these kinds of medical texts date to antiquity from the Mediterranean World to China. On the other hand, the term *pharmacopoeia* has been used to refer to the collective knowledge of medical virtues and therapeutic preparations of different substances as held by any society, culture, or group of specialists within a society or culture.[13] Such pharmacopoeias—which Pablo Gómez has called "social pharmacopoeias"—exist not just in textual form but also in the oral traditions and embodied practices of communities of healers around the globe.[14] Embracing both conceptions of the pharmacopoeia is especially helpful in the context of the early modern Atlantic World—a time and place when agents of European commercial and colonial enterprises coopted much healing knowledge from African and Native American cultures and then represented this knowledge as their own in medical and natural historical texts printed in the imperial centers of early modern Europe.[15]

This volume also seeks to broaden our thinking about the production and circulation of medical knowledge in the early modern Atlantic World by highlighting the ways in which pharmacopoeias are best understood not as finished and stable products but instead as stages in the larger processes of collecting, coopting, organizing, revising, and controlling knowledge about healing goods. While many of the producers of pharmacopoeias, especially those produced by colleges of physicians and under the authority of royal or municipal governments in Europe, viewed their texts as attempts to bring order and stability to worlds of healing goods, this volume highlights the instability of healing knowledge and of the various pharmacopoeias and "pharmacopoeic forms" that took shape in all corners of the early modern Atlantic World.[16] Pharmacopoeias—printed, manuscript, and oral—were often the subject of editing and revision just as the knowledge they represented changed over time and across space. By juxtaposing accounts of making textual pharmacopoeias at different times and places in the early modern Atlantic World, this volume seeks to call attention to the ways in which pharmacopoeias produced and were the products of colonialism, globalization, and the rise of the nation-state.

Focusing, for a moment, on pharmacopoeias as manuscripts and printed books, we note that this rich and varied set of historical sources has received surprisingly little attention beyond the specialized field of the history of pharmacy. Some studies of official pharmacopoeias—lists of approved medicinal substances and their preparation published under the auspices of municipal, national, and imperial governments—have tended to focus on particular regional pharmacopoeias or with an eye toward tracking changes in therapeutic practice or the use of specific drugs from one edition of the pharmacopoeia to the next.[17] At the same time, another group of studies in the history of pharmacy have generally analyzed pharmacopoeias in the context of the guild and professional activities of pharmacists, and to a lesser extent physicians, within specific national contexts, often during the nineteenth and twentieth centuries.[18] Taken together, the essays in this volume engage some of the foundational themes of this existing literature on the history of pharmacopoeias, including the guild activities of both pharmacists and physicians, the changes in therapeutic knowledge introduced by individual apothecaries, the efforts to monopolize therapeutic knowledge or practice, and the textual tradition of Western pharmacy as a whole. Yet this volume offers a unique opportunity to gain new insight on classic themes by juxtaposing case studies from the different times and places in the early modern Atlantic World.

Pharmacopoeias have received surprisingly little attention in the history of science and medicine more broadly, despite a robust literature that has focused on the organization and classification of natural phenomena as key forms of scientific and medical knowledge.[19] This is somewhat peculiar, especially since a strong case can be made that the organization of knowledge about medicinal substances was a vital part of scientific and colonial enterprises of Europe in the early modern world.[20] Although pharmacopoeias and the practices that produced them generally involve only a subset of natural phenomena, as opposed to broader early modern scientific projects aimed at cataloguing all flora and fauna, knowledge of medicinal substances was especially important because it helped to save lives and often led to the rise of new and valuable medical commodities, such as cinchona bark, guaiacum, and ipecac. Moreover, as evidenced by the case of Nicolas Monardes (1493–1588), a Spanish physician who published one of the early works on the medicinal uses of American flora and fauna, medicine was one of the primary lenses through which Europeans made sense of the natural phenomena of the Americas.[21] The same could be said for Africans who were forcibly relocated to the Americas.[22] Ultimately, the essays in this volume highlight the importance of pharmacopoeias and related texts as historical sources by showing how such sources can provide new insights in the entangled histories of knowledge resulting from encounters between Europeans, Africans, and Native Americans in the early modern Atlantic World. In this way, this volume places the study of pharmacopoeias in the context of scholarship emerging out of the dialogue between scholars working on the histories of science, medicine, globalization, and empire.[23]

Many of the essays in this volume also engage in significant ways with the history of drugs and pharmaceuticals. In addition to a robust field of scholarship on pharmaceuticals in the nineteenth and twentieth centuries, a growing number of historians of the early modern world have revealed important chapters in the history of medicinal substances from the sixteenth to the eighteenth centuries.[24] By focusing on pharmacopoeias and related texts rather than on the histories of individual substances, this volume enriches the early modern history of drugs by calling attention to what was known about medicinal substances, how it was known, where this knowledge was produced and circulated, why different groups took an interest in this knowledge, and how these groups negotiated their divergent ways of knowing and using these materials. In this way, the pharmacopoeias and related texts discussed in this volume provide a framework for examining early modern efforts to produce, circulate, and organize knowledge about medicinal substances as both a general category of phenomena and as contested sites where knowledge of medicinal goods was produced, organized, and sometimes lost. Pharmacopoeias and related efforts to organize the world of healing goods were therefore not just economic, political, scientific, and imperial projects—although they were, of course, all of these. Perhaps most important, they were epistemic projects that established not just *what* was known but *how* this knowledge was to be produced and to what end.

Many of the essays that follow also shift our attention away from formal networks of knowledge such as learned societies, colleges, and guilds even as several essays focus their attention on these institutions. By applying the fine-grained analytical techniques often reserved for other sources in the history of knowledge, the contributions to this volume also show how pharmacopoeias and their related texts offer insights into the more ephemeral practices of recording, organizing, and transmitting information about materia medica by artisans, colonial officials, Indigenous peoples, and others who, unlike European pharmacists and physicians, rarely had a recognized role in the production of official texts and medicines. In addition to the efforts of physicians and apothecaries, the case studies presented highlight the interactions and negotiations between a heterogeneous cast of characters engaged in the collection, organization, communication, and control of knowledge of medicinal substances, including actors such as Mohegan author Samson Occom (discussed by Kelly Wisecup in chapter 7) and Portuguese officials in Brazil and India (discussed by Timothy Walker in chapter 5). When viewed from this perspective, official pharmacopoeias published under the authority of elite actors appear as just one step in the complex and contested processes of knowing and using medicinal substances. As a result, we gain new appreciation of such texts not as an endpoint in the formalization of knowledge but as a vital part of the dynamic processes whereby natural phenomena became both knowable and useful. This volume thus highlights the way in which pharmacopoeias—broadly defined—constituted and were constituted not only by formal networks of guild knowledge but also by cross-cultural interactions, commercial exchange, imperial ambition, and other complex dynamics.

Including chapters that analyze such a diverse array of actors enriches our understanding of the historical genealogy of pharmacopoeias as tools of communication, standardization, and control. Indeed, one additional characteristic that makes pharmacopoeias so engaging as historical sources is that, in many cases, they clearly emerged out of the intersection of knowledge and the aspirations of state power. After all, the governments and medical bodies that produced official pharmacopoeias envisioned these texts first and foremost as a means of regulating and standardizing therapeutic practice. They often did so with an eye toward both the national and imperial contexts of their work. Consequently, in addition to exploring pharmacopoeias as traces of early modern epistemologies, several essays use pharmacopoeias and related texts as a means to interrogate the formal and informal contours of state power, imperial ambition, and emergent professional tensions. In particular, several of the contributions attend to the ways in which royal and municipal governments as well as professional guilds had only limited means to impose their particular vision of pharmaceutical therapeutics. Taken as a whole, this volume thus explores the development of important linkages between state power and scientific epistemology in the early modern Atlantic World. As several chapters make clear, however, such linkages were often contingent on local practices and concerns and, as a result, were not as straightforward as we might assume.

ORGANIZATION OF THE VOLUME

We have organized these essays into four parts, each dealing with a common theme or set of themes in different geographical contexts and time periods. The first section, "Pharmacopoeias and Textual Traditions," includes essays by Paula

De Vos, Emily Beck, and Matthew James Crawford that focus on the construction and function of pharmacopoeias as texts vis-à-vis other genres of medical writing. Paula De Vos (chapter 1) examines the textual traditions in Galenic pharmacy in the Mediterranean and their dissemination through pharmaceutical texts that were published in early modern Spain and circulated throughout the Spanish Atlantic World. Her essay provides an important reminder of how the conventions and traditions of European medical texts shaped pharmacopoeias even as their early modern authors sought to incorporate new substances and new information gleaned from cross-cultural interactions in the Atlantic World. In her essay (chapter 2), Emily Beck examines the development of what is generally considered to be Europe's first pharmacopoeia, the *Ricettario Fiorentino*, in the fifteenth and sixteenth centuries. By reading early editions of the *Ricettario* alongside manuscript recipe books from Florence, Beck shows how the tensions between learned and lay healers could inform the structure and function of pharmacopoeias as much as textual traditions and state aspirations. In particular, her analysis reminds us to note the disjuncture that often arose between the intended purpose of these texts and their actual use by healers. In the final essay of this part (chapter 3), Matthew James Crawford further explores the tension in pharmacopoeias between existing tradition and new materia medica in his analysis of the first edition of the *Pharmacopoeia Matritensis* published under the auspices of the Spanish Crown in 1739. While the Crown cast its new pharmacopoeia as an assertion of royal sovereignty, the *Pharmacopoeia Matritensis* also reveals the challenges to fostering uniformity in the art of pharmacy throughout Spain's vast empire. Collectively, these essays highlight a central theme: that our understanding of pharmacopoeias depends on what we define as the relevant context for analyzing these texts, whether it be the medical marketplace of an Italian city-state, the political reforms of a transatlantic empire, or the cultural traditions of the Mediterranean World.

The second section, "Pharmacopoeias and the Codification of Knowledge," includes essays that focus primarily on the complex dynamics associated with the attempts to collect and codify knowledge about medicinal substances. Here, the phrase "codification of knowledge" refers to two different, but interrelated, processes. On the one hand, the phrase refers to the efforts by individuals and early modern states to protect certain forms of knowledge through legal or other strategic means such as secrecy. On the other hand, it refers to efforts to standardize knowledge about materia medica into useful and stable taxonomies that can serve a variety of functions. These two processes sometimes operated in tension with one another but they also sometimes reinforced one another. In his essay (chapter 4), Justin Rivest examines an important instance of this dynamic in his analysis of the efforts of the Contugi family to assert their royal privilege to produce a secret remedy known as *orviétan*, a recipe for which also appeared in the *Pharmacopee royale* and several other pharmacopoeias. This case allows Rivest to probe the tensions between the codification of healing knowledge in the form of pharmacopoeias and the juridical standing of, and legal battles over, proprietary knowledge in the form of secret remedies. Whereas these pharmacopoeias in France represent a case of the codification of knowledge for the purposes of making it openly available to medical practitioners, Timothy Walker, in chapter 5, explores a case where the reverse happened. His essay examines how efforts of Portugal's imperial government treated colonial medical knowledge as a strategic resource that needed to be kept secret as it was being collected and organized. Walker also illuminates many of the challenges that the Portuguese imperial government faced in its efforts to integrate healing knowledge from different parts of the globe and the way in which secrecy was strategically used as part of the effort to do so. Unlike the example discussed by Rivest, Walker's analysis suggests that the two modes of codification reinforced one another in the Portuguese case. Of course, imperial power did not always operate in such a unified way. William Ryan's essay (chapter 6) shifts our gaze from the Portuguese world to the British Caribbean and focuses on the editorial practices of Hans Sloane while he was at the helm of the *Philosophical Transactions*. Ryan examines Sloane's efforts to organize knowledge about the various medical substances and botanical specimens that he encountered during his time in Jamaica. His chapter suggests that scientific and medical knowledge in the British Caribbean was deeply implicated in relations of colonial power and, as a result, was not as metaphysically tidy as we might think. Ryan also suggests that Sloane's editorial practices risked exposing both trade secrets and, perhaps more importantly, the limited state of colonial medical science. The codification of knowledge, in both senses of the phrase, was a halting and imperfect project.

The third section, "Pharmacopoeias and the Construction of New Worlds," includes essays that focus on the complex social, cultural, and intellectual practices that enrolled pharmacopoeias and related texts in the imaginative construction of what Europeans sometimes called the "New World." As the essays in this section further demonstrate, however, the extension of European colonial power was much more arbitrary and capricious than historians sometime assume. They also highlight the role that efforts to convert healing knowledge into pharmacopoeias and related texts shaped and were shaped by ideas about the peoples and places of the Atlantic World. Benjamin Breen uses his essay (chapter 7) to suggest that practices that we now consider "scientific" and practices that we now consider "spiritual" were actually quite close to one another, and that the seeds of the "divergence" between the two, as he terms it, were rooted in broader social and cultural assumptions and practices rather than simply the rise of scientific epistemology. Chris Parsons, in his essay (chapter 8), uses the case of Canadian maidenhair—a delicate fern that grows abundantly across eastern North America—to examine how pharmacopoeias became an important textual form in which claims about colonial space were articulated and consumed. Both of these essays deal with the extension and ultimately arbitrary nature of European colonial knowledge in the early modern world. In the final essay of this section (chapter 9), however, Kelly Wisecup describes a case that had relatively little to do with the extension of European colonial power as she discusses a unique but insightful document—a list of medicinal herbs and roots composed by Mohegan Samson Occom—that highlights the importance and vitality of Native medical practices in the early modern Atlantic World.

The chapters in the fourth section, "Pharmacopoeias and the Nation," examine the relationship between pharmacopoeias and the emergent nation-state in Britain, France, and the United States. Each of these chapters takes a different approach to examining the connections between politics, broadly understood, and the standardization and organization of knowledge about healing goods. In his essay (chapter 10), Stuart Anderson gives us detailed political history of the origins of the London, Edinburgh, and Dublin pharmacopoeias and illuminates the contingent events that shaped the origins and development of these three official pharmacopoeias. His essay also explains the conditions that gave rise to the unique situation in which Britain produced three distinct pharmacopoeias of national—and indeed international—significance. Antoine Lentacker, in his contribution (chapter 11), argues that following the revolution, the French national pharmacopoeia was one "paper technology" among many that the new nation-state used to track people and things by defining and controlling their names. Finally, in his essay on the United States pharmacopoeia (chapter 12), Joseph M. Gabriel traces the ways in which Indigenous knowledge of healing plants was both incorporated into, and eliminated from, early American medical science. Replicating elite assumptions about the nature of a democratic society, the first and subsequent editions of the *Pharmacopoeia of the United States of America* grew out of, and helped to create, a distinctive American medical science forged at the intersection of science, class, race, and nation. In doing so, Gabriel argues, the creation of a national pharmacopoeia helped create the discursive conditions for the new nation itself. Indian beliefs and practices—including the practice of keeping traditional knowledge secret—had no place in this new formulation except as an example of dangerous irrationalism.

THE ATLANTIC WORLD AND THE PLACE OF PHARMACOPOEIAS

As part of our effort to provide a more capacious understanding of pharmacopoeias and related texts, this volume embraces the geographical and historical context of the Atlantic World as its primary framework. As noted in recent scholarship, the phrase "Atlantic World" has many connotations and historians have employed the Atlantic World as a unit of analysis in many different ways.[25] Our appeal to the Atlantic World as a framework of analysis is admittedly tentative and represents a first step in thinking about how the history of pharmacopoeias engages processes that transcend the city, the nation, and the empire. At the same time, we recognize that not all elements of the history of pharmacopoeias and related texts can easily be made intelligible within this framework. We have therefore sought to juxtapose essays *about* the Atlantic World with essays that take place *in* the Atlantic World—some of which explicitly draw on the Atlantic World as a framework and some of which do not. In doing so, we seek to articulate an intellectual framework that highlights geographical scale and place as central components in the historical analysis of pharmacopoeias. The essays in this volume not only vary in their geographical and chronological focus; they also vary in the scale of their analyses and range from studies of individual texts and healers to specific cities or regions to entire empires; this range, of course, could be further extended to macrohistorical processes that operate at a truly global scale. The decision by a historian to conduct her work at any particular spatial scale of analysis depends on a wide variety of factors, including available sources, the context of the historiographical tradition in which she operates, and personal preferences. This is, at heart, an arbitrary decision, and as such must be balanced with the other decisions that one makes as a scholar. Each of the authors in this collection has thus decided to work at a scale of analysis that suits their own idiosyncratic needs and interests. Indeed, we have assembled what some readers may consider an eclectic mix of essays in this volume precisely to raise the point that the scale of analysis at which our scholarship takes place is vital to our efforts to make sense of pharmacopoeias and the broader political, economic, social, cultural, and epistemological processes that shaped them and that they often shaped in return.

PART I

Pharmacopoeias and Textual Traditions

1

PHARMACOPOEIAS AND THE TEXTUAL TRADITION IN GALENIC PHARMACY

PAULA DE VOS

IN SEPTEMBER 1775, a Mexico City notary conducted an inventory of the contents of the pharmacy of apothecary Don Jacinto de Herrera y Campos as part of a criminal investigation.[1] The inventory proceeded over several days, revealing a wide array of medicines and equipment in the pharmacy, as well as a collection of books to which Herrera presumably referred in his practice. Among the books were a series of pharmacopoeias, texts meant to standardize the types of medicines stocked in the pharmacy and the ways in which they were to be formulated. Herrera possessed copies of the *Pharmacopoeia Valentina* and the *Pharmacopoeia Augustana*, both produced in the sixteenth and early seventeenth centuries to standardize pharmaceutical practice in and around the cities of Valencia and Augsburg similar to the case of Florence's early pharmacopoeia—the *Ricettario Fiorentino*, as discussed by Emily Beck in chapter 2.[2] Herrera also possessed a worn copy of Félix Palacios's *Palestra pharmaceutica*, first published in 1706, which went on to serve as the basis for the *Pharmacopoeia Matritensis*, coordinated by the Royal College of Apothecaries in Madrid and published in 1739 as the first standard pharmacopoeia for the entire Spanish Empire, as discussed by Matthew Crawford in chapter 3.[3] The *Pharmacopoeia Matritensis* provided a "fixed and constant method by which the medicines are prepared that are in use in these Kingdoms for the cure of illness," and apothecaries were to follow it "without departing in any way whatsoever from its rules."[4] Its presence in the Herrera pharmacy was presumably no accident, as royal decrees stipulated that apothecaries throughout the Spanish Empire keep a copy of it in their pharmacies, and present it during pharmacy inspections.[5]

The various pharmacopoeias in Herrera's shop constituted part of the textual basis for Galenic pharmacy, the tradition that dominated pharmaceutical theory and practice in the West from the first century of the Common Era through the early nineteenth century. Galenic pharmacy was named after, and largely founded upon, the teachings of Galen (ca. 130–210 CE), a physician from Pergamon in the Roman Empire whose medical system guided Western medicine for almost two millennia.[6] Galenic pharmacy was brought to Mexico under the Spanish Empire and remained remarkably intact throughout the colonial period.[7] The contents of the Herrera pharmacy—its medicines, its equipment, and its books—were typical of the Galenic tradition. The pharmacopoeias it contained serve to indicate the importance of texts within this tradition and the fact that pharmacy from very early on was both a practical and a learned art, a manual craft supported by a substantial tradition of written works as well.[8] In the Galenic tradition, the pharmacopoeia played a particularly important role, not only in setting professional standards but also as a culmination of a series of different genres of pharmaceutical writing.

Studies of pharmacopoeias tend to emphasize their role in standardizing materials and formulations especially in the context of the nation-state and in this way equate them with the emergence of the modern era, as noted in Stuart Anderson's discussion in chapter 10 of David Cowen's definition of the genre. Yet it is crucial to remember that national pharmacopoeias even to the present day rest on the foundations of earlier texts and genres that go back centuries. The purpose of this essay is to trace the long and deep history of the textual tradition in Galenic pharmacy using pharmaceutical texts published in early modern Spain—and used throughout the empire—to delineate the main eras, authors, and areas of its foundations and the different genres of pharmaceutical writing that developed over time. Results indicate that the tradition dates back to antiquity, originating with Greek and Roman authors, expanding under the medieval Islamic Empires, and entering medieval Europe from the south, mainly through translation centers in Spain and Italy.

During these periods, the Galenic textual tradition developed four main genres of pharmaceutical writing that together culminated in the formation of the early modern pharmacopoeia. The first genre treated here consisted of ancient compilations of *materia medica* (medicinal materials). These were texts containing lists, descriptions, and glossaries of what were called "simples" in Galenic pharmacy—natural substances that derived from plants, animals, and minerals with known healing power. Another genre, that of the formulary, also originated in the ancient period but developed substantially under the Islamic Empires. Formularies, also referred to as antidotaries and receptaries, were compilations of recipes for "compounds," medicines made up of more than one simple. Two additional genres developed during the medieval and early modern periods: the procedural, which included technical advice and instructed practitioners in pharmaceutical operations and procedures; and the pedagogical, which included several components designed to instruct apothecaries in training. Elements of each of these types of text were brought together in the pharmacopoeia, a culminating genre of pharmaceutical writing in Galenic pharmacy that received increasing emphasis in the early modern period. In tracing this history, this essay aims to document this textual tradition and its genealogy, and to highlight its shared nature as part of Galenic pharmacy, a tradition that spread throughout the ancient and medieval Mediterranean and came to encompass the Atlantic World in the early modern period under the auspices of European imperialism.

TRACING THE TEXTUAL TRADITION

Authors of pharmacy treatises in early modern Spain understood and discussed the learned basis of their art, emphasizing the importance of literacy early on. In his 1632 publication of *Examen de boticarios*, Esteban de Villa declared that apothecaries "must have great knowledge," not only practical knowledge of plants and remedies but knowledge "of theory, or that which is found in books" as well.[9] Miguel Martínez de Leache, author of *Discurso pharmaceutico sobre los Canones de Mesue* (1652) emphasized the importance of the intellectual training for apothecaries by critiquing the empirics and charlatans "who apply medicines according to what they see only, without a more fundamental basis."[10] The practice of learned apothecaries, by contrast, rested upon the classic texts of ancient and medieval authors, through which apothecaries learned the tenets of pharmacy. Martínez de Leache declared that apothecaries must the "*studiosissimos*"— the most studious—and that "they must be most attached [*aficionadissimos*] to the study of letters because this way they come to grasp their chosen field; because thinking that they can claim to be an apothecary and understand medicine is impossible without having studied words and read books."[11] Knowledge only came, he advised, with long and arduous study, not "in one instant, all of a sudden.... No one is born from the womb of his mother already taught."[12]

These authors thus recognized the importance of a learned tradition in pharmacy and produced a corpus of early modern publications to support it (table 1.1).[13] These works also provide a means for investigating the earlier basis upon

TABLE 1.1. Corpus of Early Modern Spanish Pharmacy Texts (Sixteenth to Eighteenth Centuries)

Author	Date	Title	Language	Used in Survey?
Saladino/Tudela	1488/ trans. 1515 Salamanca	Compendio de los boticarios	Castilian (trans. from original Latin)	Yes
Benedictus Mattheo, Petrus	1521	Loculentissimi viri... Petri B[e]n[e]dicti Mathei... Liber in exame[n] apothecariorum q[uam]	Latin	Yes
Sepúlveda, Fernando de	1523	Manipulas Medicinarum	Latin	Yes
Laredo, Bernardino (1482–1545/1540)	1527	Sobre el Mesue e Nicolao: Modus facie[n]di cu[m] ordine medicandi	Castilian	
Navascués	1550	Ioannis Mesuae... Liber primus seu Methodus medicamenta purga[n]tia simplicia deligendi & castiga[n]di, theorematis quatuor absolutus	Latin	
Aguilera, Antonio de	1569	Exposicion sobre las preparaciones de Mesue	Castilian	
Laguna, Andreas de	1570	De materia medica	Castilian	
Fragoso, Juan	1572	Discurso de las cosas Aromaticas, arboles y frutales, y de otras muchas medicinas simples que se traen de la India Oriental, y sirven al uso de medicina	Castilian	Yes
Fragoso, Juan	1575	De succedaneis medicamentis	Latin	Yes
Jubera, Alonso de	1578	Dechado y reformación de todas las medicinas compuestas usuales	Castilian	Yes
Bravo, Juan	1592	De simplicium medicamentorum delectu & praeparatione libri duo: qui ars pharmacopoea dici possunt	Latin	
Velez de Arciniega, Francisco	1592	De Simpli	Latin	
Castels, Juan Antonio	1592	Theorica y Pratica de boticarios en que se trata de la arte y forma como se han de componer las confectiones ANSI interiores como Exteriores	Castilian	Yes
Oviedo, Luis de	1595	Methodo de la colección y reposicion de las medicinas simples, y de su correcion y preparación	Castilian	Yes

TABLE 1.1. (continued)

Author	Date	Title	Language	Used in Survey?
Velez de Arciniega, Francisco	1624	*Theoriae pharmaceuticae septem sectionem*	Latin	
Suárez de Figueroa, Cristóbal	1615	*Plaza universal de todas ciencias y artes*	Spanish (trans. from original Italian)	Yes
Villa, Esteban de	1632	*Examen de Boticarios compuesto por Fray Estevan de Villa Monge de S. Benito...*	Castilian	Yes
Martínez de Leache, Miguel	1652	*Discurso Pharmaceutico sobre los canones de Mesue*	Castilian	Yes
Martínez de Leache, Miguel	1662	*Tratado de las condiciones que ha de tener el boticario para ser docto en su arte*	Castilian	Yes
Fuente Pierola, Jerónimo de la	1683	*Tyrocinio pharmacopeo: método medico y chimico*	Castilian	
Martínez de Leache, Miguel	1688	*Controversias Pharmacopales a donde se explican las preparaciones y elecciones de Mesue*	Castilian	Yes
Palacios, Félix	1706	*Palestra Pharmaceutica Chymico-Galenica*	Castilian	
Assín y Palacio de Ongoz, José	1712	*Florilegio Teórico-Practico*	Castilian	
Loeches, Juan de	1728	*Tyrocinium pharmaceuticum, theorico-prácticum, galeno-chymicum*	Latin	
	1739	*Pharmacopoeia Matritensis*	Latin	
Brihuega, Francisco de	1776	*Examen farmacéutico galénico-químico, teórico-práctico extractado de las mejores farmacopeas*	Castilian	Yes
Vinaburu, Pedro de	1778	*Cartilla Pharmaceutica Chimico-Galenica*	Castilian	

which they were founded, for they regularly made reference to contemporary authors as well as earlier works in the field.[14] This practice developed out of scholastic and earlier Arabic medical traditions in which authenticity and authority in a text were established by building upon and making explicit reference to a canon of respected works.[15] I used 14 of these texts (the ones most replete with such references) to carry out a quantitative study of author referencing, in which a composite list of 77 different authors referenced in the 14 works was made, and a tally

TABLE 1.2. Major Periods of Pharmaceutical Writings in Galenic Pharmacy in Survey of Early Modern Spanish Pharmacy Texts

Place	Time Period	Number of Authors (out of 77)	% of Total Authors	Number of References in Early Modern Works (out of 415)	% of Total References
Greece	400–300 BCE	4	5.2	25	6.02
Rome	0–100 CE	6	7.8	47	11.3
Byzantium	300–600 CE	5	6.5	30	7.2
Islamic Empires	800–1100 CE	8	10.4	47	11.3
Western Europe – Late Medieval	1100–1300 CE	15	19.5	73	17.6
Western Europe – Renaissance	1400–1500 CE	39	50.6	193	46.5
Total		77	100	415	100

kept of how many of the authors were referenced by how many of the 14 books, for a total of 415 references (table 1.2). From this tabulation, I was able to organize authors and references according to place and time and to determine the most widely cited (and presumably the most important and influential) authors over time. These results were then corroborated with book lists provided by several early modern Spanish pharmacy authors who listed "the most necessary works in the pharmacy," discussed further below.

Preliminary results show a clear pattern of works produced in times and places that corresponded to the foundation and spread of Galenic pharmacy from ancient Greece and the Hellenistic world to the Roman and Byzantine Empires, and from the medieval Islamic world to medieval and early modern Europe. They also indicate that although Galen provided the namesake for the tradition, its foundations began in the classical period and went on to involve many contributors over time. Among the references to earlier authors and texts in the early modern Spanish pharmacy treatises, there were references to 4 authors from classical Greece that were referred to a total of 25 times in the 14 sources, making up a total of 6% of the 415 citations recorded. There were 6 different Greco-Roman authors from the first century CE who were referred to 47 times, making up 11.3% of the citations. Byzantine authors numbered 5, with 7% of the citations (30 of 415). European works were referred to the most, with medieval authors numbering 15 with 73 references for a total of 17.6%, while the 39 Renaissance authors, with 193 references, made up almost half, or 46.5%, of the total citations.

One of the clearest characteristics of the textual tradition in Galenic phar-

TABLE 1.3. Top Ten Most Commonly Cited Authors in Survey of Early Modern Spanish Pharmacy Texts

AUTHOR	PERIOD AND REGION	NUMBER OF WORKS THAT REFER TO AUTHOR/TEXT (OUT OF 14)
Mesue, John, Yuhanna ibn Masawaih ("pseudo-Mesue") (Common Era - dates unknown)	Medieval Islamic Empires/ Medieval Europe	14
Dioscorides, Pedanius (ca. 40–90)	Roman Empire	13
Galen, Claudius (129–200)	Roman Empire	12
Platerius, Mattheus (1120–1161)	Medieval Europe – Salerno	12
Pliny the Elder (23–79)	Roman Empire	11
Matthiolo, Pietro Andreas (1501–1577)	Early Modern Europe – Italy	11
Avicenna, Ibn Sina (980–1037)	Medieval Islamic Empires	10
Juan Serapion, Yahya ibn Sarafyun (9th century)	Medieval Islamic Empires	9
Villanova, Arnald de (ca. 1240/1235–1311)	Medieval Europe – Spain, France	9
Sylvaticus, Mattheus (1285–1342)	Medieval Europe – Salerno	9

macy is that it developed in and around the ancient and medieval Mediterranean, undoubtedly aided by the long history of intense cross-cultural interaction across these waters that has taken place since the development of the first seafaring societies along its shores. Indeed, nine of the top ten authors cited in the survey (tables 1.3 and 1.4) wrote in the ancient or medieval period, including Dioscorides and Pliny, who wrote encyclopedic works on simples; and Galen, who authored a number of pharmaceutical and pharmacological works. They were all widely cited in the early modern texts, as were the medieval authors from the rich Islamic pharmaceutical tradition, including Serapion (Ibn Wafid), Avicenna (Ibn Sina), and possibly John Mesue (the anglicized version of Yuhanna ibn Masawaih). Mesue, in particular, produced three works, a book of simples, a formulary, and a set of theorems or canons regarding the selection and preparation of simples that went on to multiple editions in the age of print.[16] Medieval European authors Arnald de Villanova, Mattheus Plateraius, and Mattheus Sylvaticus wrote important antidotaries and books of simples that reflected advancements in pharmaceutical knowledge through the translation of Arabic works into Latin in Toledo and Salerno, and through the establishment of medical schools in the universities of Montpellier, Padua, and Bologna.

The importance of these early works is further highlighted by the fact that

TABLE 1.4. Ancient and Medieval Authors Cited in Early Modern Spanish Pharmacy Texts

Place	Time Period	Number of Authors	Author Names
Classical Greece	400–300 BCE	4	Hippocrates Theophrastus Aristotle Plato
Roman Empire	0–100 CE	6	Dioscorides Galen Pliny Celsus Themistius Strabo
Byzantine Empire	300–600 CE	5	Paul of Aegina Nicholas Myrepsus Oribasius Aetius of Amida Hesychius of Miletus
Islamic Empires	800–1100 CE	8	Avicenna Mesue (?) Serapion Al-Razi Averroes Haly Abbas Avenzoar Maimonides
Western Europe	1100–1300 CE	15	Nicholas Salernitanus Bernard de Gordon Arnald de Villanova Mundinus de Liuzzi Simon Genuense Juan de Abano Gentiles de Fulgineo Gilberto Anglico (Salerno) Mattheus Platerius (Salerno) Mattheus Sylvaticus (Salerno) Nicolaus Praepositus Albertus Magnus Jean de St. Amand Benardi de Gordonio Guy de Chauliac

TABLE 1.5. Major Periods of Pharmaceutical Writings as Cited in Early Modern Spanish Pharmacy Texts (Altering Categories to Include Commentaries and Translations while Removing Early Modern Western Europe)

Place	Time Period	Number of Authors	% of Total Authors	Number of References in Early Modern Works	% of Total References
Greece	400–300 BCE	4	7.7	25	8.7
Rome	0–100 CE	6	11.5	47	16.4
Commentaries on Dioscorides		5	9.6	29	10.1
Total Roman References (Including Later Commentaries)		11	21.2	76	26.5
Byzantium	300s–600s CE	5	9.6	30	10.5
Commentaries on Nicholas Myrepsus		1	1.9	3	1.0
Commentaries on Paul of Aegina		1	1.9	2	0
Total Byzantine References (Including Later Commentaries)		7	13.6	35	12.1
Arabia	800–1100 CE	8	15.4	47	16.4
Commentaries on Mesue		9	17.3	39	13.6
Total Arabic References (Including Later Commentaries)		17	32	86	29.9
Western Europe – Late Medieval (Without Commentaries)	1100–1300 CE	13	25	65	22.6
Total		52	100	287	100

a number of the early modern authors referenced in the survey had published commentaries of some of these works, including five translations and commentaries of Dioscorides's *De materia medica* and no less than nine commentaries of Mesue's works. Indeed, these two authors stand out as two of the most important, foundational authors in Galenic pharmacy, its namesake notwithstanding.[17] Taking these commentaries into account and placing them within the time period of the original author, a picture of the Mediterranean, and particularly Arabic, foundations of Galenic pharmacy's textual tradition emerges (table 1.5). Classical

27

Greek authors accounted for 7.7% of the total of 77, with 8.7% of the 415 references; imperial Rome accounted for 21% of authors and 27% of references; Byzantine authors (mainly Oribasius, Aetius of Amida, and Paul of Aegina, whose works were largely encyclopedic compilations of Galen's corpus)[18] accounted for 14% of authors and 12% of the references; and Arabic authors made up 32% of the total authors and 30% of the references. Thus, medieval Arabic texts (counting commentaries as well) made up the largest influence on the Spanish textual tradition and, together with Rome, made up fully half of the authors and references in the survey.

This Galenic tradition entered Western Europe from the south, where the areas of Southern Europe that bordered the Mediterranean stand out as the major producers of pharmaceutical texts, with France producing about one-quarter of the medieval works, Iberia about one-fifth, and the Italian states almost one-half (table 1.6).[19] This trend began to change, however, in the Renaissance and early modern period, which witnessed the gradual growth in northern influence on publishing in pharmacy. Authors of early modern Northern European works in the survey amounted to 8, or 20.5% of the total of 39, while Southern Europeans made up 79.5% of total early modern authors, down from 93% in the medieval period. Of southern works, it is notable that Spanish sources increased substantially, from 2 medieval authors cited to 9, or 23% of authors of early modern works. Thus, despite growing evidence of Northern European influence, Southern Europe continued to dominate in the spread of Galenic pharmacy and its texts.[20]

Other writings corroborate these findings.[21] Certain early modern Spanish pharmacy texts listed authors whose works were considered essential reading for apothecaries in the Galenic tradition, essentially setting up a standardized textual canon of core, foundational works for the profession. The Greek, Roman, Arabic, and Southern European texts to which they referred reflect the findings of the survey. According to Saladino da Ascoli's 1488 *Compendium aromatarium*, there were eight books "necessary to any apothecary," including that of Dioscorides from Rome; the Arabic works of Abulcasis (al-Zahrawi), Avicenna (Ibn Sina), and Serapion (Ibn Wafid); and the *Antidotario Nicolao*, Platearius's *Circa Instans*, and the glossary *Libro de Sinonimas*, by Simon of Genoa, all from the medieval Italian medical school of Salerno.[22] A few decades later, Pedro Benedicto Mateo's 1521 publication *Liber in Examen Apothecariorum* expanded on this list by including thirty books "which the apothecary must have," including those of Dioscorides, Pliny, and Galen; Avicenna, Mesue, and Serapion; and Arnald de Villanova, Simon of Genoa, Platearius and the *Antidotarium Nicolao*.[23,24] For Cristobal Suarez de Figueroa in 1615, the "most common" books that the apothecaries needed to consult were those of Dioscorides, Galen, Pliny, Celsus, Nicolao Proposito, Nicholas Mirepsus, and Mesue, along with a host of "modern"

TABLE 1.6. Medieval and Renaissance Authors (Northern versus Southern Europe) in Early Modern Spanish Pharmacy Texts

Place	Late Medieval Authors	Number	% of Total (15)	Renaissance Authors	Number	% of Total (39)
England	None	0	0	None	0	0
Germany	None	0	0	Bernard von Kronnemburg Sessen Leonard Fuchs Valerius Cordus Juan Placotomo	4	10.2
Low Countries	None	0	0	Quiricus de Augustis Matthaeus Lobelius	2	5.1
Switzerland	None	0	0	Johannes Jacobus Wecker Konrad Gesner	2	5.1
Total Northern/Central Europe	None	0	0		8	20.5
Italy	Mundino de Liuzzi Simon Genuense Pietro de Abano Gentiles de Fulgineo Gilberto Anglico (Salerno) Mattheus Platerius (Salerno) Mattheus Sylvaticus (Salerno) Nicolaus Praepositus (Salerno)	8	53.3	Saladino da Ascoli Giovanni Manardi Johannes Jacobus Manlius de Bosco Ioannes Costeus Antonio Musa Brasavola Girolamo Savonarola Christophorus de Honestis Paulo Suardo Antonius Guainerius Giovanni Matteo Ferrari da Grado, Hermolao Barbaro Pierio Valeriano	13	33.3
Portugal	Petrus Hispanus, Pope Jon XXI	1	6.7	Rodrigues de Castelo Branco, "Amatus Lusitanis"	1	2.6
Spain	Arnald de Villanova Maimonides	2	13.3	Alonso de Jubera Juan de Vigo Petrus Benedictus Mattheo Antonio Castells Juan Fragoso Juan Navascues Sanguesano Hernando Sepúlveda Andrés de Laguna Bernardino de Laredo	9	23.1

TABLE 1.6. (continued)

Place	Late Medieval Authors	Number	% of Total (15)	Renaissance Authors	Number	% of Total (39
France	Albertus Magnus Jean de St. Amand Bernard de Gordon Guy de Chauliac	4	26.7	Symphorien Champier Laurent Joubert Guillaume Rondelet Brice Bauderon Ioannes Tagaucius Jean Fernel Jean Ruelle Jacobus Sylvius, Jacques Dubois	8	20.5
Total Southern Europe/ Mediterranean		15	100		31	79.5

authors. Finally, Esteban de Villa's *Examen de boticarios* of 1632 listed the books "that are written on this art," choosing "the most curious and useful that I have been able to find by many different authors," dividing them between "Greeks," of which he listed 6, "Arabs" of which there were 7, and "Latins" which included 70 of his fellow Europeans of the medieval and early modern periods.[25] Despite the preponderance of the works of Villa's contemporaries, however, they were not the works he would most recommend as being essential to the apothecary. For this he turned to the classics. For "the books that the apothecary must have," Villa stated that "of all the books listed above, it remains to know which ones are very necessary to the apothecary . . . so that the apothecary is not, as they say, without the weapons of his art, having everything pertaining to it."[26] These works included the same eight referred to in the *Compendium aromatarium*, with the caveat that practitioners employ an up-to-date commentary of *De materia medica*. Villa also recommended the "famous" work of Luis de Oviedo, a treatise published in 1581 on the selection and processing of simples that was one of the hallmarks of the procedural genre.

MAJOR GENRES OF THE PHARMACEUTICAL TEXTUAL TRADITION

Together, these authors formed the basis for a variety of pharmaceutical genres that evolved over time. The secondary literature on genres of medical writing has generally identified two categories of pharmaceutical writing: books of simple medicines, or "herbals," and books of compound medicines, or formularies.[27] There is general agreement that medical and scientific writing changed relatively little over time, particularly before the age of print, largely due to the

Scholastic practices referred to above in which authors relied heavily on their predecessors for themes, discursive style, argument structure, and textual strategies, thus making changes only incrementally.[28] The most pronounced changes in genre, it is said, occurred during the Scientific Revolution, when there was a transition from "Scholasticism to empiricism and then to rationalism" in medical writings.[29] An examination of early modern Spanish pharmaceutical texts and their antecedents, however, reveals a line of development that appears to have little to do with the Scientific Revolution. Instead, its major developments, as shown previously, correspond to the spread of Galenic pharmacy throughout the Mediterranean and into Europe. It also shows a wider range of genres that culminated in that of the pharmacopoeia. In addition to the herbals and formularies, which go back to antiquity, were glossaries of drug names and substitutes, procedural texts that described pharmaceutical operations and techniques in detail, and pedagogical texts designed for training pharmacists. Whereas glossaries appear in conjunction with the herbals and formularies, the procedural and pedagogical texts largely developed over the medieval and early modern periods, as pharmacy grew into an increasingly specialized medical profession. Elements of these genres then came together to form the early modern pharmacopoeia that eventually entered the Atlantic World.

Books of Simples

Books of simples treated the raw materials and natural resources of Galenic materia medica, or collections of medicinal substances. Simples were made up of plant, animal, and mineral substances, though plants dominated, making up 80–90% of all simples. Presumably this is why treatises on simples came to be known in the medieval and early modern periods as "books of herbs," or "herbals," and formed the basis for early botanical studies.[30] They were also referred to as dictionaries, particularly when organized alphabetically. Books of simples usually included lists of plant simples with descriptions of their leaves, flowers, or berries as well as their healing powers, or "virtues"—how they worked and the conditions for which they were indicated. In Galenic pharmacy, the simples described in the herbals were often assigned a primary quality (hot, cold, wet, or dry) and a degree of intensity (on a scale of 1 through 3 or 1 through 4) that was part of Galenic humoral theory, which taught that the body was composed of 4 "humors" that each had a dominant characteristic (being hot, cold, wet, or dry). Disease resulted from the imbalance of these humors, with medicines applied of opposite qualities to counteract them and restore balance.

Books of simples had been in use throughout the Mediterranean for centuries, building on the base established largely by the writings of Dioscorides, but with important contributions, as described above, from Galen and Pliny. *De materia medica* was translated into Arabic under the Abbassids (750–1258 CE), and excerpts of it remained influential in Western Europe after the fall of Rome. Arabic authors also produced books that greatly expanded the number of simples thanks to the widespread reach and cosmopolitan nature of the Islamic Empires, which gave them access to Persian, Babylonian, and Indian medical traditions (among others), in addition to the Greek.[31]

The knowledge imparted from these works reached medieval Europe through the translation movement, and Dioscorides' contribution received renewed impetus in the sixteenth century with the publication of several new translations and annotations of *De materia medica,* including those of Amato Lusitano in Portugal, Andrés de Laguna in Spain, and Pietro Matthiolo in Italy, which was "consulted by scholars from Cambridge to Cracow."[32] Investigations of medicines from the Americas and Asia lent new impetus to this tradition beginning in the sixteenth century, with publications of such authors as Garcia d'Orta, Nicholas Monardes, Francisco Hernandez, and Juan Fragoso, describing simples from these areas that were new to the European pharmacopoeia. These works were similar in structure to the earlier books of simples from the ancient and medieval Mediterranean, including botanical descriptions, qualities and degrees, and virtues.

In addition to the traditional herbal were different types of reference works on simples that arose over time and added to the Mediterranean textual tradition. As early as the writing of books of simples were works on drug substitutions, also called *succedanea,* or lists of "quid pro quo" that itemized possible safe and effective alternatives for the simples called for that might not be locally or readily available, or might be too expensive, as in the case of recipes that called for precious stones.[33] Although Galen is usually identified as the first in the Western medical written tradition to provide a list of drug substitutes, tablets with two-columned lists of drugs have been identified as such in ancient Assyria during the time of King Sardanapalus, 668–626 BCE, and from ancient Egyptian papyri as well.[34]

Another type of work that developed as early as the first century CE were "synonyms," or glossaries of simples in which a variety of names in different languages were given for the same simple, due to the fact that the identification of simples, and particularly plants, "presented enormous difficulties" in general and especially upon translation.[35] The need for such glossaries arose when Roman authors sought to make sense of earlier Greek writings, in which transliteration of the name of the simple would not necessarily be understood, as plants are notorious for having local names.[36] Moving into the Middle Ages, as Greek works were translated into Arabic and then Latin, the need for these glossaries continued, with famous synonyms produced in which the name of the simple was given in

Greek, Arabic, and Latin. This medieval subgenre was especially crucial for clearing up confusion and for standardization of terms when transliteration of a local name would not necessarily have any meaning in another language.[37] The need for these works was an indication of the extent of trade and exchange among world systems of the early Mediterranean, with the important additions made to the Greek and Roman materia medica by Arabic scholars and practitioners. This genre continued to be produced in the early modern period. Juan Fragoso's 1575 *De succedaneis medicamentis*, for example, discussed possible substitutes for 250 different simples, as well as alternate recipes for many compounds.

The Formulary

Next, perhaps the most widely recognized pharmaceutical genre was the formulary, which consisted of recipes for compound medicines with brief instructions as to how to formulate them. In the Arabic world they were referred to as "dispensatories," or *aqrabadhin*, which signified "an antidotary given by the grace of God" and derived from the Greek *graphidion*, meaning "list" or "registry."[38] In the Middle Ages, these works were often referred to as antidotaries or receptaries. They were generally encyclopedic in character and were meant to serve as easy-to-use reference works for apothecaries. One of the earliest extant texts specifically devoted to compounding was *On Compounds* (*De Compositiones*) of Scribonius Largus (ca. 1–ca. 50 CE), which included recipes for over two hundred compounds organized in a "head to toe" (*a capite ad calcem*, "from head to foot") arrangement according to the part of the body they treated (though certain sections focused on certain medical preparations).[39] Galen followed in the next century with two treatises on compound medicines, *On the Composition of Drugs according to Kinds*, generally arranged by type of remedy, and *On the Composition of Drugs according to Places* (hereafter referred to as *Kinds* and *Places*, respectively), the latter arranged like Scribonius Largus's work, according to the place of the body it healed.[40] These works are recognized as major milestones in the history of compounding and of the formulary genre. However, subsequent Arabic formularies incorporated and standardized so many new elements that, as I have argued elsewhere, they effectively invented the pharmacopoeia that eventually became the standard in early modern Europe.[41] Whereas Galen's books were mainly organized according to disease, the Arabic formularies were organized by the type of compound, and included several types not used in Hellenistic medicine. Hellenistic medicine, for example, recognized a series of compounds categorized by the way they were applied to the part of the body they were meant to cure (gargles, eye washes, incense, perfumes, sneeze inducers, toothpastes, enemas, pessaries, and suppositories). There were also a series of compounds categorized by their function or overall effect on the body (purgatives, emollients, astringents, abstergents, cathartics, and emetics). A third type of compound, "method-based" compounds, could be classified by the materials and methods used to make them. These included, for Galen, the honey-based confections of theriac[42] and the bitter hieras, as well as ointments, oils, liniments, poultices, pastilles, and plasters.

Arabic formularies went on to add a host of new sugar-based compounds to those already in use, including electuaries and confections, jams, marmalades, and preserves, lambatives (thick, viscous medicines ingested by licking), and various classes of syrups, which were arguably the most important and certainly the most numerous type of compound in Arabic—and later European—pharmacopoeias.[43] These additions were certainly important, but it was the structural elements of Arabic pharmacopoeia that effectively formed the foundation of the genre. For the most part, Arabic formularies were divided by chapters, with each chapter representing a different type of compound. Each entry in the various chapters followed a standard structure that included the compound name, ingredients (often with specific measurements), a short set of instructions, indications, and dosage. In addition, pharmacopoeias increasingly emphasized method-based compounds, which made up the majority of their chapters.

Later, European pharmacopoeias followed this same organization and emphasis, and the categories of compound medicines grew increasingly standardized, largely based upon a formulary that appeared in the late thirteenth century in northern Italy and whose significance for European pharmacy is difficult to overemphasize. That work was the *Grabadin* of John Mesue, which was divided into twelve chapters, all of which fell into the method-based categorization.[44] This increasing focus on procedure and operation (rather than application or effect) that the focus on method-based compounds indicated reflected the increasing specialization and professionalization of pharmacy in the late medieval period, a trend that continued into the early modern era.[45] Indeed, European books of compounds tended to follow Mesue's categorizations, with pharmacopoeias becoming increasingly streamlined over the course of the sixteenth and seventeenth centuries. Spanish pharmaceutical treatises that contained formularies, for example, all followed (with some but little variation) Mesue's categories of method-based compounds, with very few of the application- or effect-based compounds. In this way, Mesue's *Grabadin*, which owed its formulation to Arabic precedents, arguably formed the basis for the modern pharmacopoeia.

Procedural Texts

In addition to these genres were two other types of pharmaceutical writing in Galenic pharmacy that developed in the medieval and early modern periods as part of the growth of pharmacy as a separate profession within the field of medicine. The first, which I have termed "procedural," was a type of treatise that first appeared in the medieval Arabic world. Procedural texts included instructions as to how to perform the operations needed to prepare both simple and compound medicines, and in their truest form were a collection of general rules and instructions as to the behavior of medicinal matter according to its substance and the ways to prepare medicines. In this way, despite their practical nature, procedural texts were the most theoretical of the pharmaceutical subgenres and led to intimate knowledge and understanding of the natural matter with which the apothecaries worked.[46]

Like the mixing of medicines, technical advice and know-how had certainly been part of the earliest stages of Galenic pharmacy, but it tended to constitute "tacit knowledge" among practitioners and was rarely written down or treated in a systematic way. Thus, the earliest identifiable procedural texts did not appear until the later medieval period, with such works as the *Liber servitoris*, a remarkable Arabic treatise produced around the year 1000 CE that described in detail the processing of minerals, plants, and animals for use as medicines. The treatise was written by Abū al-Qāsim Khalaf ibn al-'Abbās al-Zahrāwī (936–1013 CE), Latinized to Abulcasis, a prominent physician in Cordoba. It was the twenty-eighth chapter of a larger pharmaceutical treatise, the *Kitab al-Tasrif*. In it, al-Zahrawi made direct reference to Dioscorides and Galen, showing their influence, but overall the work was highly original, written in clear language with detailed instructions as to procedures, techniques, and materials. According to the author, the overall purpose of the *Liber servitoris* was to explain how to prepare simples for inclusion into compound medicines, there being "many compound medicines, the many simples for which have need of preparation" before being incorporated into them.[47]

The *Liber servitoris* was divided into three parts or chapters, the first having to do with inorganic materials, or "the preparation of stones and minerals only," the second treating herbs, and the third animals and animal parts. In part one, al-Zahrawi described the various ways to prepare metals, natural salts, and stones for inclusion in pharmaceutical recipes. The main operations employed in their processing included washing, burning, and sublimation (a procedure involving the distillation of solid materials). He described, for example, various ways to make, burn, or wash the "dross," or ores, of impure metallic compounds like iron, litharge of lead, and gold to be used in ointments. Part one also explained how to sublimate yellow arsenic and mercury and how to make lime by bleaching ash with eggshells, seashells, or white marble.[48]

The second part of the *Liber servitoris* dealt with "roots and plants" and various ways to process them, explaining how best to obtain their juices and mucilaginous parts; how to extract starch from grains; how to peel and core nuts and seeds; how to prepare purgative herbs; how, when, and where to best collect herbs; and how to make medicinal taffy. Al-Zahrawi also treated several thermochemical processes, including how to distill oils and vinegar, how to prepare distilled medicinal waters, and how to burn branches and other plant parts (roots of trees, seeds, resins, herbs, wine dregs) to obtain ash.[49] In part three, al-Zahrawi focused on processing animal parts and products, explaining how to obtain blood from live animals, how to whiten beeswax, how to make medicine from infant urine, and how to obtain and preserve bile. As with mineral and plant materials, the *Liber servitoris* also described several thermochemical processes using animal materials—explaining how to burn seashells, oyster shells, eggshells, crabs, scorpions, and snakes, as well as silk and wool.[50] In this way, the *Liber servitoris* described many basic, necessary processes the apothecaries would use to prepare natural materials, or simples, for inclusion into compound medicines. It was well known to the early modern European medical community, with nine editions of the work published before 1501.[51]

The next major procedural text, which had an even wider impact on the development of European pharmacy, was Mesue's *Canons*. The *Canons* were essentially a set of pharmacological rules that provided general directions, or from the title of the work, laws as to how to choose, prepare, and apply simple and compound remedies. The *Canons* were divided into four sections, called "theorems" in later editions, that dealt, respectively, with the selection, preparation, application, and effects of simples. The most valuable sections of the *Canons* for apothecaries were the first two, the first treating the "election," or gathering, storage, and evaluation of simples, and the second giving directions as to their "correction" in order to prepare them for inclusion in a compound medicine and/or application to the human body. In the first two *Canons*, there was virtually no discussion of health or disease; they consist of relatively simple instructions as to how to evaluate and prepare simples and compounds in the most effective way possible. The focus was on pharmaceutical technique, not medical theory: Galenic humoral medicine plays a very minor role, and simples and their properties, or "virtues," were conceived of in materialist terms as substances that could be released or dissipated depending on the pharmaceutical operations applied. In this way the *Canons* were meant for practicing apothecaries who were becoming increasingly professionalized and, as with the *Grabadin*, more and more focused on the methods of formulating medicine.

The first two sections of the *Canons*, in fact, were so important that they were copied, annotated, and explained over and over in edition after edition of pharmaceutical works published throughout Europe in the fifteenth through eighteenth centuries. Early published editions of Mesue's works outpaced those of Dioscorides, Pliny, Avicenna, al-Zahrawi, and Arnald de Villanova (to name a few), and the information presented in the *Canons* went on to provide much of the basic vocabulary, themes, and organization for subsequent pharmaceutical publication. In the first *Canon*, Mesue provided a general rule or set of rules by which to judge and classify a simple and its powers according to its substance; its qualities and degree; its texture, flavor, odor, and color; the time in which it was harvested and stored; and the place it came from.

The second section, or theorem, of the *Canons* had to do with the "correction" of medicines, or the different operations that a practitioner could perform in order to render a medicine safe and effective. Here, Mesue outlined the basis of pharmaceutical technology by identifying four types of operations that the apothecary could employ in order to enhance or alter a medicine's virtue.[52] In Spanish pharmacy, these operations came to represent the main components of the apothecary's work and served to provide the very definition of pharmacy itself. The four operations of Mesue included decoction, or the act of cooking simples through the application of heat; infusion, or the immersion of a simple in boiling liquid; lavation, or washing, of a simple; and trituration, or the division of a simple into smaller parts usually through grinding or crushing. For Mesue, the preparation of all simples was necessary in order to render them safe and effective for human use. To do so, the practitioner needed to know four main things about the simple to be prepared: the nature of its substance or density; the nature of its virtue, or healing property; whether it was strong or weak; and whether it "worked" or released easily or with difficulty. The nature of these characteristics would determine the vigor with which these operations should take place, with the end result always to produce and preserve from the simple a virtue of optimum strength.

Mesue's *Canons* arguably had the greatest impact on pharmaceutical concepts and procedures in the early modern era, but another work in the Spanish tradition, Luis de Oviedo's *Método de la colección y reposición de las medicinas simples y de su corrección y preparación* (1581) stands out in its efforts to elaborate upon these procedures. Part of Oviedo's purpose was to provide specific information to practitioners as to how to prepare medicines to fill in gaps left in the ancient works. For Oviedo, it was crucial that medicines be prepared properly; otherwise, the very medicines meant to cure an ill would themselves cause more harm, so that "instead of giving health, they remove it, and instead of freeing us from the illnesses that afflict us, they make them worse."[53] Yet the ancients did not leave clear instructions as to how to do so, so it was up to Oviedo to include the detailed information that practitioners needed to know in order to proceed: "It being the case that Galen and Dioscorides and other ancient doctors did not give enough consideration to simple medications; in writing about the manner of their preparation they did not provide a straightforward [method] which deprives us of knowing [how to prepare] the remedy. For this reason we have found it very necessary to add . . . the way to prepare them, so that those repressed by illness are not left unaided."⁵⁴ The influence of the *Canons* is ubiquitous throughout the *Método*. Not only does Oviedo refer to Mesue and his commentators, but the procedures he describes that make up the core of the work are the four operations identified in the *Canons*, with the aim at all times to correct and preserve the simple's virtue in its most optimal, efficacious form. Oviedo's work was, in this way, of great import to the profession as well. It went through three more editions after its initial 1581 publication (1595, 1522, and 1692), and it was one of only two sixteenth-century works still used in Mexican pharmacies two centuries later.

The Pedagogical Text

The final genre, the "pedagogical" text, arose, as did the procedural, with the increasing professionalization of pharmacy.⁵⁵ These texts were a product of early modern Europe, the first appearing in Italy in the late fifteenth century, the *Compendium aromatarium* by Saladino da Ascoli. They tended to have one or more of several different components that evolved over time and built upon the earlier Arabic tradition: the establishment of a definition of pharmacy and the duties and responsibilities of the apothecary; the increasing use of the vernacular; the arrangement of procedural and theoretical information in the form of an examination or dialogue that became increasingly standardized and formulaic over time; and the inclusion of several different subgenres within one text (i.e., pedagogical elements combined with lists of simples, compounds, synonyms, and drug substitutes; as well as instruction on technique, operations, and procedures). These works were written primarily by apothecaries and intended for apothecaries. Aguilar (1569), for example, stated that his work was meant for "learned apothecaries, the true ministers of the art."⁵⁶ Similarly, Jubera wrote in 1578 "for the benefit of those of my profession and art of pharmacy."⁵⁷ Overall, these texts were not concerned with explicating disease nor with humoral theory, beyond the effect a medicine would have on a particular humor. Rather, they were most concerned with enumerating and explicating the responsibilities of the apothecary, giving precise directions as to how to prepare medicines, setting out a canon of works that the pharmacist needed to know, and making sure that their writings were not obscured by arcane language or even by the use of Latin.

The *Buen Boticario*

The first characteristic of the pedagogical genre was its emphasis on the "*buen boticario*," or the morality and good character of the apothecary so that he would fulfill his responsibilities in an ethical and conscientious manner, as noted also by Antoine Lentacker (chapter 12) in his chapter in this volume. The *Compendium aromatarium*, for example, included a discussion of the definition and moral attributes of the apothecary that would become a theme of later works, and was based upon Arabic precedents, whose professionalizing trends included emphasis on the morality of the apothecary and the ethical practice of medicine.[58] In this work, a hypothetical doctor asked an apothecary: "What is the responsibility of the apothecary?"[59] The apothecary replied that he had two main responsibilities: to know how to "grind, clean, infuse, cook, and distill" substances in order to prepare them well, and once prepared, to know how best to preserve them. The second question that Saladino's physician would ask, "What must the apothecary be like?"[60] required an answer that delineated both the moral qualities that an apothecary must demonstrate, and the ethics of his practice. It stipulated that the apothecary be God-fearing, pious, serious, and mature; a practitioner who took care to prepare only the freshest medicine of the best quality, and only those ordered by the physician, without any unauthorized substitutions. The apothecary, as discussed above, was also meant to be learned and to have mastered the corpus of the major ancient and medieval works in the field.

Use of the Vernacular

A second characteristic of the pedagogical text was the use of the vernacular. Although the use of Latin did not become obsolete, the publication in Spain of pharmacy books in Latin diminished over the course of the early modern period. In the sixteenth century, the number of major pharmaceutical texts written in Latin (6) almost equaled those written in Castilian (7). In the seventeenth and eighteenth centuries, however, Latin publications diminished: only one of six major seventeenth-century treatises was written in Latin, and two out of six in the eighteenth century (see table 1.7).[61] Authors of Spanish texts were highly conscious of the reason they wrote in Castilian: to appeal to the professional group of apothecaries who were not (according to the authors) always well versed in Latin, licensing requirements notwithstanding. By writing books accessible to this group, the authors believed themselves to be making the fruits of centuries of knowledge available to the apothecaries, and through the apothecaries to the general public. Several sixteenth-century vernacular works sought to justify the use of Castilian in this way. Alonso de Tudela translated Saladino da Ascoli's *Compendium aromatarium* in 1515, for example, so that apothecaries could "understand all the things pertaining to their art." Many had been prevented from doing so "because the majority of the apothecaries of these kingdoms lack [knowledge of] Latin and could not benefit from such a beneficial book, [so] it seemed to me a very useful and even necessary thing to translate it into Castilian so that

TABLE 1.7. Works of Early Modern Spanish Pharmacy

Century	Books in Latin	Books in Castilian	Total
16th	6	7	13
17th	1	5	6
18th	2	4	6
Total	9	16	25

they could realize the fruits of Doctor Saladino's work in composing it."[62] In 1569, apothecary and author of *Exposicion sobre las preparaciones de Mesue*, Antonio de Aguilera, expressed similar sentiments: realizing the "great lack" of Latin among many of his counterparts, he says, "I was induced by the begging of many of my apothecary friends to bring to light this present work, such a necessary declaration of [pharmaceutical] doctrine, in our clear Castilian romance language."[63] In 1578, Alonso de Jubera stated that his book was in Castilian so that "those who have not studied [Latin] can enjoy it more easily."[64] Similarly, Luis de Oviedo wrote his book in 1581 "so that those who know Latin as well as those who do not (of which there are many) are able to benefit from it."[65] A few years later, Antonio Castells concurred, stating, "I wished to take on this work for the public good, principally for the apothecaries who do not know Latin well, so that with this brief treatise . . . they can understand the theory as well as the practice that is involved in the method [of making medicines], which will ensure their work and make up for the deficiencies caused by their lack of education."[66] Such a step was necessary to make their work as effective as possible "in order to benefit the sick, who put their lives in our hands."[67] As late as 1778, approval for Pedro Vinaburu's *Cartilla Pharmaceutica* derived in part because his understanding of medieval Latin texts allowed him to express and explain it in Castilian "for the well-being and use of all."[68]

The Use of Dialogue

Another noticeable element of the pedagogical genre that developed in the early modern period was the use of dialogue in setting out the most fundamental principles of the profession. Scholars have examined the use of this literary form in works of Renaissance natural philosophy, most notably in those of Galileo, Giordano Bruno, and Leibniz, arguing that it served an important rhetorical and pedagogical purpose.[69] Not only did it expose the logic of the argument in a clear manner but it allowed the author to anticipate objections and highlight the logical flaws within them. Dialogue also set up a dialectical dynamic in which the "learner" was essentially coached by the "knower" to arrive at the proper conclusion. Such was the case for the pharmacy writing, in which dialogue served a clear pedagogical purpose: to initiate pharmaceutical neophytes into the fundamental principles and practices of the profession in a way that invited their participation.

The first instance of the dialogue comes in a very rudimentary form in the *Compendium aromatarium* through the questions, discussed above, that the hypothetical physician would ask the apothecary. This pattern was followed closely in the *Examen apothecariorum* of Pedro Benedicto Mateo (1521), who began the work by presenting three questions: "What is an apothecary? ... What is a [pharmacy] examination? ... [And what was] most necessary to know" about pharmaceutical theory and practice? The respondent then went on to answer each question beginning with "Dico," or "I say."[70] This literary device grew more common over the sixteenth century, in which the dialogue would continue for several pages and, in some cases, throughout the entire work. The participants in the dialogue, furthermore, were given clear identities and roles that evolved over time, but that always represented a hierarchical pairing of a "knower" who was questioning, coaching, and teaching a "learner."

In Antonio de Aguilera's explanations of Mesue's *Canons* (1569), for example, the dialogue continued throughout the entire work and played an explicit role in the text. As Aguilera explained, he had "put this work together in the form of a dialogue in which the chapters are divided by question and answer so that in this way the doctrine will be clearly understood."[71] In this way, Aguilera believed that Mesue's doctrine would be "explained literally and to the letter of the text."[72] For the dialogue, Aguilera chose two telling characters, the first being a physician named Apollo, a man of "great knowledge and wisdom" whose name harkened back to the Greek god "who invented medicine and was the first to identify the workings and virtues of herbs."[73] Such a character would have complete knowledge of the doctrine, which would allow him "to propose and ask" the appropriate questions to clarify it.[74] The other participant in the dialogue was an apothecary named Curio, a name chosen explicitly because "it conjures and means a man who is curious, solicitous, and an expert in his art and office." And indeed, the text follows the dialogue form faithfully from beginning to end, with a presentation of Mesue's *Canons* in Latin, to which the two participants respond in Castilian, switching roles within the lines of text.

Another sixteenth-century work presented in the form of a dialogue was Alonso de Jubera's *Dechado y reformacion de todas las medicinas compuetas usuales* (1578), a work that was meant to standardize the recipes given for compounds. The dialogue also continues throughout the work, and like Aguilera's, was inserted within the lines of text. It takes place between a father and son, the father seeking to teach his son the principles of pharmacy and the best way to make certain compounds. The work begins with eight short chapters that explain the main categories of plants and their healing abilities, the different parts of plants used in medicine, and the various operations discussed by Mesue by which to prepare medicines—decoction, infusion, washing, and grinding. In each case, the son asks the father for definitions and explanations, and the father responds largely by quoting ancient and medieval authorities.

Three works in the later seventeenth and eighteenth centuries show a final evolution of the dialogue from physician/apothecary and father/son pairings to the more expressly pedagogical and schematic pairing of teacher and student. These works have many elements in common, indicating an increasing standardization of the form and the larger genre. Fuente Pierola's *Tyrocinio pharmacopeo methodo medico y chimico* (1660) and Félix Palacios's *Palestra pharmaceutica* (1706) both use "Pregunta/Respuesta—Question/Answer" (immediately abbreviated to "P" and "R") between, one would assume, a hypothetical teacher and student, while Juan de Loeches's *Tyrocinium pharmaceuticum* (1728) begins with a dialogue between the "Magister" and the "Discipulo" (immediately abbreviated to "M" and "D" in the text). In all three of these works, the dialogue forms only the first part of the text in which the general definition and principles of pharmacy are presented. The characters or markers are not inserted within the text, but are rather justified along the left margin of the page or column and in this way dominate the style of this first section. The questions and answers are more succinct and formulaic, and deal with one aspect of pharmacy at a time, showing that the form and the content have become standardized parts of the text. The dialogue first sets out to define pharmacy, then to define what a medicine is, followed by a series of questions about how to choose, prepare, and compound medicines. Though each work has its own particular characteristics, they all follow this increasingly standardized format.

Pharmacopoeias as Culmination

The final textual genre discussed here, the pharmacopoeia, was in effect the culmination of earlier ones, because it brought together elements from these genres into one comprehensive work. *Pharmacopoeia*, deriving from the Greek "to make drugs," was a term first used by the Greek writer Diogenes Laertius in the second or third century and later adopted by European authors beginning in the sixteenth century.[75] Different pharmacopoeias included different elements and genres, but they generally included several different sections within the volume. These sections comprised the newer pedagogical elements—dialogue, best practices, and ethics, increasingly written in the vernacular—as well as elements of the older genres, including materia medica (sometimes with botanical information, sometimes without), glossaries, recipes for compounds, and information on techniques for selection and processing. Over time, they developed into official, legally enforced regional, national, and imperial standards, and continue to be a genre of great significance to modern pharmacy.[76]

The *Compendium aromatarium* represents an early example of this kind of work. It consisted of seven parts, the first consisting of a dialogue defining and outlining the apothecary's major responsibilities, followed by a commentary on Mesue; a list of the compound medicines from the *Antidotario Nicolao*; a brief treatise on weights and measures; instructions as to how to collect, prepare, and preserve simples; and a final section outlining in detail "the way to organize a pharmacy with the all things in it that it needs."[77]

If the *Compendium aromatarium* constitutes an early example of the early modern pharmacopoeia, then it reached its culmination with Félix Palacios's *Palestra pharmaceutica* (1706), which served as the basis for the standard formulary for the Spanish Empire.[78] Palacios's work is often identified with the emergence of chemical medicine in the late seventeenth century, but it was firmly an amalgamation of traditional Galenic pharmacy and chemical pharmacy—part of the "chemico-Galenic compromise" typical of many pharmacopoeias of the period.[79] Indeed, the traditional elements of the work are just as important as the chemical, and the *Palestra* was in many ways the culmination of two thousand years of a developing pharmaceutical literature that resulted from and, in turn, had a major impact upon, the increasing specialization and professionalization of pharmacy. The text begins with the usual dialogue defining pharmacy, followed by a series of questions and answers concerning the selection and preparation of medicines. Next is a series of lists of mineral, animal, and plant simples, each in Latin with a Spanish translation followed by descriptions of how to prepare them according to Mesue's operations. The first part of the book concludes with a description of the different instruments used within the pharmacy and the ways to moderate applied heat. The second part of the book discusses drug substitutions, weights, measures, and chapters describing an array of different compounds—not only the method-based compounds but earlier application-based ones as well, from syrups, plasters, and ointments to gargles, chewables, potions, lotions, eye washes, enemas, suppositories, perfumes, and incense. Each chapter begins with a short treatise describing the compound, its history and etymology, and techniques for preparation, followed by individual recipes that list ingredients and instructions, as well as what they cure and the dose to give. In this way, the *Palestra* combined older genres—lists of simples and compounds as well as glossaries and drug substitutions—with newer elements of the procedural genre, in its detailed discussion of operations, and of the pedagogical genre, in its use of the vernacular and use of dialogue to define and describe pharmacy.

CONCLUSION

In this way, it is possible to see the development of the textual tradition in Galenic pharmacy, from its origins in ancient Greece and Rome, through the evolution of new genres in the medieval Arabic world, to their adoption and expansion in early modern Europe, especially Spain. An overview of this tradition, based upon tabulation of references to authors in early modern Spanish pharmacy texts, reveals the unparalleled influence of Dioscorides and Mesue especially in the incorporation of their methods and materials into later texts. It also serves to identify the range of genres within the textual tradition, including the herbals (books of simples), glossaries, and formularies that developed and grew increasingly standardized from the medieval to the early modern period. In addition to these genres was the introduction of two more types of texts from the late medieval and early modern periods—the procedural text, which outlined and elaborated upon techniques and practice; and the pedagogical text, which brought together several new elements designed to aid the preparation of apothecaries in training. These elements included discussions of the ethics and morality of the apothecary, lists of books the apothecary needed to read and keep in his shop for reference, use of the vernacular, and use of the technique of educational dialogue between expert and neophyte. These various elements came together in the pharmacopoeia, a genre that ran to several parts, beginning with Saladino da Ascoli's *Compendium Aromatarium* in the late fifteenth century and culminating with Félix Palacios's encyclopedic *Palestra pharmaceutica*, published in 1706. In this way, the textual tradition in Galenic pharmacy spread from the ancient Mediterranean to medieval Europe through the Islamic Empires, and then moved on to the Atlantic World, as evident in the pharmacopoeias kept in Herrera's Mexico City shop.

2

AUTHORITY, AUTHORSHIP, AND COPYING
⁕

The *Ricettario Fiorentino* and Manuscript
Recipe Culture in Sixteenth-Century Florence

EMILY BECK

IN 1498, the College of Physicians of Florence offered up the *Nuovo Receptario Fiorentino* as a concise, easily understood compilation of reliable recipes for College-approved medicines.[1] The recipes in that first edition came mostly from such antique and medieval Arabic authorities as Mesue (777–857 CE), Niccholao (ca. twelfth century CE), Avicenna (980–1037 CE), and Rhazes (854–925 CE).[2] More contemporary authors would win places in the volume in subsequent editions, but even as the College shifted from Arabic to classical Greek and Roman authors, the foundation of the *Ricettario* was a medical canon that had been established centuries previously. Although scholars debate whether it was the first European pharmacopoeia, the *Ricettario Fiorentino* served as a model for later regional and national pharmacopoeias across Europe and the New World, which echoed both its form and content.[3] In addition, the authors that the Florentine College of Physicians cited largely continued to form the base of official pharmacy across Europe and in the Americas during the first few centuries of contact, as discussed by Paula De Vos (chapter 1).

The authors of the first edition of the *Ricettario Fiorentino*, the Florentine College of Physicians, directed their advice toward *speziali* (apothecaries).[4] By the mid-sixteenth century, apothecaries' shops and stock had become subject to regulation by the College of Physicians. All of the rules about the physical characteristics of shops, pieces of advice about collecting and conserving ingredients, and recipes were meant to ensure that apothecaries would be the official providers of medicines, making specific preparations that would be both effective and safe, generally to be dispensed only on the prescription of a physician.[5]

It is clear, however, that many medicines were not made in this period by apothecaries acting at the behest of physicians. Sixteenth-century manuscript medical recipe books support the general view that family, friends, and members of the community mostly made their own medicines and cared for themselves because of cost, convenience, and habit.[6] These handwritten reference volumes of recipes taken from all kinds of sources also make it clear that apothecaries were not the only individuals compounding complex medicines and that doctors were not the only ones prescribing them. Although the *Ricettario*'s authors were concerned about the effect of these unregulated medical practices that appear in manuscript recipe books, they only sparsely mentioned those healers who were not part of the official medical hierarchy.

The *Ricettario* and the legal goals of the College of Physicians raise questions about how closely aligned the practices of different groups of professional and lay practitioners were and whether unlicensed healers took notice of all of these official pronouncements. Although it may be tempting to take the *Ricettario*'s recipes and assume that they were common to practitioners throughout the medical marketplace, Nancy Siraisi has argued that "medical texts [of all kinds] are essentially prescriptive; consequently, they are unreliable and inadequate sources of information about actual medical activity and its social context."[7] Whether lay healers used the *Ricettario Fiorentino* is unclear, although it is very probable that it was not a standard pharmaceutical reference text in informal medical spaces beyond apothecary shops. In the words of James Shaw and Evelyn Welch, "One of the fundamental problems in the history of medicine is to bridge the gap between written works of theory and the reality of daily practice."[8] Although manuscript medical recipe books are medical books that were prescriptive, they were also personal archives of practical information and thus provided a way of seeing beyond official recommendations to individual habits.

The medical practices of lay healers in the early modern period are, for a variety of reasons, less evident in the historical record even though informal treatments from family and friends constituted the most common form of healthcare for early modern Europeans. While they assuredly took advantage of the multitude of published vernacular sources of information, handwritten recipes were a fundamental part of lay healers' practices. Recipes for all kinds of medicines were traded among friends and colleagues as well as purchased from experts. Establishing the exact mode of transmission of individual entries in medical manuals and recipe collections is often problematic, but it is clear that recipes frequently circulated in handwritten, printed, and oral forms. The collection of recipes over a person's lifetime led to stacks of individual sheets and scraps of paper, amassing in folders and boxes. Recipe compilers sometimes rewrote their recipes in bound books to organize them and keep them in one place for future use, molding their practices (and ideas of their work) into the paper genre of "the recipe book." Recipe books are strong evidence that laypeople routinely practiced medicine, often with attention to the same theories and with similar ingredients that academic doctors promoted.

Authors of both manuscript and printed books occasionally cited the authors of the medical recipes they recommended to their readers. Like references in modern cookbooks to celebrity chefs, medical authors in sixteenth-century Florence referred to famous medical authorities, from ancient Greece to their contemporaries. They used these references to bolster their own authority: the prestige of the original author transferred to the recipe, and to the sixteenth-century author, assuring contemporary readers that their recipes were carefully chosen and would be effective. Although the recipes in manuscript recipe books and in the printed *Ricettario Fiorentino* volumes appear to be very similar in many cases, the authors of these two types of pharmaceutical texts prioritized very different aspects of medical authority in order to reach their distinctive goals. On the one hand, the College of Physicians regularly called upon ancient and medieval canonical authorities. On the other, manuscript recipe book authors much more often referred to contemporaries. Although printed pharmacopoeias are fundamentally local, reflecting the authority structures and preferences of a given site, the manuscript recipe books examined here are products of personal relationships in specific medical marketplaces, where knowledge of contemporary practitioners seems to have held more weight than deep knowledge of the works of individuals like Avicenna and Mesue.

Although, in many ways, a comparison of official and lay medicine is tricky to perform, it is important to examine diverse document types together in order to visualize the ways that different practitioner groups may have interacted. This chapter is an exercise in this type of assessment, placing two editions of the official *Ricettario Fiorentino* (1498 and 1567) in conversation with five Tuscan, sixteenth-century manuscript medical recipe books from the collections of the Biblioteca Riccardiana in Florence. The comparison lends itself both to understanding how the College of Physicians changed its intellectual preferences over time and to visualizing the different ways in which official and lay healers used their pharmaceutical reference volumes to present themselves as knowledgeable and trustworthy in the medical marketplace. Approaching the writing practices of these manuscript authors alongside the official documents of the College exposes the complex differences in the citation preferences, as well as the striking similarities in terms of the way practice was composed.

THE *RICETTARIO FIORENTINO* AND MEDICAL AUTHORITY IN FLORENCE

The Florentine College of Doctors and Apothecaries was founded in 1218 and experienced varying levels of success in controlling the medical practitioners of the city-state throughout the medieval and early modern period.[9] It was not until 1498 that the College printed its first recipe book to which apothecaries were bound, the *Nuovo Receptario Fiorentino* (later retitled *Ricettario Fiorentino*). Although it would not gain legal control over apothecaries and medical practice in general in Florence until the mid-sixteenth century, the College offered this book as the common recipe source for apothecaries and physicians to ensure shared knowledge of the composition of medicines.

The authors of the first edition began their book by arguing that people in the region were in danger because of the variety of medical recipes in use. Not only was this miscellany of practices dangerous for patients, it was also bad for physicians because it undermined their authority. They insisted that standardization of recipes and medical decision-making in a volume written by physicians was the key to avoiding the mistakes that came from apothecaries taking liberties with the composition of medicines and the ills that came of them (both for patients and physicians). To make the book user-friendly, they would follow the simple style of previous authors like Mesue, Niccholao, Avicenna, and Galen and put forth a book that contained only what was necessary, nothing superfluous.[10] Although many changes were introduced to the *Ricettario Fiorentino* over the course of sixteenth-century editions, the introduction always emphasized the importance of a standardized volume for apothecaries.[11] By introducing a new, standard reference book of medical recipes that relied on canonical authorities but was written in the local Tuscan dialect, the authors implicitly insisted upon the careful integration of apothecaries—who were not required to know Latin—into their official medical hierarchy.[12]

Followed by the preface and a table of contents, the 1498 *Nuovo Receptario Fiorentino* began with several "doctrines," sections of varying length with recommendations of what apothecaries needed to know in order to practice their craft. The first described the physical space of the apothecary shop and insisted that it be "free from wind, dust, humidity, and smoke."[13] The second listed the necessary books to keep on hand, including those of authors like Avicenna and Mesue, as well as specific books like the *Almansore* of Rhazes and the "antidotary of Niccholao" (the *Antidotarium Nicolai*), many of which continued to be standard pharmaceutical reference texts well into the seventeenth century.[14] The series of doctrines were followed by information about collecting plants in different months, descriptions of ingredients and how to conserve them, and compounding instructions for diverse medicines.

Additions to the *Ricettario Fiorentino* between the 1498 and 1567 editions reflected the professionalization of the College and supported its claim that its members were learned and superior to other practitioners in the medical marketplace. The entry on aloe, for example, was a mere forty-two words in 1498 but by 1567 the entry had grown to over two hundred words and included additional information about where the plant was found and its characteristics according to Dioscorides.[15] The increasing professionalization of the College is evident in this expansion: by referencing different authors and making specific recommendations about the origins and characteristics of pharmaceutical ingredients, the compilers asserted their knowledge and authority over the multitude of choices available to apothecaries through an ever-increasing (and often ever-suspicious) global trade network.[16] Many entries, for example, specifically identified from where the ingredient should be sourced, and some warned against false versions. Adulteration was a concern in this era, as was the confusion about plant identification and naming practices.

Although the authors of the *Ricettario* were mostly traditionalists in their inclusion of canonical doctors' recipes, they included three important American ingredients in the third (1567) edition. They asserted that the celebrated sixteenth-century cure for syphilis, "Legno Guaiaco" (guaiac) came from the "west Indies" and that the guaiac that came from Isla Beata, an island off the southern coast of the Dominican Republic, was the most renowned. The authors noted that readers should know that the bark could be adulterated with that of "the ash tree, or of mulberry, or other similar trees: you can identify these from their taste and smell."[17] The second ingredient was "Balsamo Occidentale." Although Balsam of Peru was well known in Europe, the *Ricettario* simply identifies the West Indies as the origin of this ingredient. The last American ingredient, sarsaparilla, was also to be found in the West Indies, specifically in "Isola della punta del mare del Sur."[18] The majority of the ingredients in the *Ricettario Fiorentino* could be found growing around Florence, further afield in Italy, or in apothecary shops having been imported from places like Crete, India, and the Middle East. Since Italy was not a major player in New World explorations, it is perhaps not surprising that these late fifteenth- and early sixteenth-century versions of the *Ricettario Fiorentino* did not include more American ingredients.[19]

According to Sali Morgenstern, the 1567 edition of the *Ricettario* was "thoroughly corrected and expurgated," and was "no longer the work of a more or less independent body of physicians and apothecaries," since they were to be selected by the duke rather than elected by the members of the guild.[20] The development of legal authority with the backing of the duke likely encouraged the *Ricettario*'s expansion into a more authoritative, comprehensive volume that would act not simply as a reference book but also as a legally binding text with well-researched, trustworthy information of all sorts.[21] The authors of the *Ricettario* made substantial changes to the volume between the first (1498) and third (1567) editions. As the book took on a new role in the medico-legal landscape of sixteenth-century Florence, the authors expanded the book as a whole and began to shift their interests to increasingly diverse source material, although they still primarily focused on canonical rather than more local authorities, such as highly regarded physicians from Florence or the surrounding region.[22]

PROFESSIONAL PHYSICIANS, LAY HEALERS, AND CITATION PRACTICES

Apothecaries were not the only individuals making medicines, of course. Unlicensed individuals continued to make their own remedies as they always had.[23] Manuscript medical recipe books are often some of the only remaining sources that provide a window into lay medical practice in the early modern period. These handwritten notebooks make it clear that the ingredients and instructions for drug compounding to which nonprofessional practitioners ascribed were not so different from the recipes provided in the *Ricettario Fiorentino*. Manuscript medical recipe books provide a link between the prescribed knowledge of Florentine medical authority and the day-to-day practice of medicine among unlicensed healers. The comparison between manuscript recipe books and the *Ricettario Fiorentino* offers evidence of the convergences in theory and practice and divergences in authority between professional and lay healers in the Florentine sixteenth-century medical marketplace.

The similarities and differences in the genres of early printed pharmacopoeias and manuscript recipe books are significant and demonstrative of the deviations in knowledge practices in medicine in the early modern period. The Tuscan manuscript recipe books from the Biblioteca Riccardiana in Florence that will serve as counterpoints to the printed, official *Ricettario Fiorentino* editions were written in the same approximate time period that the College of Physicians created its official recipe book.[24] Riccardiana MS 3059 and MS 3049 are both fifteenth-century manuscripts. Part of MS 3059 is dated July 20, 1460, but there are no dates in MS 3049, although the hand is fifteenth century. Riccardiana MS 3057, 3044, and 2376 are all primarily sixteenth century.[25]

The construction of individual recipes and organizational tools is indicative of the different kinds of readers that print and manuscript recipe book authors expected. The titles of recipes, for example, are particularly telling. The *Ricettario*

Fiorentino rarely included information in recipe titles about the use of a recipe and, in general, there are no recommendations for how recipes or ingredients should be used by patients. More commonly, the titles are specific to each entry, such as "An Imperial Purgative According to Niccolao."[26] The recipes in the manuscripts, however, often have descriptive, symptom- or illness-specific titles, occasionally with citations. For example, the title "Cerotto da Crepati" in one manuscript describes the recipe as a plaster for cracks.[27] Authors of manuscript recipe books expected readers to approach the volumes with particular illnesses or complaints in mind. The *Ricettario*, on the other hand, was accessible to many readers because it was written in Tuscan, but its authors maintained significant professional distance between their expertise and the lack of training of other practitioners. Many medicines could be adjusted "according to the fantasy of the doctor," but the authors of the *Ricettario* worried about apothecaries who would adjust medicines on their own without adequate knowledge and the direction of a physician, leading to "endless scandal." Therefore, they wrote: "In this, our present recipe book, we have not indicated anything about what each [recipe] is appropriate for, because we hope that he who has [this book already] knows how to use it, and that he who does not know, will learn, and will learn how to use it canonically."[28]

Aside from titles, the content of the manuscript recipes reinforces the idea of a less-academic manuscript reader, since recipes in manuscripts more often than not included compounding instructions, but the *Ricettario* often separated compounding instructions into different, general sections. Neither the manuscripts nor the *Ricettario* regularly include information about dosage or for whom a recipe was best suited. This may be explained in part for the *Ricettario* by the simple fact that apothecaries were meant only to fill prescriptions from physicians who would have, presumably, already communicated that information to their patients.

Provenance notations, the part of the recipe entry that described the origin of the recipe, were included in many, though not all, recipe entries and are important details because they represent the bidirectionality of the strategic uses of authority in medical writing. Provenance notations like the names of specific authors and geographic details reflect both the author's mastery of particular theories as well as his or her (or their) membership in certain medical communities. In their study of seventeenth-century English manuscript recipe books, Elaine Leong and Sara Pennell found that "over one-third of recipes came with the name of a donor or 'author,'" many of which were family members or personal acquaintances.[29] They categorized recipes and other medical information as a type of currency, commodities "which flowed between people, and the authority and reliability of which was inflected by the circumstances of that movement."

Leong and Pennell argued that citation practices in the process of recipe book compilations could reinforce the authority of individual recipe entries and allow compilers to be legitimate participants in the medical marketplace.[30]

Provenance notations in the Italian recipe books examined here, however, suggest that Italian manuscript medical communities were somewhat different from their English counterparts. Rather than serving as receipts of social interaction, Italian manuscript authors largely skipped over references to family members and personal relationships, citing other authorities in the process of creating a social and "professional" identity via expressed mastery of particular kinds of knowledge. Importantly, only one recipe author across the manuscripts examined here contains a reference to a family affiliation: "Benedetto, my uncle," in MS 2376.[31] Provenance notations were included alongside recipes in both manuscript and print recipe books, notably in the *Ricettario Fiorentino*. Like print authors, manuscript compilers made decisions about whose recipes to include in their volumes based on efficacy, trust, and authority.

As already mentioned, the *Ricettario Fiorentino* was greatly expanded between the first and third editions of 1498 and 1567. In addition to the more detailed ingredient entries, over one hundred new recipe entries were included in the 1567 *Ricettario*. Although the number of recipe author references remained essentially the same regardless of the additional entries, there was a significant shift in which authors were included in the volumes.[32] Numerous first edition authors are not present in the third edition (55.5%), and 84.2% of third edition authors are not present in the first edition.[33] Importantly, the majority of referenced authors only appear once in each volume. The change in representation of the few authors who actually make up the majority of references is most noteworthy, however. Table 2.1 provides a breakdown of the recipe authors with multiple references in the 1498 and 1567 editions.

Recent histories of Italian medicine have characterized the medical marketplace as one that was extremely diverse, with all sorts of practitioners, from physicians to charlatans, vying for patients.[34] Those who were chosen to contribute medical recipes to print and manuscript volumes represent the theoretical bases and individual practitioners in whom the most trust was placed. Even with the differences in the distribution of recipe author references in the two editions of the *Ricettario*, it is key not to lose sight of the fact that the authors with multiple references listed in table 2.1 represent over 90% of all of the recipe author references in the volumes. The major meaningful trend that appears between the first and third editions is the increasing preference for classical and local authorities over the medieval Arabic authorities whose work had made up the European medical canon for centuries.[35] For example, in 1498, Mesue, said to be a highly influential physician who lived and worked in Baghdad in the ninth century,

TABLE 2.1. Recipe Authors with Multiple References in the 1498 and 1567 Editions of the *Ricettario Fiorentino*

AUTHOR	NUMBER OF CITATIONS IN 1498 EDITION	NUMBER OF CITATIONS IN 1567 EDITION
Mesue	274	183
Galen	16	90
Niccholao	73	31
Avicenna	21	12
Niccolao Alessandrino	0	18
Rhazes/Almansore	17	7
Paulo	0	13
Andromaco	3	10
Gentile da Fuligno	8	2
Alexandro / Alessandro	6	3
Guglielmo Piacentino	5	0
Giovanni da Vico	0	5
Serapione	3	0
Aezio	0	4
Damocrate	0	4
Carpi	0	3
Azaranio Albuchasi	2	0
Christofano di Giorgio	2	0
Figliuolo di Zaccheria	2	0
Niccholo Falucci	2	0
Asclepiade	0	2
Lucio Cathagate	0	2
Montagna	0	2

accounted for almost 60% of recipe references.[36] In 1567, however, his contributions had dropped to 44.6%. On the other hand, Galen, the ancient Greek physician whose system of humors was fundamental to early modern medical theory, rose from 3.5% in 1498 to 22% in 1567.

These changes in focus in the *Ricettario* editions are reflective of medical opinion more generally in much of Europe in the mid- and late sixteenth-century. For example, the German natural philosopher Leonhart Fuchs asserted his mistrust of medieval Arabic authorities in his several botanical studies. Since the age of Dioscorides, plant names had become so confused that natural philosophers and physicians alike were increasingly concerned about whether patients actually received the medicines they were prescribed. Issues of naming were not resolved for many years despite the efforts of many authors and their illustrators. *De Historia Stirpium*, Fuchs's landmark 1542 herbal, was largely based on the opinions of Dioscorides, Pliny, and Galen, a deliberate choice to move botanical medical knowledge away from Arabic writers toward classical authorities.[37] More broadly, the transition within the *Ricettario* reflects the Renaissance trend of rejecting Arabic translations and returning to classical sources. With this analysis, the *Ricettario* demonstrates the College's responsiveness to changes within the local medical community, making the volume uniquely Florentine and distinct from other areas in Europe. Paula De Vos, for example, has shown the opposite shift away from classical sources and toward Arabic authorities among Spanish pharmaceutical texts in the same period (chapter 1).

As the Florentine College of Physicians chose a new intellectual lineage for themselves in the early sixteenth century, lay medical practitioners may not have experienced a similar drive to assert their connections to classical authorities. Although making a comparison between manuscript and printed recipe books is difficult, the exercise reveals the similarities and differences in practice and scholarly positioning of diverse groups of people. The comparison shows how lay healers diverged from official physicians by focusing more attention on local connections rather than ascribing to humanistic trends.

As mentioned above, these individual manuscripts were written by different people and were likely produced in Tuscany around the same period as the first and third editions of the *Ricettario Fiorentino*. Although they are anonymous, as a group they can function as a window into lay medical practice at the same time the College produced the *Ricettario Fiorentino*. MS 3044 has no recipe citations, MS 3049 has 5, MS 3059 has 11, MS 3057 has 31, and MS 2376 has 189 (table 2.2). Of the over one hundred recipe authors mentioned in the three volumes of the *Ricettario Fiorentino* and Riccardiana manuscripts, there are only eleven authors who appear both in a printed and manuscript volume: Mesue, Galen, Niccholao, Rhazes/Almansore, Andromaco, Gentile da Fuligno, Serapione, Antonio della Scarperia, Marsilio di Sancta Sophia, Maestro Bonino, and Dino di Firenze. It should not be surprising that these authors together account for 70% of the recipe author references across all volumes—both manuscript and print—consulted in this chapter. More significantly, this percentage breaks down to only 11% of the manuscript recipe author references and 81% of the *Ricettario Fiorentino* recipe author references.[38] Put plainly, although there was some overlap, manuscript authors and the College of Physicians generally turned to different author-

TABLE 2.2. Recipe Authors with Multiple References in Manuscript Volumes

MS RICC. 2376		MS RICC. 3049		MS RICC. 3057		MS RICC. 3059	
AUTHOR	No.	AUTHOR	No.	AUTHOR	No.	AUTHOR	No.
A.M.S. in S[anta]. M[aria]. N[ovella].	45	Maestro Anselmo da Genova	2	Niccholao	4	Galen	2
Piu persone	23			Rhazes / Almansore	3	Maestro Bonino	2
Ben[edet]to mio Zio	21			Dino / Dyno [del Garbo] di Firenze	3		
m.o Fabio Moriconi d'Amelia	19			Mesue	2		
Giuliano Cerusio in Pisa	18						
Marcho C.M. Agostino Dottori in S[an]ta M[ari]a N[ovell]a	15						
Maggio Barranti da Monte Varchi Medico Fisco e Cerus[ic]o	10						
S.A.S.	4						
M.o Maggio in S[an].ta M.[ri]a N.[ovell]a	3						
m. Tommaso ligniani, cerusio a' Citta di Castello; Mattiolo; Troncone	2						

ities when constructing their pharmaceutical reference books: where the College focused first on Arabic authorities and then on classical ones, the lay authors of these manuscripts included references to more local individuals.

The inclusion of geographic locations in some provenance notations in the manuscript recipe books allows us to map parts of the authors' knowledge networks. It is likely, for example, that the abbreviations "S.M.N." and "S.ta M.a N.a" in MS 2376 refer to Santa Maria Nuova (today, Novella), one of the main cathedrals in Florence, located today near the main train station.[39] Santa Maria Novella has one of the longest operating church pharmacies in the world. Santa Maria Novella also operated a large hospital in the medieval and early modern period, although today it produces primarily luxury bath and body products like lotions and perfumes instead of medicines. Since the author included sixty-four recipes with references to the church, it was evidently an important center of medical knowledge for this author. This author also included recipes from surgeons in Pisa and Città di Castello and the ambassador from Lucca (all cities in central Italy).[40] Even though manuscript authors also referred to some of the same canonical authorities that the College of Physicians did, these geographic notes demonstrate that the knowledge networks of manuscript authors could also be relatively local. Unlike the authors of the *Ricettario Fiorentino*, who rarely included information about where their recipe authors were from, including indications of geographic proximity seems to have been a strategy of manuscript recipe book authors for demonstrating knowledge of and connections in the local medical marketplace. Although the percentage of individual recipe authors with associated geographic locations is similar in manuscript and printed volumes, significantly more individual recipes have authors with geographic information in manuscripts than in printed volumes (tables 2.1 and 2.2).

Rather than include recipe authors associated with a variety of places to demonstrate their medical network, the authors of the *Ricettario Fiorentino* wrote in the first person to emphasize their ownership over medical practice in Florence itself. For example, the authors frequently noted which medicines were or were not made "here." In the 1498 edition, "Lactovaro di Re," a recipe by Mesue, for example, was "not used."[41] Because the phrasing of the titles is vague, it is unclear whether the authors of the volumes were warning their readers against these recipes or whether they were simply uncommonly prescribed but that apothecaries should know how to make them just in case. Considering the general importance of the authors of unused recipes within the *Ricettario*, it is possible that the authors simply felt they needed to include the recipes because they were part of the canon. It is also possible that the authors themselves did not recommend the recipes but knew that some practitioners preferred them. The recipe for "Magisterial Plaster for the Spleen Sickness" shows that authors at least occasionally included recipes simply because many other people used them and thought they were useful: "We do not use this [remedy]; but, because others use this [recipe] and the others [written above], and [their users] are notable, we put this [recipe] here [anyway]."[42]

Two recipes in the 1498 edition provide additional insight to the College's rationale for including recipes that they did not actually recommend using and suggest that mid-sixteenth-century humanism that drove the official medical community to begin rejecting Arabic sources was apparent already in this first edition of the *Ricettario*. The recipe for "Dyarodo" includes a long justification for not using the recipe and is followed by an alternate suggestion: "We have found this electuary described by Nicholao and Mesue. So, we put forth [here] the one that Mesue described in his antidotary that we do not use. Although we do not use this electuary, because Gentile da Fuligno used it and since [he found that] it was very useful, we put it here. We included the recipe we use below, but we do not use this one."[43] The following entry, "Dyradon Abatis," begins, "This is the electuary of Niccholao that is commonly used."[44] From this case, it seems likely that many of the recipes that were not in use were included because they were authored by canonical authors (like Mesue) and/or the compounding apothecary needed to be able to make them because they were still used by some practitioners, even though the official authority did not find them as good for one reason or another. By the 1567 edition, these recipes that were "not used" had been deleted from the volume, a decision likely related to the fact that the College had gained legal control over apothecaries and did not see a need to continue to include recipes that they did not recommend be in the repertoire of apothecaries.

In the 1567 edition, the *Ricettario Fiorentino*'s authors continued to reference Florentine medical practice in order to assert their control over the marketplace by asserting that some recipes were made in particular ways in Florence. For example, the recipe for "Honeysuckle Unguent according to Carpi" says to "grind the honeysuckle (we use those with yellow flowers)."[45] Similarly, the recipes for "Apostles Unguent according to Avicenna" and "Diapipereos according to Galeno" say specifically that "we" make the recipe "in the following way."[46] Another instance of local preferences is in the recipe "Diamoron made by Galeno," in which the authors say, "Although Galen uses apples in his [recipe for] Diamoron, we use loaf sugar."[47] The *Ricettario*'s recipe for theriac also includes the regional preference for "galangal, which we choose in place of Jerusalem rose."[48]

Examining manuscript recipe books and the printed *Ricettario Fiorentino* editions side-by-side illustrates the differences in the ways that both groups included information about place to reinforce their goals. Where the manuscript authors referenced medical authorities linked to particular locations to show their participation in a knowledge network, the authors of the *Ricettario Fiorentino* discussed Florentine pharmaceutical preferences to emphasize their authority claims over the marketplace. Since the College used first person and local preferences to assert control, they could continue reaching further afield for pharmaceutical knowledge, either to authorities across Europe, North Africa, and the Middle East or to the new pharmaceutical ingredients entering Europe from New World expeditions.

COMMON RECIPES ACROSS SOURCE TYPES

Delving deeper into both manuscript and print, the recipes themselves become powerful points of comparison. Recalling that there were only a few authors in common between our manuscript and printed volumes, it should be unsurprising that the manuscripts and the *Ricettario Fiorentino* only have a few recipes in common. The fact that any recipes are the same, however, is important because it is a signal of a pharmaceutical knowledge base that was at least somewhat overlapping. Three recipes (tables 2.3, 2.4, and 2.5) from MS 3057 serve as our examples here.

Because recipes as a discourse type are highly variable according to personal preference (reordering ingredients, using local names or different spellings, etc.), it is difficult to attribute too much significance to similarities between disparate sources. In the following examples, the ingredients and ingredient amounts are similar enough to make it clear that the recipes were essentially the same between manuscript and print even though they are not exactly the same. Rather than being copied directly from the *Ricettario* or from manuscript into the *Ricettario*, it is likely that these recipes circulated in various other forms, both print and manuscript, and came to each author via diverse channels. Considering that the authors of these recipes are Mesue, Niccolao, and della Scarperia, all authors whose recipes circulated among medical practitioners in a variety of ways, it is not surprising that there might have been some degree of recipe standardization.

Recipes for "Pills of Rhubarb according to Mesue" appear in MS 3057 and in both editions of the *Ricettario Fiorentino* (table 2.3). What is interesting from our standpoint of considering knowledge transfer is the order in which the ingredients appear and the similarity of language across recipes. It seems very likely, for example, that the author of the manuscript recipe had a similar source as the authors of both *Ricettario* volumes. In the recipe for the pills of Scarperia, the similarities are striking and the differences seem to stem from individual authorial preference rather than a true deviation from a common source (table 2.4). Similarly, the major difference between the manuscript and print version of the "Diacimino di Nicolao" is the lack of compounding instructions in the manuscript version (table 2.5).

Because there is no information about the manuscript authors, it is difficult to speculate as to why MS 3057 has more recipes in common with the *Ricettario* than the other manuscripts studied here. This author evidently had access to other printed or manuscript medical sources, since the beginning of the manuscript is taken up with a copy of Antonio della Scarperia's treatise on fevers. While not exactly the same as the *Ricettario* versions, the ingredients are very similar and likely reflect a common recipe source that was modified through

TABLE 2.3. Comparison of Recipes for "Pillole di Reubarbero secondo Mesue"

Source	MS 3057 (16th century)	Ricettario Fiorentino (1498)	Ricettario Fiorentino (1567)
Title of Recipe	Pillole di Reubarbero secondo Mesue	Pillole di Reubarbero di Mesue et Usansi	Pillole di rhabarbaro di Mesue
Recipe	Rx. Raued. 3.iii. Sugo di regolita, Sugo di assentio, Mastice an. 3.1, Mirabolani citrine. 3.iii ½. Seme di appio, Seme di finochio. an. 3 ½. Torcisci diarodon 3.iii ½. Gliempigra 3x. & si faccino pillole.	Recipe reubarbero fine dr. iii, sugho di regolita, sugho di assentio, ana dr. I; mastice dr. I, mirabolani citrine dr. ii ½; seme di appio, seme di finocchio, ana dr. ½; trocisci diarodon dr iii ½; spetie di yerapigra semplice on. 1 dr ii. Con acqua di finocchio fa' pillole.	Pillole di rhabarbaro di Mesue. Rx. Rhabarbaro fine 3.iii. Sugo di logorizia d'assenzio, Mastice an. 3.1. Mirabolani citrine 3.iii ½. Seme d'Appio di Finocchio an 3 ½. Trocisci diarhodon 3.iii ½. Hiera picra 3.x. Acqua di Finocchio q.b. Fa pillole.

TABLE 2.4. Comparison of Recipes for "Diacimino di Nicolao"

Source	MS 3057 (16th century)	Ricettario Fiorentino (1597)
Title of Recipe	Diacimino di Nicolao	Diacymino di Niccolao Alessand.
Recipe	Rx. Cinamomo, Gherofani, an. 3.ii ½ Zinziberi et Melano piperi, an. 3.ii g. Galange, cimbre, calomento, an. 3.1. e ii Ameol, Leuistico, an. 3.1.g xviii Macro piperi 3.1. Nardo, Noce moscada, an. e. ii ½ & in altro e 1 ½ Mele che basti	Rx. Cymino tenuto infuse per un dii Avanti nell'aceto, & rasciutto 3. viii.e. i Cinnamomo, Gherofani, an. 3. ii ½ Gengiovo, Pepe nero, an. 3.ii ½ Galanga, Santoreggia, Calamento, an. 3.1 e. ii. Ammi, Leuitico, an. 3.i.g.xviii Pepe lungo Nardo indica, Cardamomo, Noci moscade, an. e. ii ½ Pesta ogni cosa, & fa spezie & componsi con Mele stiumato quanto basta, Et usasi hoggi farne in piaster con zucchero, mettendo once una e mezzo di spezie per lib. di zucchero.

TABLE 2.5. Comparison of Recipes for "Pillole di Maestro Antonio dalla Scarperia"

Source	Ms. 3057 (16th century)	Ricettario Fiorentino (1498)	Ricettario Fiorentino (1567 and 1597)
Title of Recipe	Pillole di Maestro Antonio dalla Scarperia	Pillole di Maestro Antonio della Scharperia sono simile alle sopra scripte et usansi	Pillole di Hiera con agarico di maestro Antonio da Scarperia.
Recipe	Rx. Cenamomo, Zafferano, Spigaceltica, Squinanti, Bachera, Cassia lignea, Spigo nardi, Xilobalsamo, Carpobalsimo, An. 3. 1 Viuole, Epittimo, Rose rosse, Coloquintida, Mastice, an. 3.[?] Argario, Turbiti, an. 3.ii Reubarbero, Scamonea, corretta, an. 3.iii Aloe epatico, 3.xv. Fa pillole triata [?] vino bianco.	Recipe cennamomo, croco, spignonardi, squinanti, bacchera, mastice, cassia lignea, xilobalsamo, carpobalsamo, viole, epithamo, rose rosse, coloquintida interior, ana dr. 1; agario fine, turbitti fini; ana dr. ii; reubarbaro fine, diagridii, ana dr. iii; aloe pathico dr. xvi. Componi con trebbiano et triacha.	Rx. Aloè epatico, Cinanamomo, Nardo indica, Zafferano, Schinantho, Bacchera, Mastice, Casia, Silobalsamo, Carpobalsamo, Viole, Rose rosse, Epithimo, Coloquinthida, an. 3.1; Agarico eletto, Turbith fini, an. 3.ii; Rhabarbaro 3.iii Scamonea preparata, 3.iiii Theriaca onc.ii Trebbiano quanto basta. Fa pillole

various transmissions as well as each author's preference. The lack of similarity between the *Ricettario* and a manuscript is not an indicator of a lack of interaction in the marketplace, however. For example, the author of MS 2376 organized his or her manuscript in sections of recipes given by individual people. For example, "Giuliano Cerusico in Pisa" gave eighteen recipes and "Maggio Barranti da mote varchi" gave ten.[49]

CONCLUSIONS

The variations in cited authors, as well as in the representations of local culture in each volume, reflect the different goals of manuscript authors and the authors of the *Ricettario Fiorentino*. Scholarship on the *Ricettario Fiorentino*, and on the history of pharmacy and natural history in general, has tended to relegate lay practice to a different and often unknowable location when compared with official practice. Although, as Cristina Bellorini puts it, "the multifarious world of empirics, quacks, and wise women ... would not provide an easy comparison with an official pharmacopoeia," it is important to examine these practices next to each other in order to determine several key concepts.[50] The comparison of these Riccardiana manuscripts and the *Ricettario Fiorentino* volumes emphasizes that lay practitioners operated in different terms than official healers, but that both their realms of practice and their intellectual contexts were overlapping. The idea of the transfer of medical knowledge as a unidirectional stream, or even a bidirectional stream, is not particularly useful when we are confronted with documents that highlight the nebulous nature of these historiographical categories.

When applied to manuscript recipe books, the concept of "epistemic genres," or genres of cognition, becomes a powerful way of explaining how recipe book types functioned within and across social groups. Gianna Pomata argues, "As shared textual conventions, genres are intrinsically social. Contributing to a genre means consciously joining a community."[51] As an epistemic genre, manuscript recipe books were firmly situated within both writer and reader cultures that had clear format conventions that allowed all parties to recognize and take advantage of diverse knowledge. The similarities between manuscript recipe books and printed pharmacopoeias, thus highlight the uniformity of writer and reader conventions across textual formats and genres. This resemblance is important considering the diversity of knowledge bases and social situations across the range of literate medical practitioners who read and wrote and made recipes in the sixteenth century.

Although the five Riccardiana manuscripts under investigation here are anonymous, their authors were not necessarily very different from the people writing or using the *Ricettario Fiorentino*. Literacy was valued for both genders in a variety of social statuses, so anyone from a priest to a merchant might have written the volumes. Official practitioners composed many of the extant manuscript recipe books written in Italian, so it is clear that some healers used official texts and also maintained their own stores of medical knowledge in separate notebooks. At the same time, there is no evidence that the authors of these manuscripts were official practitioners. Although the manuscripts examined here show a noticeable preference for local authors, common authorities exist across document types. And regardless of official interests in new kinds of medicine, the *Ricettario* was a document that was very slow to reflect the manuscript preference for non-canonical authors and non-Arabic authorities.[52]

As a product of the College of Physicians, the *Ricettario Fiorentino* was explicitly outward facing: the textual evidence of a body of practitioners attempting to craft a public image that would be authoritative and encourage (and later, require) followers because of a new and growing understanding of the importance of municipal control over health. Citations in the *Ricettario Fiorentino* editions emphasize that the College's medieval focus on Arabic authorities began to wane by the mid-sixteenth century and were replaced by classical and Italian authors. Although they maintained attention toward official, approved authors, they adjusted the recipes to reflect their local medical customs. The end goal was to document all approved, official medical customs in the region in an effort to standardize practice for the health of patients and status of practitioners.

Manuscript recipe books are, on the other hand, textual evidence of individual practitioners deciding whose advice they trusted and whose name would reflect the most authority and authenticity back on the manuscript author. These manuscripts are notably recalcitrant: their stories are hidden between lines of text that do not overtly say very much at all. However, the decisions that manuscript authors made about whose advice to include in their books allude to communities that had different geographic and temporal foci than official practitioners.

The provenance records in both manuscript and print serve the same essential purpose: to be a record of knowledge origin and of the writers' interactions with members of his or her knowledge community. Rather than using canonical knowledge, manuscript authors used their recipe books as repositories for evidence of interactions. Although lay practice is still not easily defined, the trope of the unknowable lay practitioner is one that needs reevaluation. The practices of laypeople can be placed in comparison with professional recommendations. The differences within recipe genres that become apparent upon comparison are important examples to help further historical studies of the interactions within and across knowledge communities. Accepting that manuscript recipe books were both public and private documents and evidence of practice allows us to recognize that their authors crafted them in a calculated manner not so dissimilar from the College of Physicians. Provenance notations, evidence of local interactions, and subtle differences in textual organization (such as in titles and compounding instructions), expose the confluences and divergences of manuscript and print medical communities in sixteenth-century Florence.

3

AN IMPERIAL PHARMACOPOEIA?

The *Pharmacopoeia Matritensis* and *Materia Medica* in the Eighteenth-Century Spanish Atlantic World

MATTHEW JAMES CRAWFORD

IN 1739, a new pharmacopoeia was published in Europe under the title *Pharmacopoeia Matritensis*. It was just one among many in a crowded field of pharmacopoeias published by royal governments, cities, and colleges of physicians throughout eighteenth-century Europe, as described by Stuart Anderson (chapter 10) in this volume.[1] This new pharmacopoeia was produced by the Royal Protomedicato, the body responsible for regulating the medical professions in Spain, and was promulgated under the auspices of the Spanish Crown. A copy of the "Decree of the Royal Protomedicato," which accompanied the pharmacopoeia, emphasized the "great utility to Public Health that would come from having a fixed and regular method for making the Medicaments that are used in these lands for the curing of diseases."[2] The decree continued with an exhortation to officials in "all the Provinces and subject Kingdoms" that "all Pharmacists in their respective districts should have a copy of the book entitled *Pharmacopoeia Matritensis* and should follow all the rules and methods for the production of Medicaments that are in this [book]."[3] Such language indicated that this new pharmacopoeia represented the Crown's official policy regarding the preparation of medicaments and was enforceable throughout Spain's territories, not just in

Spain but also in the Americas. Finally, the decree explained that pharmacists who failed to adhere to the "rules and methods" listed in the *Pharmacopoeia Matritensis* would be fined "200 ducats" and lose their license to practice pharmacy.[4]

As an index of the medicinal substances and the techniques for preparing medicaments, the *Pharmacopoeia Matritensis* offers some insight into the collection, certification, and circulation of healing knowledge in an imperial context in the early modern Atlantic World. The Madrid pharmacopoeia is especially interesting because it was published during the era of the Bourbon Reforms, a period in the history of the Spanish Atlantic in which the Crown and its advisors sought to revitalize Spain's ailing imperial enterprise through a variety of reforms aimed, in part, at consolidating royal power.[5] As a result, this pharmacopoeia is useful for exploring the medical and epistemological dimensions of Spanish imperial reform under the Bourbons, who ascended to the Spanish throne in the eighteenth century.

Although pharmacopoeias have received scant attention as primary sources outside of the history of pharmacy, such an approach to the *Pharmacopoeia Matritensis* is not without precedent. Indeed, in an article published several decades ago, George Urdang, then director of the Institute for the History of Pharmacy at the University of Wisconsin–Madison, emphasized the ways in which pharmacopoeias might illuminate broader historical developments beyond pharmacy and medicine.[6] He characterized pharmacopoeias as "witnesses of world history" and urged historians to consider the ways in which the development of pharmacopoeias reflected key historical developments. Urdang was especially interested in the ways in which pharmacopoeias "became gradually a matter of national ambition, a part and a proof of national sovereignty and unity."[7] His main reason for exploring these texts as barometers of the growth of "national sovereignty" and "nationalist ideology" was to provide a parallel to his own time in the mid-twentieth century—a period that he described as "a new period in world history" in which "greater unification" was the trend.[8] He predicted that the promulgation of a universal pharmacopoeia in the later decades of the twentieth century would emerge out of what he saw as a trend toward global unification in twentieth-century politics and culture, just as earlier pharmacopoeias had reflected the rise of the nation-state in early modern Europe.

At first glance, Urdang's characterization of pharmacopoeias as "witnesses of world history" seems prescient of the recent global turn in the history of science and medicine.[9] Yet, in spite of the grander aspirations suggested by this phrase, Urdang's article only examined the case of the emergence and development of pharmacopoeias in Europe starting with the *Nuovo Receptario* issued by the College of Physicians in Florence in 1498, an episode discussed by Emily Beck in her essay (chapter 2). While taking its cue from Urdang's suggestion of thinking about pharmacopoeias as historical sources emerging out of the intersection of ways of knowing and ways of governing, my essay seeks to reconceptualize the relationship between pharmacopoeias and the contexts out of which they emerged. Whereas Urdang chose to characterize pharmacopoeias as "witnesses of world history," we should recognize that pharmacopoeias were not just passive mirrors of the worlds they inhabited but also played an active role in constituting the different social, cultural, and natural worlds in which they were embedded.[10] In addition, whereas Urdang's analysis focused on the relationship between pharmacopoeias and the nation-state, it is worth exploring the relationship between pharmacopoeias and a different kind of political unit: empire.[11]

This essay argues that the history of the first official pharmacopoeia of the Spanish Empire is best understood as part of a larger process of the globalization of healing knowledge and the hybridization of Indigenous and European healing knowledges in the Atlantic World. While the Crown and its advisors envisioned the *Pharmacopoeia Matritensis* as a tool for asserting royal sovereignty and making pharmaceutical therapeutics uniform throughout the empire, a closer reading shows that the pharmacopoeia's role as a tool of empire is ambivalent at best. Whereas much existing scholarship has focused on the cross-cultural exchange of medical and natural knowledge that took place in the Americas, this essay explores the extent to which echoes of these processes reverberated in the pages of an official pharmacopoeia printed in early eighteenth-century Europe.[12]

THE GENRE OF PHARMACOPOEIAS IN EARLY MODERN EUROPE

Although they varied quite a bit in content, form, and structure, pharmacopoeias became an identifiable genre of medical text in early modern Europe with certain shared characteristics.[13] The primary function of pharmacopoeias was to list the plant, animal, and mineral substances used medicinally and to provide the recipes for preparing these materials for therapeutic use. These substances were known collectively as materia medica and some of them were also "simples," a term for substances that could be used therapeutically without any additional preparation or manipulation. Of course, pharmacopoeias were not the only type of text to include this kind of information. As noted by Emily Beck (chapter 2), the genre of pharmacopoeias shared characteristics with many other kinds of texts such as natural histories, dispensatories, and recipe books.

Pharmacopoeias varied in terms of how much information they provided on individual materia medica. Some gave a description of the substance, the botanical, animal, or mineral source of the substance, its geographical origins, and sometimes its medicinal uses as in the case of the *Pharmacopoeia Augustana*.[14] Some just listed the substances by their name, as in the case of early editions of the *Pharmacopoeia Londinensis* and the *Codex Medicamentarius* of Paris.[15] The other common sections of pharmacopoeias described the techniques for preparing plant, animal, and mineral substances for therapeutic use including recipes for compound medicines in which several individual substances were mixed together. As described by Paula De Vos in this volume (chapter 1), these components of early modern pharmacopoeias in Europe developed from several genres of ancient and medieval medical writing.

Many, but not all, pharmacopoeias declared their allegiance to a particular theory or natural philosophy of medicine. In Europe in the early eighteenth century, there were two main approaches to pharmacy and the preparation of medicaments—"galenical pharmacy," which followed the methods first articulated by ancient Greek and Roman physicians and pharmacists, and "chemical pharmacy," which followed the methods of sixteenth-century physician Paracelusus and his followers.[16] "Galenical"—or more commonly "Galenic"— pharmacy was based primarily on a pharmaceutical treatise by Roman physician Galen that emphasized the use of botanical substances as medicaments with minimal processing of these substances into oils, syrups, and ointments. "Chemical pharmacy" emphasized the use of mineral or chemical substances, such as mercury, and the processing of these materials through more intensive chemical techniques, such as distillation.[17] Some pharmacopoeias specialized in either Galenic or chemical pharmacy. Other pharmacopoeias included medicaments and methods from both types of pharmacy, as in the case of the *Pharmacopoée royale galenique et chemique* (1676), by Moyse Charas (1619–1698), and the *Palestra Pharmaceutica Chymico-Galenica* (1706), by Félix Palacios (1677–1737).[18]

One way to make sense of the variety of pharmacopoeias produced in early modern Europe is to focus on authorship and patronage. Some pharmacopoeias were the product of an individual physician, pharmacist, or natural philosopher who developed their own compendia of effective remedies, such as *Pharmacopoeia regia*, by Johann Zwelfer (1618–1668).[19] Examples of these kinds of pharmacopoeias include those that were advertised as a collection of remedies by a specific physician, but published after that physician had died, such as in the case of the *Pharmacopoeia Bateana*, which was attributed to physician George Bate (1608–1669) even though the first edition did not appear until 1691.[20]

Other pharmacopoeias were the product of a corporate entity, such as a college of physicians or society of apothecaries, with patronage from the state. These pharmacopoeias were often associated with a particular city and some received patronage from municipalities, as in the case of the first pharmacopoeia, *Ricettario Fiorentino* of Florence, as discussed by Emily Beck in this volume (chapter 2), or the *Pharmacopoeia Augustana* of Augsburg. Others were patronized by the royalty as in the case of the *Pharmacopoeia Londinensis*, as discussed by Stuart Anderson in this volume (chapter 10), or by a national government, as discussed by Antoine Lentacker (chapter 11) and Joseph Gabriel (chapter 12). One characteristic of official pharmacopoeias that makes them compelling sources is that they were legally enforceable, at least in theory, and represent important state interventions into knowledge of the natural world, as noted by Antoine Lentacker in chapter 11. Pharmacists and apothecaries licensed in a particular jurisdiction were obliged to use the materia medica and the methods for preparing and mixing them as listed in the official pharmacopoeia. The *Pharmacopoeia Matritensis*, which was developed under the auspices of the Spanish Crown, was a product of this early modern European tradition of official pharmacopoeias dating back to the late fifteenth century, as discussed by Emily Beck (chapter 2).

Various political entities and governing bodies—from monarchies and city councils to associations of physicians—hoped to use official pharmacopoeias as a means to standardize therapeutics, combat fraud, and increase oversight of the preparation of medicaments by apothecaries. Like their modern counterparts, early modern official pharmacopoeias attempted to draw a firmer line between orthodox medical practice and quackery, as described for the case of France in Justin Rivest's essay (chapter 4). Yet they often did so by implication rather than the explicit prohibition of specific substances or practices. When it came to materia medica or medicinal simples, those plant, animal, and mineral substances that appeared in the pharmacopoeia were the ones recognized by the state and its medical advisors as legitimate. Those that did not appear were not considered legitimate by the official authorities and, as a result, the medicinal use of such substances was discouraged even if not explicitly prohibited. Similarly, at a time when there could be many different recipes for the same preparations or compound medicines, pharmacopoeias imposed order by designating just one version of a recipe as the official version, a function described further by William Ryan in his account of Hans Sloane's efforts to collect and curate healing knowledge from the Caribbean (chapter 6).[21]

PHARMACOPOEIAS AND THE REGULATION OF PHARMACY IN EARLY MODERN SPAIN

While governments and rulers throughout Europe had employed official pharmacopoeias for several centuries, the *Pharmacopoeia Matritensis* did not appear until the eighteenth century. In some ways, it is surprising that it took so long for Spain to produce an official pharmacopoeia. After all, the Spanish Crown had taken an interest in regulating medicine and the medical professions in the fifteenth and sixteenth centuries, including several policies such as the establishment of the Tribunal del Real Protomedicato in 1477 and a royal order issued by Philip II in 1593 that called for the development of a "general pharmacopoeia."[22] In addition, the influx of new materia medica from the Americas raised questions about the utility of these new materials in medicine and pharmacy. As a result, the Crown and its Council of the Indies, which oversaw governance of Spanish America, had turned their attention to the investigation of American materia medica since the earliest decades of the sixteenth century.[23] Furthermore, official pharmacopoeias already existed in some Spanish cities, such as the *Concordia Pharmacopolarum Barchinonensium* (first published in Barcelona in 1511) and the *Officina Medicamentorum* (first published by the College of Apothecaries in Valencia between 1601 and 1603).[24] Yet it was only in 1739, in the midst of the Bourbon Reforms, a political movement that had the strengthening of royal power as one of its central goals, that the Crown and the Real Tribunal del Protomedicato finally published the *Pharmacopoeia Matritensis*. Although the title associated this project explicitly with Madrid—the political capital of Spain and its empire—the pharmacopoeia applied to all licensed pharmacists throughout the Spanish Empire as indicated in the "Decree of the Royal Protomedicato" that accompanied the pharmacopoeia.

While imperial reform may have provided the impetus for the development of an official pharmacopoeia by the Spanish Crown, the content of the *Pharmacopoeia Matritensis* had its origins in other official pharmacopoeias, such as the *Pharmacopoeia Augustana*, the format of which the *Pharmacopoeia Matritensis* followed closely, and a pharmacological handbook entitled *Palestra Pharmaceutica Chymico-Galenica*, published by Félix Palacios (1677–1737), a pharmacist based in Madrid and member of the Royal Society of Medicine and Other Sciences in Seville. Palacios's text was quite successful, with nine editions of the work appearing between 1706 and 1792.[25] In the history of Spanish pharmacy, Palacios is most remembered for his efforts to introduce and promote the use of chemical medicines in Spain, as reflected in his decision to include both chemical and Galenic pharmaceutical techniques in his popular handbook. The *Pharmacopoeia Matritensis* borrowed heavily from Palacios's text—especially his descriptions of the practices and processes for selecting and preparing botanical, animal, and chemical medicaments—as well as for producing a number of preparations including infusions, wines, decoctions, clysters, cataplasms, extracts, electuaries, pills, and many others. The *Pharmacopoeia Matritensis* also drew much of its list of medical simples from those included in various chapters describing the most "common" botanical, animal, and mineral materia medica in Palacios's *Palestra Pharmaceutica*.[26]

THE DEVELOPMENT AND STRUCTURE OF THE *PHARMACOPOEIA MATRITENSIS*

The *Pharmacopoeia Matritensis* first appeared in 1739, more than a century after Philip II (r. 1556–1598) had ordered its creation. According to Antonio González Bueno, the use of the adjective "matritensis" was likely a reflection of the centralizing impulse of reformers in the Bourbon government, as well as a continuation of the practice of naming pharmacopoeias after their city of origin (figure 3.1).[27] The second and third editions of the *Pharmacopoeia Matritensis* appeared in 1762 and 1775, respectively. In 1780, the Spanish Crown issued a royal order that divided the Real Tribunal del Protomedicato (Royal Protomedicato), the government body responsible for regulating the medical professions in Spain, into three sections—one for medicine, one for pharmacy, and one for surgery. Under the auspices of the new pharmacy section of the Royal Protomedicato, a new edition of the official Spanish pharmacopoeia was published under the title *Pharmacopoeia Hispana* in 1794, with a second edition appearing in 1797 and multiple editions appearing in the nineteenth century.[28]

The first edition of the *Pharmacopoeia Matritensis* that appeared in 1739 was divided into six main parts. The first part provided an introduction to pharmacy with chapters on the "subject and object" of pharmacy, on medical simples, on the instruments used in pharmacy, and on "common pharmaceutical operations."[29] The second part included more than thirty chapters describing different kinds of compositions and "medical mixtures," while the third, fourth, and fifth parts described recipes for different types of prepared medicines, including extracts, conserves, juleps, syrups, powders, pills, oils, and others. Finally, the sixth and seventh parts dealt with "chemical operations," including the distillation of waters, spirits, oils, and other chemical preparations.[30]

For the purposes of thinking about the *Pharmacopoeia Matritensis* as a tool of empire, let us consider the chapters from the first part that focus on the "simples" of the pharmacopoeia. In these chapters, the lists of "simples" included those plant, animal, and mineral substances that could be used therapeutically on their own as well as other substances that were not necessarily medically active, but were commonly used in the preparation of compound medicines. These lists of materia medica that appeared as the opening chapters of official pharmacopoeias served as a kind of master ingredients list that included any and all substances that a pharmacist might need to prepare the various remedies in the pharmacopoeia. Given that the *Pharmacopoeia Matritensis* was the official pharmacopoeia of the entire Spanish Empire, its lists of materia medica are useful for assessing the extent to which Spanish pharmacy had assimilated medicinal substances from all regions of Spain's vast territories, especially the Americas.

De la Botica de la Comp.ª de Jhs.

PHARMACOPOEIA MATRITENSIS

Regii, ac Supremi Hispaniarum

PROTOMEDICATUS

AUCTORITATE, JUSSU ATQUE AUSPICIIS

Nunc primùm elaborata.

Quæ non prosunt singula, multa juvant.

MATRITI

E TYPOGRAPHIA REGIA

D. MICHAELIS RODRIGUEZ.

MDCCXXXIX.

FIGURE 3.1. Title page of the *Pharmacopoeia Matritensis* (1739). The handwritten notes indicate the names of previous owners of this volume, including a pharmacy run by the Society of Jesus. Courtesy of the Lloyd Library and Museum.

AN IMPERIAL PHARMACOPOEIA?

One notable feature of *Pharmacopoeia Matritensis* relative to other European pharmacopoeias of the early eighteenth century was that it included two chapters on "simples"—one on "official simples" and another on "exotic simples."[31] The chapter on "official simples" included a series of lists of the names of substances under several different subsections, including a subsection on the "vegetable kingdom" divided into "herbs and leaves most used," "flowers," "seeds," "fruits," "woods," "roots," "barks," "fungi," "liquid juices," and "gums, resins, balsams and insipid juices"; a subsection of medicinal substances from animals divided into "animals, their parts and excretions" and "fats"; a subsection of "metals, stones and minerals"; and finally a subsection of "marine objects" (*marina*). Overall, the list of official simples included 481 items (table 3.1). Of these, 378 (78.6%) were botanical substances; 45 (9.4%) were animal substances; 48 (10%) were mineral substances; and 10 (2.1%) were marine substances.[32] While materia medica from the Americas were not excluded entirely, identifiably American materia medica (such as cacao, ipecac, and guaiacum) account for only a small portion of all the materia medica (18 out of 481, or 3.75%).

The subsequent chapter of the *Pharmacopoeia Matritensis* focused on "exotic simples" and provided much more information on each item. In addition to listing each substance by name, the chapter included a short paragraph describing the medical "virtues" and geographical origins of the substance. It discussed 164 medicinal substances and, of these, 137 (83.5%) also appear in the list of "official simples," while 27 (16.5%) were unique to the list of "exotic" simples. For the purposes of analyzing the geographical distribution of the origins of these "exotic" simples, we can group the materials into larger geographical regions based on their identified geographical origins as coming from the following areas: Asia (including India), the Middle East, Africa, Europe (including one item from Russia), and the Americas (table 3.2).[33] Thirty items, representing 18.3% of the "exotic simples," had no geographic origin identified. Several items were identified as coming from two or more geographic regions and, in those cases, the items were counted for each region. This survey of the geographic origins of these "exotic simples" reveals that 48 (29.3%) of the materials in the chapter came from a region in Europe, which suggests that the label "exotic" refers to the rarity of the substance rather than to non-European origins. In addition, the data reveal that the largest portion—59 substances (36%)—of the "exotic simples" were identified as originating from a region or place in Asia (table 3.2). Meanwhile, 25 (15.2%) of the "exotic simples" were identified as originating from a region or place in the Americas, followed by 19 (11.6%) identified as originating from a region or place in the Middle East, and 18 (11.2%) identified as originating from a region or place in Africa (table 3.2).

The presence of these "exotic simples" in the *Pharmacopoeia Matritensis* high-

TABLE 3.1. Percentage of Botanical, Mineral, and Animal Substances in Part I, Chapter II ("De Simplicibus officinalibus") of *Pharmacopoeia Matritensis* (1739)

Type of Substance	Number	% of Total Number of Substances
Botanical	378	78.6
Mineral	48	10.0
Animal	45	9.4
Marine	10	2.1
Total	481	100

TABLE 3.2. Regional Distribution of Substances listed in Part I, Chapter III ("De Simplicibus Exoticis") of the *Pharmacopoeia Matritensis* (1739)[1]

Region	Number	% of Total Number of Substances
Asia	59	36.0
Europe (including Russia)	48	29.3
Americas	25	15.2
Middle East	19	11.6
Africa	18	11.0
No place of origin indicated	30	18.3

1. Some substances were listed as originating from more than one region. For example, if a substance was listed as originating from Ethiopia and India, then it was counted as a substance for both "Africa" and "Asia." It is for this reason that the total number of substances (199) is greater than the total number of substances (164) listed in Part I, Chapter III, "De Simplicibus Exoticis," and why the total percentage is greater than 100%.

lights the connection between the practice of European pharmacy and the emerging global drugs trade in the early modern world. Some "exotic" substances had been known and used by pharmacists in Europe and the Mediterranean World since antiquity, while other substances such as those from the Americas were new additions to European pharmacopoeias. At the same time, it is important to recognize that non-European substances represented only a small portion of the total materia medica in the pharmacopoeia. If we subtract the "exotic simples" that originated from Europe, the remaining 105 substances represent 21.9% of the total number (481) of medicinal substances included in the list of "official simples" in chapter two of the *Pharmacopoeia Matritensis*. A similar trend is seen for medicinal substances from the Americas, which represented only 15.2% of the "exotic simples" and only 5.2% of all the "official simples" listed in the pharmacopoeia. Such numbers help to put into perspective the presence of non-European exotic materia medica in this official Spanish pharmacopoeia.

Moving beyond the numbers, it is worth pausing here to recall that the Spanish Crown and its Royal Protomedicato in Madrid intended for the *Pharmacopoeia Matritensis* to serve as the official pharmacopoeia for the entire Spanish Empire, including its American kingdoms as outlined in the "Decree" included in the front matter of the pharmacopoeia. In practice, such a policy meant that pharmacists were obliged to have a copy of the pharmacopoeia on hand and also to stock the substances and ingredients listed therein.[34] Let us imagine the experience of a pharmacist in Mexico City or Lima or one of the smaller cities in Spanish America in light of such a policy. Would they have had the same kind of access to materia medica from around the globe as pharmacists in Madrid? In light of the preponderance of materia medica from Europe in the pharmacopoeia, we might ask: How much difficulty did pharmacists in Spanish America face in acquiring medicinal substances from Europe?[35] Ultimately, a close reading of the materia medica in the Spanish Empire's official pharmacopoeia, in combination with attention to the geographic origins of medicinal substances, gives us a sense of how challenging it must have been to practice imperial pharmacy in the American territories of the Spanish Empire.

In effect, the official pharmacopoeia's requirement that pharmacists use specific substances meant that pharmacists in Spanish America faced some challenges that their counterparts in Spain did not. The lists of required simples in the *Pharmacopoeia Matritensis* were enforceable by law, and so pharmacists throughout the empire were required to have these materials on hand. As shown by the work of Paula De Vos on pharmacy and the drugs trade in colonial Mexico, demand from Mexican pharmacists actually supported a robust trade in medical simples from Europe that reached Mexico via Spain, and a recent study by Linda Newson shows a similar situation for early colonial Lima.[36] But what about the quality of these materials? A pharmacopoeia published in Mexico City in 1615 provides a clue. In a note to the reader in his book, *Quatro Libros. De la Naturaleza y virtudes de las plantas, y animales que estan recevidos en el uso de Medicina de la Nueva España*, Francisco Ximénez, a Dominican missionary, observed that "the medicines that are brought from Spain, having crossed the vastness of the ocean, lose their [medical] virtue."[37] By categorizing some materia medica (including all of the American materia medica) as "exotic," which is, after all, a relative term, the *Pharmacopoeia Matritensis* espoused and endorsed an unmistakably European perspective on pharmacy and asked pharmacists throughout Spain's vast empire to do the same.

It is difficult to know how pharmacists in colonial Spanish America reacted to this situation due to a lack of sources. It is certainly possible that some of these pharmacists may have considered European materia medica better than medicinal substances from the Americas. After all, according to some prevailing medical and environmental theories developed mostly by Europeans, the American environment had a degenerative effect on local fauna.[38] So, it would not have taken that much imagination to apply this logic to the local American flora as well. In addition, we might imagine how pharmacists in colonial Spanish America, especially those born and trained in the Americas, reacted when they found that their local materia medica was labeled "exotic" in the empire's official pharmacopoeia—the pharmacopoeia that they were bound by law to follow. As a result, some pharmacists in colonial Spanish America would have found ways to subvert or simply ignore the requirements of the official pharmacopoeia. After all, medical inspectors were not peering over the shoulders of pharmacists at all times. Nonetheless, the production and dissemination of the *Pharmacopoeia Matritensis* as the official pharmacopoeia of the Spanish Empire meant that the art of pharmacy had now become a matter of empire. It also meant that a distinctly European vision of healing was enshrined in the empire's official policy regarding the regulation of pharmacists, as reflected most prominently in the ratio of New World materia medica to Old World materia medica.

Similar to Benjamin Breen's question in chapter 7 about the apparent bias against African medicines in European pharmacopoeias, we might ask, "Was the emphasis on European and Old World medicaments necessarily a product of an imperial vision of pharmacy?" Other evidence suggests that the story is a bit more complicated. Here, a comparison with two texts that influenced the *Pharmacopoeia Matritensis* is useful. The first is the *Palestra Pharmaceutica Chymico-Galenica*, first published by Félix Palacios in 1706 with subsequent editions appearing throughout the eighteenth century. It is likely that the *Pharmacopoeia Matritensis* was based partly on the third or fourth edition of Palacios's text (which appeared in 1730 and 1737, respectively, and expanded significantly on the content of the first edition).[39] Like the *Pharmacopoeia Matritensis*, Palacios's pharmacopoeia included a list of the common simples used in pharmacy spread across three chapters—one that focused on animals, one that focused on plants, and one that focused on minerals. The substances in these three chapters came from all corners of the globe, including the Americas, as did the materials in the *Pharmacopoeia Matritensis*, although their geographic origins were not explicitly identified. Notably, Palacios included a chapter on the "simples that are absolutely necessary for recipes" that described those substances most commonly used in compound medicines.[40] In this chapter, Palacios listed sixty-three items—plant and animal materials as well as minerals—but not a single substance from the Americas.[41]

Similar trends are evident in a key text that influenced Félix Palacios: a book by the French chemist and pharmacist Nicolas Lémery (1645–1715) titled *Cours de Chymie*, which first appeared in 1675.[42] Although the title would seem to suggest that it was a work of chemistry, Lémery's text focused mainly on the chemical manipulations of the most common plant, animal, and mineral sources of materia medica. Lémery's text was quite popular in its time and he published ten editions between 1675 and 1713. In addition, in 1703, Palacios produced and published a Spanish translation of Lémery's work—a project that may have provided the impetus for Palacios to publish his own pharmacopoeia a few years later.

As in the case of the *Pharmacopoeia Matritensis* and Palacios's *Palestra Pharmaceutica*, only a small number of American medicaments appear in Lémery's list of common materia medica. For example, of the forty-four substances in the first edition, only three (7%) were from the Americas, including jalap, guaiacum, and tobacco. In the fourth edition, published in 1681, a fourth American medicament was added: *quinquina*, the popular name in France for the bark of the cinchona tree, a medicament for treating intermittent fevers that was gaining in popularity in the late seventeenth-century Atlantic World.[43] Such evidence from Lémery's text further suggests that the small proportion of American materia medica in Palacios's *Palestra Pharmaceutica* and the *Pharmacopoea Matritensis* was not unusual. In addition, this evidence shows that American materia medica were a much more significant presence in *Pharmacopoeia Matritensis* relative to these other texts. In many ways, this feature of these texts makes sense if we remember that Lémery's *Cours de Chymie* and Palacios's *Palestra Pharmaceutica* were produced for European pharmacists, and so probably reflect what materials were most commonly available and familiar to them. In other words, the primarily European perspective of the text that would become the official pharmacopoeia of the Spanish Empire may not necessarily have been born of some grand design to impose European therapeutics and repress American materia medica. Instead, it may have simply been the result of an emphasis on the familiar and the force of tradition, especially since many official pharmacopoeias were modeled on official pharmacopoeias that preceded them and were published elsewhere.[44]

Additional insight into the matter comes from reconsidering the lists of medical simples required and approved for therapeutic use in Spain's official pharmacopoeia. It turns out that the *Pharmacopoeia Matritensis*, following the lead of Palacios, gave pharmacists some latitude to diverge from the officially sanctioned ingredients and recipes. It did so by permitting the practice of substituting one medical simple for another. Both Palacios's pharmacopoeia and the *Pharmacopoeia Matritensis* provided a list of materials that could be used *in place of* the required materia medica. This feature of the text was typical of the genre that also appeared in seventeenth-century editions of the *Pharmacopoeia Augustana*, one of the model texts for Spain's pharmacopoeia. At the same time, the decision to include this section in the *Pharmacopoeia Matritensis* suggests that the members of Spain's Royal Protomedicato were sensitive to the realities of pharmaceutical practice in imperial contexts: in the far-flung corners of the empire, including even some regions of the Iberian Peninsula, some medicinal substances were going to be scarce due to the contingencies of trade and the offerings of the local environment. In other words, at the same time that the *Pharmacopoeia Matritensis* espoused a European vision of pharmacy, this text also recognized the limits of this vision. Pharmacy, in practice, was a much more heterogeneous enterprise than it was on paper. Moreover, the technique of substitution provided a warrant, of sorts, for pharmacists in Spanish America to use local or readily available materia medica in their treatments. At the very least, the presence of a chapter on substitutions shows that the official pharmacopoeia of the Spanish Empire was more flexible than it might seem at first.

In addition to this technique of substitution, it is also important to pay closer attention to *which* American medicaments became part of European pharmacopoeias. Consider the case of quina, or cinchona bark. In the decades before the publication of the *Pharmacopoeia Matritensis* in 1739, quina had become one of the most important medical commodities in the Atlantic World in terms of both the volume and value of the quina trade relative to the trade in other medicinal drugs. Official records of imports to Cádiz, Spain's main Atlantic port, indicate that between 1718 and 1724 merchants imported approximately 15,000 pounds of the bark annually and by the 1750s that number was up to approximately 300,000 pounds of bark imported annually.[45] Similarly, studies by Patrick Wallis of drug imports to London in the seventeenth and eighteenth centuries show that drug wholesalers in England were importing nearly 100,000 pounds of bark annually in the early 1750s.[46] Demand for the bark was so great because many in Europe had come to recognize quina as an effective treatment for the intermittent fevers that plagued many populations in Europe and other parts of the Atlantic World.

In addition to the value and volume of the trade in the bark, the case of quina is also notable in that it shows how the inclusion of even a small number of new remedies from the Americas in the official pharmacopoeias of Europe could have significant effects on the theory and practice of therapeutics of European physicians and pharmacists. As it turns out, this Andean wonder drug presented a challenge to European medical theory: it was difficult for European physicians to explain *why* or *how* the bark worked.[47] From the perspective of Galenic medicine that was the dominant paradigm in early modern European pharmacy and medicine, the bitterness of cinchona bark meant that it should be classified as a "hot" remedy. As a result, it made no sense that the bark proved efficacious in the treatment of fever—a condition which would usually call for a "cold" remedy to counteract the excess of heat in the patient. In this instance, the case of quina shows how we get a better sense of the effects of materia medica from the Americas by looking beyond the relatively small numbers of American medicaments incorporated into the official European pharmacopoeias, and looking more closely at the impact of individual substances.

CONCLUSION

Let us consider how closer attention to the presence of American materia medica in the *Pharmacopoeia Matritensis* might encourage us to think differently about European pharmacopoeias in the early modern Atlantic World. A useful concept is the notion of "disturbance" as used by geographer Robert Voeks in his study of the pharmacopoeias of folk healers in the tropical rainforest of the Americas.[48] In his article, Voeks challenges the long-standing myth that the knowledge and practices of Indigenous or folk healers in the Americas derive from some kind of ancient wisdom of pristine tropical landscapes that predated and resisted the influence of European colonialism. Instead, he argues that the use of medicinal plants by rural peoples in the tropical Americas are better understood as what he calls "disturbance pharmacopoeias." Such pharmacopoeias include a suite of indigenous or traditional medicaments of American folk healers, "enhanced by new medicinal foods, ornamentals and weeds," that came in the wake of European colonization. In this way, Voeks's notion of Indigenous pharmacopoeias as "disturbance pharmacopoeias" recognizes and recasts folk healing traditions as dynamic and historical enterprises rather than storehouses of timeless knowledge.

The idea of a "disturbance pharmacopoeia" also seems useful for dispelling some of the myths about the pharmacopoeias—official and unofficial—developed by European healers, colleges of physicians, and governments in conjunction with imperial and commercial expansion in the early modern world. As reflected in the case of some European physicians' concern about the use of cinchona bark, European healing traditions were also disturbed, so to speak, by encounters and exchanges with Indigenous materia medica, just as Indigenous healers and healing traditions in the Americas were disturbed by their encounters with Europeans (to say nothing of other kinds of disturbances). In addition, the pharmacopoeias promulgated by early modern European states and associations of physicians were themselves already the product of previous disturbances, if we keep in mind the long-distance exchanges of goods and medical knowledge between Europe, Africa, and Asia in the centuries before 1492.

Reframing early modern European pharmacopoeias under the rubric of "disturbance pharmacopoeias" further shows how George Urdang's notion of "pharmacopoeias as witnesses to world history" is worth both resurrection and revision. To read European pharmacopoeias as the products of disturbance resulting from cross-cultural encounters is to treat this important group of early modern medical sources not as witnesses to the ossification of medical knowledge in the name of a protonationalism, as suggested by Urdang, but as witnesses to and participants in the complex process of globalization that shaped medical knowledge and practice in an increasingly interconnected early modern world. Finally, it is important to emphasize that characterizing pharmacopoeias as mere witnesses is to overlook the important and dynamic role that this set of texts played in fostering and promoting the exchanges and development of healing knowledge across time and space. In short, pharmacopoeias were as much agents of world history as witnesses to it.[49]

Another way to assess pharmacopoeias and their impact on healing knowledge is to focus on pharmacopoeias as a genre of medical writing. As noted in this essay, existing trends and traditions in the format and structure of early modern European official pharmacopoeias exerted significant influence on the *Pharmacopoeia Matritensis*. For the most part, Spain's first official pharmacopoeia grouped its materia medica and compounded medicines under section and chapter headings that were commonplace in early modern Europe. The same goes for the content of the pharmacopoeia in that the majority of simple and compound medicines were those that had appeared in other official pharmacopoeias. Yet, at the same time, the *Pharmacopoeia Matritensis* did introduce some structural novelties—such as having an explicit chapter on "exotic" simples—and included materia medica and compound medicines that were unique to (or at least more popular in) Spanish and Iberian medical traditions.

Finally, by recognizing that the pharmacopoeias in our own time are descendants—even if only distant—of early modern European pharmacopoeias, we are encouraged to consider the ways in which current and ongoing attempts to formalize and codify healing knowledge are also "disturbance pharmacopoeias" of a sort, not just in the sense of being the products of long-term historical processes of disturbance, exchange, and cooptation but also in the sense of being witnesses and agents of disturbance as our pharmacopoeias continue to shape and be shaped by the interactions between Western medicine and myriad bodies of dynamic healing knowledge that persist in societies and cultures around the globe.

PART II

Pharmacopoeias and the Codification of Knowledge

4

BEYOND THE PHARMACOPOEIA?

Secret Remedies, Exclusive Privileges, and
Trademarks in Early Modern France

JUSTIN RIVEST

WHILE PHARMACOPOEIAS like the Parisian *Codex medicamentarius* governed the world of apothecaries in ancien régime France, there also existed a parallel and more loosely organized tradition whereby pharmaceutical innovators were granted exclusive royal privileges for the sale of novel therapeutic preparations, called "secret remedies." These privileges in effect granted legal monopolies over the sale of a given drug. Many privilege holders held no formal medical credentials and were not members of the medical corporations—the faculties and colleges of physicians, or the guilds of surgeons and apothecaries. As such, while their monopolies were narrowly circumscribed to the production and sale of a single drug, they were often resented by apothecaries, who saw them as encroaching on their own corporate privileges to compound and dispense all drugs within a given jurisdiction.

Published pharmacopoeias and secret remedy privileges each offered different modes of organizing the world of healing goods. One aimed to standardize the recipes of established drugs, the other to reward therapeutic innovation. These two opposing modes coexisted throughout the ancien régime and were in some ways mutually dependent upon one another for their definition. The

1728 protocols of France's earliest secret remedies commission demonstrate that pharmacopoeias played an instrumental role in defining pharmaceutical novelty: commissioners compared prospective secret remedies to established pharmacopoeia preparations in order to ensure that they were not simply disguised or lightly modified versions of existing drugs. To put it simply, if a drug could be found in a pharmacopoeia, then it was not novel, and was therefore unworthy of a privilege.

But what happened when the recipe of a secret remedy was published in a pharmacopoeia? In this essay I explore this question through the example of *orviétan*, a secret remedy that the Contugi family sought to monopolize through a royal privilege. The family patriarch, Christophe Contugi, began his career as an *opérateur*, or more pejoratively, a "charlatan" in the classical sense of the term: an itinerant vendor who peddled drugs to the public with an accompanying stage-show in the market square.[1] By the end of his life, however, he was a respectable bourgeois and his successors sold the drug from an established boutique. Although successive generations of the Contugi family held a royal privilege for the exclusive sale of orviétan in France from 1647 onward, by 1682 no fewer than eight European pharmacopoeias, two of which had been published in France, included a range of different recipes purporting to produce orviétan. The matter came to a head in that year, when the royal apothecary Antoine Boulogne took advantage of this situation by selling his own orviétan from a boutique located directly across the road from that of the Contugis at the foot of the Pont Neuf, one of Paris's busiest thoroughfares. Was orviétan a hereditary medical secret of the Contugi family? Or was it a widely known drug, closely related to theriac, whose secret recipe had already been disclosed in pharmacopoeias?

This essay explores the use of pharmacopoeias in answering these questions. From the perspective of state authorities, pharmacopoeias came to be closely intertwined with the world of secret remedies by providing a criterion for assessing whether or not a drug was novel. Although specific procedures were never rigidly codified, it is clear that pharmacopoeias played a formal role in granting monopoly privileges by 1728 and were probably being used informally for this purpose in the preceding decades. The essay begins by examining the legal context of pharmaceutical privileges in ancien régime France, comparing "secret remedies" to those compounded by apothecaries. It then provides a background of the Contugi family orviétan monopoly and details the legal battles between the family and the royal apothecaries in 1682–1685. This case demonstrates that, well before 1728, pharmacopoeias furnished apothecaries with arguments in judicial debates, serving as instruments in their larger economic struggle with familial monopolies like that of the Contugis. The Contugis for their part responded not only by insisting that the pharmacopoeia preparations had not accurately captured their secret recipe but also by redefining their monopoly around a trademark ("the sign of the sun") and a privileged point of sale at the foot of Pont Neuf.

DRUG CATEGORIES: OFFICINAL, MAGISTRAL, SECRET, AND SPECIFIC

University-trained physicians regularly denigrated operators like the Contugis as "charlatans" and "empirics," but they and others like them might more justly be called "privileged vendors" of so-called secret remedies (*remèdes secrets*).[2] The terminology surrounding their drugs, and those compounded by apothecaries, provides a useful entry into the question of what role pharmacopoeias played in establishing these distinctions. When applied to remedies or medicines, the terms "secret," "proprietary," and "patent" are sometimes used interchangeably, but each places emphasis on a different dimension of the phenomenon of medical monopoly privileges. The notion of remèdes secrets, for instance, emphasized the medical secret of a drug, understood to include its recipe, preparation, and the identity of its main ingredients. "Proprietary," by contrast, emphasizes the attempt to monopolize a remedy by a proprietor, which can take several forms: a drug can be monopolized through a legal privilege, through trade secrecy, or by both in conjunction. Many secret remedies in early modern France were protected by royal monopoly privileges, and in most cases their recipes were also concealed—at least initially—by the shroud of trade secrecy. In both senses, then, they can be called "proprietary." The terms "secret" and "proprietary" also converge insofar as both legal monopolies and medical secrets could be sold, inherited, and otherwise transmitted as property in the ancien régime.[3]

The term "patent medicines" poses a series of problems which are particularly acute to anglophone readers. "Patent" originally referred to the letters patent that granted a privilege—a right that belonged to a corporation or even to a single individual—to the exclusion of others. European monarchs granted a variety of privileges, including the establishment of monopolies in medicine and other economic spheres. By the nineteenth century, however, many "patent medicines" were not protected by patents registered with government patent offices; the term became pejorative, a tool of distinction used by an "ethical," scientific pharmaceutical industry, particularly in the United States, that sought to distinguish itself from supposedly unscrupulous and unscientific commercial competitors.[4] I avoid using the term "patent medicines" for this reason, and prefer "proprietary" or "secret remedies." Likewise, doing so avoids conflating ancien régime privileges with modern notions of intellectual property. Privileges did not require a public disclosure (via a government patent office) of the full details of their invention—called "specification"—in exchange for the legal and financial protections of an explicitly temporary monopoly. Vendors generally strove to keep their recipes secret indefinitely as a further bulwark against counterfeiters. Further still, a royal privilege was not guaranteed by a "right" to intellectual property in the modern sense; rather, it was the gift of a benevolent sovereign, which aimed to reward inventors for adding to the glory of France. It would protect their investment from interlopers, and ensure the quality of the product for the general public against counterfeiters and adulterators.[5] Finally, we should recall that ancien régime privileges were enmeshed in the normal systems of patronage that characterized the culture of the royal court. Indeed, modern intellectual property emerged explicitly from critiques of this system, which opponents portrayed as being arbitrary, secretive, and open to favoritism and venality.

The term *remèdes spécifiques* was often used interchangeably with *remèdes secrets* in the eighteenth century. In pharmacological terms, "specific remedies" (or simply "specifics") were drugs that targeted a defined disease or condition in a localized way—for instance, intermittent fevers, venereal disease, or dysentery. The reason for conflating the two is that so many privileged secret remedies were also "specifics," pharmacologically speaking. Rather than merely provoking a purge or some other intervention aimed at rebalancing the patient's humors, "specifics" were supposed to respond to a given disease regardless of the peculiarities of an individual's personal temperament. This made them unusual in the normal medicine of the period, which treated humoral imbalances within the whole body. Indeed, the "one-size-fits-all" approach of "specifics" smacked of medical empiricism to Galenic physicians, but found advocates among chymical physicians and in localized medical contexts, notably military medicine, where large patient volumes made individualized treatment and calibrated regimens impracticable. Harold J. Cook has also connected this "deindividualization" to the growing commercialization of proprietary drugs, which were touted as "specifics" so that they could be marketed to the largest possible number of patients.[6]

Because they were protected by legal monopolies and medical secrecy, proprietary "secret" or "specific" remedies stood apart from the drugs normally compounded by apothecaries. Apothecaries' drugs came in two general types: they were either *officinal*, meaning they were prepared according to a standard, publicly available formula from an established pharmacopoeia, and often stocked ready-made in apothecaries' boutiques; or they were *magistral*, tailored to fit an individual case from the personal prescription of a physician.[7] Secret remedies differed from officinal compounds in that they were supposed to be new, rather than being based on standardized recipes. Further, secret remedies differed from magistral prescriptions insofar as they were not usually tailored to fit the individual circumstance of a patient—every patient received the same drug.

As we shall see in the following sections, these categories could easily blur into one another in practice. From the perspective of regulatory officials, pharmacopoeias provided a criterion for assessing whether or not the secret remedies proposed by inventors were actually novel. Conversely, the recipes of ostensibly secret remedies could also be disclosed in pharmacopoeias, turning them into officinal preparations. They could thus offer a foundation for the claims of apothecaries that such remedies could not be monopolized within families and should instead be incorporated into their corporate monopoly over drug production.

THE ROLE OF PHARMACOPOEIAS IN EVALUATING SECRET REMEDIES BEFORE AND AFTER 1728

On July 3, 1728, the State Council of Louis XV promulgated an *arrêt* to bring order to the medical anarchy then reigning in the capital.[8] At first glance the text of this arrêt reads much like any other early modern denunciation of illicit medical practice. But the producers and vendors of these pernicious remedies were no ordinary empirics, as the operative portion of the edict demonstrates. The state council and the king's first physician, Claude Jean-Baptiste Dodart (1650–1730), were, in fact, targeting secret remedies vendors who already held royal privileges. They were ordered to deposit their privilege documents, along with samples of their drugs, to the office of the lieutenant general of police within two months, so that they could be reexamined.[9]

This attempt to reexamine and, if necessary, revoke existing drug privileges was a response to the fact that a veritable industry of privileged vendors had emerged by the early eighteenth century. This arrêt was the first of a series that would provide increasingly articulate legislation regulating the trade in proprietary remedies in France from 1728 onward. But the need for all hitherto granted privileges to be reexamined and for the medications they licensed to be tested anew suggests that the practice of granting individual privileges for proprietary remedies had somehow run off the rails in the first decades of the eighteenth century.

In his 1762 *Jurisprudence de la médecine*, the most extensive treatise on medical regulation in ancien régime France, the physician and jurist Jean Verdier provides some notion of how the 1728 arrêt came into being, and what abuses it was intended to rectify. Verdier observed that the most famous empirics had long known that the best way of evading challenges from the faculties and other medical corporations in France was to secure a royal privilege. And since kings would never grant such privileges without medical counsel, the approbation of specific remedies came through custom to constitute an established right (*droit établi*) of the king's first physician (*Premier médecin du roi*).[10] The drugs were sometimes tested on patients—usually in relatively small numbers—under the supervision of the royal first physician or other court physicians.[11] Its secret recipe would also be disclosed to the first physician, who would judge whether or not it was worthy of a privilege, which would then be issued by the secretary of state for the king's house (*Secrétaire d'État de la Maison du roi*). As Verdier puts it, "Such was the order followed in recent centuries, by which means empiricism has been tolerated."[12]

Verdier admits that the licensing of proprietary remedies was in some cases desirable, despite the fact that many vendors had no formal medical qualifications. Pure chance and even overt charlatanism had sometimes enriched medicine with new remedies, in his view, therefore empiricism could not be rejected wholesale.[13] This was particularly important in the case of "specifics," whose occult virtues could not be determined through any reasoned causal account of their sensible qualities, but could only be learned through the trial-and-error associated with the practice of so-called "empirics." Writing at the height of the Enlightenment, Verdier saw in empirics a potential wellspring for medical innovations. He believed that a careful regime of privilege granting, kept within narrow bounds, could ensure that the true fruits of empiricism would be safely reaped while carefully avoiding the abuses that might come with it.

According to Verdier, important steps toward producing such a balanced situation came in 1728 following the remonstrances of the royal first physician Dodart. In his account of the reforms, Verdier provides an image of a first physician who was constantly imposed upon by various court interests to gratify their respective medical clients by licensing their medications—regardless of the fact that many were dangerous, veritable "weapons against humankind," in his words.[14] Sharing this "embarrassing" authority with other commissioners offered Dodart a means not only of examining the drugs more formally but also of diffusing blame for any rejections.

Precious little has survived documenting the work of the 1728 commission. Indeed, the earliest surviving institutional archive for French drug regulation dates to 1778, when the Société Royale de Médecine was placed in charge of assessing new drugs for monopoly privileges.[15] Fragmentary evidence suggests that a growing collection of drugs and their vendors' accompanying parchment privileges were deposited with the lieutenant general of police in 1728. But when the commissionaires convened to analyze them, what was their modus operandi? One short *mémoire*, dated October 16, 1729, and signed by René Hérault (1691–1740), lieutenant general of police, has survived and provides some hints as to their procedures. The commissioners were expected to distinguish between three types of remedy: those that are dangerous, those that are salutary, and, finally, the pragmatic category of "indifferent remedies." In the latter case, the mémoire observes that the commission should not abolish those privileges that it has pleased the king to grant for "indifferent" remedies that cause no ill effects. Their vendors were permitted to continue holding privileges to sell such "indifferent" remedies, although—a subsequent note clarifies—they should no longer hold the *exclusive* rights to do so: "In this case the exclusion must be abolished, and the apothecaries must be given the liberty to compound these remedies."[16] This clause likely served as a compromise, insulating the commission from the wrath of former monopolists whose privileges might have been granted through the intercession of powerful patrons at court, including even the king himself.

The remainder of the mémoire expresses the two principal goals of the commission; namely, to confirm the novelty of the drug and to carefully specify the circumstances in which it should be used. The first of these has direct bearing on the role of pharmacopoeias within its examination proceedings: "In the future, [the commission will] no longer grant privileges for remedies under any pretext whatsoever unless they have been analyzed beforehand, and that it be evident that the remedy is distinctive, not present in any Pharmacopeia, and that it is salutary."[17] The second goal was to ensure the careful specification of the circumstances and conditions in which an approved drug should be used—presumably in an effort to prevent vendors from selling their drugs as panaceas or cure-alls. It is interesting to note that while the mémoire leaves the precise method of assessing whether a drug is "salutary," "indifferent," or "harmful" in the hands of the commissioners—it merely specifies that these are to be determined "par l'experience." "Experience" in this case meant the drug's use on small numbers of patients, personally witnessed by the commissioners themselves. Secondly, "if it is judged necessary, by analysis *(par l'analyse)*" in the sense of chemical analysis.[18] No specific procedures are mandated in either case—no number or type of patients is specified, for example; nor are there any specific instructions for the chemical analysis. Therapeutic novelty, however, is given a more precise criterion: it is to be confirmed by the absence of the drug from "any pharmacopoeia." In this way, pharmacopoeias played a crucial role in defining privilege-worthy innovation in the procedures of the 1728 commission.

This raises the question of whether pharmacopoeias and other works of *materia medica* were checked prior to 1728. No detailed documentation survives regarding the earlier procedures, but anecdotal evidence suggests that the commission may simply have codified existing practices. Take, for example, the following anecdote from Bernard Le Bovier de Fontenelle's *éloge* of Guy-Crescent Fagon (1638–1718) for the Académie Royale de Sciences. Fagon was royal first physician from 1693 to 1715, and was responsible for recommending many drugs during his tenure. Implicitly, the 1728 reforms may have been intended to reexamine privileges that were granted under his watch. According to Fontenelle, however, Fagon was "no friend to empirics." He was interested in medical secrets, and had bought several on behalf of the king, but "he wanted these to be true secrets, that is to say, unknown up until that point, and consistently useful," and had often shown would-be privilege holders "who believed they possessed a treasure" that their secret had, in fact, already been disclosed publicly in "books" or "mémoires," of which pharmacopoeias were doubtless the most important.[19] These comments suggest that the notion of using the recipes from printed sources as a criterion to establish the novelty of a pharmaceutical invention had already been an established practice under Fagon. Both before and after 1728, it seems that, at least in theory, the absence of a remedy from the pharmacopoeia played some role in establishing its novelty and defined it as a "secret remedy" worthy of a privilege.

THE INHERITED SECRET OF THE CONTUGI FAMILY ORVIÉTAN MONOPOLY

It seems clear then that pharmacopoeias played an important role in adjudicating which drugs came to be protected by monopoly privileges and which did not. But to return to the question that opens this essay, what happened when the recipe of a privileged secret remedy was disclosed in print? Orviétan offers a useful case study. For nearly a century, from 1647–1741, four successive generations of the Contugi family fought to extend and later preserve their exclusive privilege to sell orviétan in the kingdom of France.[20] Touted primarily as a cure for all poisons—from the venom of vipers and rabid dogs to human poisons like arsenic—and pestilential diseases like plague and smallpox, orviétan eventually came to be used for any number of lesser discomforts such as colic and digestion problems. The Contugi family's claim to the orviétan monopoly was contested from the beginning, and not just by apothecaries: other itinerant operators likewise claimed to be the sole possessors of the drug's secret, but the Contugis were unique in securing royal endorsement in the form of their monopoly privilege. Their efforts to preserve this privilege, first granted to the family patriarch, Christophe Contugi, in 1647, were closely tied to their efforts to defend the notion that orviétan was a hereditary medical secret known only to members of the family.

The first pharmacopoeia to include a recipe for orviétan was the fourth edition of Johannes Schröder's *Pharmacopoeia medico-chymica* (1655).[21] The recipe, titled "Electuarium Orvietanum," makes no special note of this fact, and orviétan does not appear in any other pharmacopoeias until after 1665, following the publication of two different recipes in Bordeaux and Paris; by Thomas Riollet on the one hand, and by the physician and traveler Pierre-Martin de la Martinière (1634–1676) on the other.[22] Because Riollet has been explored elsewhere, I will focus here on La Martinière, whose comments also reveal interesting details about the origins and circulation of orviétan before the Contugis sought to monopolize it.[23] In keeping with the "professor of secrets" tradition, La Martinière takes explicit aim at the keepers of medical secrets, and that of orviétan in particular, censuring those who would conceal and monopolize cures "under the shadow of a little lucre."[24] La Martinière was an author of popular medical works: throughout his voluminous writings, he reproached charlatans and alchemists as occasional fraudsters, all the while recognizing their often useful medical innovations and casting himself as a charitable discloser of their secrets for the benefit of the common people.[25] In his 1665 *Traitté*, after describing the composition of the ancient antidotes theriac and mithridatium, he launches into a critical investigation of the origins of orviétan, which he portrays as a latter-day successor of these two drugs.[26]

La Martinière reports the "fable" (as he qualifies it) given by most orviétan vendors; namely, that the drug originated in late sixteenth-century Italy, where it was invented by a shepherd from the town of Orvieto (hence the name) and then transmitted through a complicated chain of a physician, a cardinal, and finally a succession of apothecaries and their apprentices. But the Contugis were not the only ones who claimed to be the exclusive heirs of the orviétan secret in the middle decades of the seventeenth century. Even before orviétan recipes began to appear in pharmacopoeias and before they were challenged by the apothecaries, the Contugis faced challenges from rival charlatans selling their own varieties of orviétan. These challenges were especially pronounced when Christophe Contugi left Paris and "went on the road" to sell orviétan in the provinces, endeavoring to exercise the full kingdom-wide extent of his exclusive privilege to sell the drug. His privilege provided him with legal grounds for pursuing rivals as counterfeiters and adulterators, but he still had to register it in each jurisdiction, and its enforcement required active litigation. On several occasions he encountered rival vendors who likewise claimed to be the sole inheritors of the original orviétan secret. These rivals often bolstered their claims by securing permissions from local jurisdictions, such as Christophe Poloni, who was endorsed by the Estates General of Languedoc.[27] In 1656, for instance, Contugi's efforts to register his royal privilege in provincial jurisdictions triggered lawsuits with rival orviétan operators in Toulouse and Bordeaux.[28] In reference to these sorts of disputes over exclusivity, La Martinière jibes that the proliferation of so many contrary claims and convoluted chains of transmission from the original inventor tends to discredit all of the orviétan operators: "One [operator] claims to be the grandson of this physician, another says the physician was his great-grandfather, another claims he was the grandfather of his father-in-law, and that in the lineage of this father-in-law, he was the only one to inherit the secret, getting it through marriage to his wife; and almost all of these operators say something similar, to such an extent that, to take them at their word, this Orviétanalized physician deflowered more women than Hercules to have so many bastards, for they all carry different names."²⁹ La Martinière's comment testifies to the variety of claims then in circulation, most of which relied on marriage or kinship links. A critical reappraisal of the origins and transmission lineage of the orviétan secret(s) lies beyond the scope of the present contribution, but for our purposes here, it suffices to say that Contugi's claim to the secret went through his wife, who had been widowed by an earlier orviétan vendor: "Clarice Vitraria, sole and unique inheritor of Jean Vitrario, physician, who married Clarice, the widow of Hierosme Fioranti the first to be called *l'Orviétan*."³⁰

Whatever the status of his claims to exclusive inheritance of the secret, Contugi secured his first letters patent to sell orviétan on April 9, 1647. Royal privilege in hand, he sold it on the Pont Neuf and elsewhere for over thirty years and fought a number of legal battles to preserve his exclusive rights to do so from counterfeiters and rival operators. In 1681, he died in comfortable circumstances, as an honorable bourgeois of Paris. Clarice's marriage with Contugi appears to have been childless, and Le Paulmier conjectures that she was still alive in 1658 but had certainly passed away by September 9, 1659. On that date, Contugi married a young actress from his troupe, Roberte Richard, who played the role of Florinde in their theater skits opposite Contugi's Capitaine Spacamont.³¹ Two decades later, she too would find herself widowed, but the privilege and secret of orviétan would pass to her eldest son, Louis-Anne.³²

The medical secret of orviétan was thus transmitted to Contugi by the widow of a previous vendor. Of course, other operators made rival claims to this inheritance, but the secret seems to have been transmitted through familial and marital channels in the Contugi family over successive generations. In their case, these claims of exclusive inheritance were bolstered by an ostensibly kingdom-wide monopoly privilege. But as the recipe became available in pharmacopoeias, the apothecaries saw an opportunity to challenge interlopers on drug production like the Contugis.

CONTUGI ORVIÉTAN VERSUS PHARMACOPOEIA ORVIÉTAN, 1682–1685

Despite their persistent troubles with rival charlatans in the provinces, the publication of orviétan recipes in the pharmacopoeias did not cause legal problems between the Contugis and the apothecaries until 1682. In the fall of that year, a new apothecary's boutique opened opposite that of the Contugis, on the rue Dauphine facing the Pont Neuf. Contugi's widow, Roberte Richard, reported that the new boutique's attendant, one Jean Regnault, was replicating their display and containers (*montre et boëttes*) and selling his own drug under the name of orviétan. Her son, Louis-Anne, was the legal holder of the orviétan privilege but he happened to be out of town, and so Roberte, unable to tolerate such a flagrant violation of her family's privileges, brought her grievance to the Châtelet, the civil and criminal court for the city of Paris and the seat of the city's *prévôt*. Roberte made the request under her own name—not that of her son—and on December 18, 1682, the authorities seized all merchandise in the offending shop. Before long, however, Roberte found herself the defendant in a countersuit: the plaintiff was the royal apothecary Antoine Boulogne, then living in Versailles, who made it known that the boutique opposite that of the Contugis belonged to him and that Regnault was simply his store clerk (*garçon de boutique*), selling the orviétan on his behalf. The Contugi–Boulogne conflict would drag on for over two years, and would raise the pivotal question of whether the Contugis could continue to hold an exclusive privilege to sell orviétan when the secret of its recipe had been "divulged" in multiple pharmacopoeias.[33]

The two parties began by jockeying with one another to secure the most advantageous tribunal to hear the case. Contugi had also carefully renewed his letters patent on July 16, 1683. These jurisdictional conflicts tied up the case for over a year, but on March 27, 1684, an arrêt placed it under the jurisdiction of the royal Conseil privé. To argue his case, Boulogne had a *Requeste servant de Factum* printed. In it he argued that the 1647 privileges that Roberte Richard invoked in the initial seizure were made out personally to Christophe Contugi and were nonhereditary, since they mentioned nothing about his widow or heirs, and thus were extinguished upon his death. Likewise, he argued that these letters had not given Christophe, much less his widow, the right to run a boutique in Paris, but rather had allotted him only the right to sell it itinerantly (*une faculté ambulante*). Consequently, the initial seizure was without legal foundation. As such, Boulogne requested that the Contugi boutique be closed and that they be fined three thousand livres for damages against him, pointing to the undue force of the seizure, alleging that the Contugi widow "closed his boutique down, broke everything that was found within it, and damaged and seized his sign," not to mention "the loss of all the *orviétan* removed by the Contugis, the production of which had cost a considerable sum."[34] He also argued that the letters patent that Louis-Anne Contugi had secured in the intervening period were not valid because he had not provided the requisite demonstration before civil magistrates, and furthermore, that he registered them at the Châtelet in prejudice of their ongoing case before the Conseil privé.

Most of these points were aimed at invalidating the initial seizure of goods ordered by Roberte Richard, but the response to the fifth question—as to whether the Contugis could, in any case, prohibit apothecaries from selling orviétan—strikes at the heart of the issue of pharmaceutical secrecy and the prerogatives of apothecaries. Boulogne argues that orviétan is not a medical secret at all, but a publicly known antidote described in various pharmacopoeias. As examples, he first cites Schröeder's *Pharmacopoeia medico-chymica*, mentioned above, along with "the Roman pharmacopeias for more than four centuries," the pharmacopoeias of Brussels, Antwerp, Lyon, Augsburg, Venice, Naples, and, crucially, the recent royally sponsored pharmacopoeia of Moyse Charas. As Boulogne would have it, physicians daily prescribed orviétan using any one of these various formulae. Boulogne granted that these recipes might not be the same as that used by the Contugis, but this difference was not due to their recipe being a hereditary medical secret but rather to the fact that the Contugis used discounted, nearly expired ingredients. Quality, then, is what distinguishes the orviétan of the apothecaries from the orviétan of the Contugis, and this quality is rooted in the skill of the practitioner. Boulogne asked rhetorically, "Who could convince themselves that a woman without experience, and her son, without having any tincture of medicine or pharmacy, would know how to make mixtures, preparations, coctions, and settle everything that goes into the composition of a remedy which apothecaries only attempt after having undertaken much work, long study, repeated public examinations, and the production of a masterpiece—in sum, after having provided evidence of their capacity to be admitted [to the guild] and be granted the faculty to exercise this art?"[35] Boulogne thus argues that his orviétan, produced with all the assurances of the corporate medical world in training and examination, is quite simply better than anything Contugi and his mother Roberte could ever hope to produce. Boulogne clinches his case with a volley of arguments against the notion that orviétan was a hereditary family secret: not only was orviétan in the pharmacopoeias but it had also been produced publicly, even in Paris, in a demonstration organized by Henry Rouvière, the syndic of the apothecaries of the royal households—Boulogne's own corporation—before the lieutenant general of police and the Paris Medical Faculty. In closing, he argues that if orviétan were the special preserve of the Contugis, it would "dismember" pharmacy and alienate apothecaries from their proper function; as such, it would run contrary to the order of arts and crafts that the king intended for his subjects.[36]

Boulogne's argument rests upon two foundations: the necessary superiority of the training and consequently skills of corporate apothecaries, and the assertion that orviétan is not itself a hereditary medical secret. The second claim has an important basis, as we have already seen: by 1685, orviétan was indeed included in several pharmacopoeias. The most recent and prestigious of these, the *Pharmacopée royale* of Moyse Charas, had criticized those "charlatans" who claimed to own the true secret of orviétan in much the same terms as Boulogne and La Martinière. Charas argued that "the good effects which well-prepared Orviétan has hitherto produced have given occasion for various affronters to employ all sorts of methods to make people believe that they or their predecessors were the sole inventors, and that they were the only ones to have the true recipe."[37] These arguments anticipate those that would be made by the American medical profession in the nineteenth century, as Joseph Gabriel's essay in this volume shows (chapter 12): itinerant operators and "Indian doctors" were likewise denounced on these grounds by incorporated professionals, who used pharmacopoeias as tools to challenge the proprietary nature of their medical secrets. Whether or not such claims to having unique knowledge of the recipe for the "one true *orviétan*" were true, the question remains: Did Charas or any of the other pharmacopoeias have the familial recipe of the Contugi family?

It is virtually impossible to say. Patrizia Catellani and Renzo Console have demonstrated that recipes of orviétan were extremely variable, and astutely point out that this variability was in fact a product of the conditions of competition and secrecy under which orviétan was originally disseminated. Secrecy encouraged a multiplication of recipes, as different charlatans, apothecaries, and physicians sought to replicate, imitate, counterfeit, or publicize the secret. Under these conditions, the number of ingredients expanded and recipes became exceedingly complex.[38] Throughout the later sixteenth and early seventeenth centuries, orviétan does not appear in printed pharmacopoeias: at least initially, members of the corporate medical community seem to have sought to keep their distance from the orviétan "charlatans."[39] Beginning with Schröder's *Pharmacopoeia medico-chymica* in 1655, divergent recipes for orviétan were readily available in the pharmacopoeias of Prévost (1666), Kratzman (1667), Herford (1667), Hoffmann (1675), and Charas (1676); the official pharmacopoeias of Rome (1668) and Lyon (1676); and in La Martinière's 1665 treatise on theriac, mithridatium, and orviétan.[40] The variant recipes in these pharmacopoeias no doubt manifested the existing diversity of orviétan recipes then in circulation, but they also likely contributed to this diversity as time went on, with individual apothecaries working to simplify, modify, or otherwise "perfect" their own orviétan. The most notable instance of innovation would eventually produce a new branch of orviétan, the so-called *orvietanum praestantius*, an opiate and narcotic analgesic, quite distinct from the original antidote in pharmacological terms.[41]

Boulogne was thus correct in pointing out that recipes for orviétan existed in numerous pharmacopoeias—even if this did not guarantee that any one of them was the Contugi recipe. Some of his other claims are more dubious, however. Orviétan was not included in the Roman pharmacopoeia until 1668, a far cry from the "four centuries" that Boulogne alleges. But Boulogne's hyperbole can be explained by another of his claims; namely, that public preparations of orviétan had recently been undertaken by Rouvière, the royal apothecaries' syndic. This almost certainly refers to Rouvière's recent public preparation of theriac, described in the 1685 *Journal des Sçavans*.[42] Beforehand, Rouvière secured all of the rare ingredients in the Galenic recipe for the theriac of Andromachus in large quantities, including exotic opobalsam, and on the day of his demonstration, he presented the Paris Faculty, apothecaries, and the interested public with the spectacle of fifty-eight dozen live vipers, pronounced a speech, and then cooked the vipers with only two spoons of water in a massive bain-marie. He then kneaded the resultant juices with the other ingredients into *trochisques* (bread lozenges), which preserved the volatile salts to which theriac's virtues as an antidote were credited.[43] The royal apothecaries' preparation followed an earlier public preparation of theriac, organized the previous year (1684) by the urban apothecaries Matthieu-François Geoffroy, Antoine Josson, and Simon Boulduc.[44]

This reference to Rouvière's public theriac demonstration thus allows us to understand some of Boulogne's more puzzling statements, namely his claim that the secret of orviétan has been included in pharmacopoeias for centuries: he seems to be saying that orviétan is largely indistinguishable from theriac. To what extent can orviétan be seen as a species of theriac? No recipe for the Contugi family orviétan appears to have survived, but the question can be answered by looking at the variety of other orviétan recipes that have come down to us. Catellani and Console have examined thirty-five different recipes and determined that the number of ingredients varies between nine and fifty-seven, with an average of twenty-six.[45] Of these, the three most frequent are angelica root, honey foam, and "aged" theriac.[46] So, it can safely be said that most pharmacopoeia recipes of orviétan contained prepared theriac, and that those that did not contained theriac's most conspicuous ingredient, viper flesh.[47] Therefore, even if orviétan cannot be reduced to theriac—as it might contain dozens of other ingredients—it seems clear that the two drugs were closely related. La Martinière even describes orviétan as having emerged from an effort to devise a "double theriac,"[48] and David Gentilcore has suggested that orviétan in its initial Italian context "was being offered as an accessible, if not 'poor man's,' theriac."[49] Indeed, it is probably not a coincidence that the renewed interest in theriac, most dramatically illustrated by public preparations like the one described in the *Journal des Sçavans*, matched up quite closely with the appearance of orviétan in the pharmacopoeia.

By the time of the lawsuit between Boulogne and Contugi, then, a whole panoply of divergent orviétan recipes were available in various pharmacopoeias, and all of these linked it back to theriac, a drug that had a considerable pedigree and whose public preparation was being used by the apothecaries to elevate their own professional status. Though this formed a critical part of Boulogne and the apothecaries' case, it seems to have drawn little response to the Contugis, who continued to hold their royal privilege and never conceded that their secret had been revealed.

By the summer of 1685, both parties found that their case was in deadlock.

Recognizing the expenses that continued litigation would entail, the Contugis and Boulogne resolved that, in order to "foster peace and friendship between them," they would follow the "good counsel" of their friends and agree to return their affairs as close as possible to the status quo that existed before the initial seizure. They settled the matter out of court, signing a notarized cessation of hostilities on August 14, 1685.[50] Interestingly, Rouvière, the royal apothecary who led the theriac demonstration, appears as a cosigner in the contract because he was serving as syndic for the corporation of the apothecaries of the royal households, of which Boulogne was a member.

The compromise they reached was to bifurcate the identity of orviétan—distinguishing between pharmacopoeia orviétan recipes and the version produced by the Contugis—and ensuring that this distinction would be clear to the drug-consuming public. The settlement prohibits Boulogne and any other royal apothecaries from selling orviétan out of any boutique on the Pont Neuf or from replicating any of the material aspects of the Contugi orviétan brand, including their displays, handbills, containers, signage, and mark, "nor to post signs with the name of *Orviétan* outside of their boutiques, as he does, nor in the roads, crossroads, and public squares of Paris."[51] The contract does not, however, prohibit them from otherwise making or selling "their *orviétan*," carefully distinguished from that of the Contugis, as the orviétan "whose composition is taught by the authors of the pharmacopoeias."[52] The Contugis thus recognized that they might still retain their cachet in advertising, branding, and sales location. So long as their clientele placed more credit in their preparation, and their rivals did not interfere with these exclusive features, then apothecary production and sale of what might be called "pharmacopei orviétan" could be tolerated. In legal terms, the settlement stressed the Contugi orviétan more as a brand rather than as a medical secret—although again, they never concede that any of the pharmacopoeia recipes described their "true" orviétan.

What was the Contugi brand in concrete terms? Answering this question takes the discussion from legal debates and into print advertising and the material culture of trademarks, in which proprietary medicines played a pioneering role.[53] The central iconographic feature of Contugi branding was the so-called sign of the sun. It appears prominently at the top of a surviving Contugi broadsheet as a sun with a human face, surrounded by the motto "ut sol solus ut sal salus" (figure 4.1).[54] The sign is surrounded by coats of arms, including those of the pope and the king of France. The rest of the border is made up of snakes, mushrooms, frogs, fish, spiders, scorpions, snails, and rabid dogs, all sources of poisons to which orviétan served as an antidote. These images may also have been present on the larger painted displays (*tableaux* and *montes*) that are mentioned in the settlement.

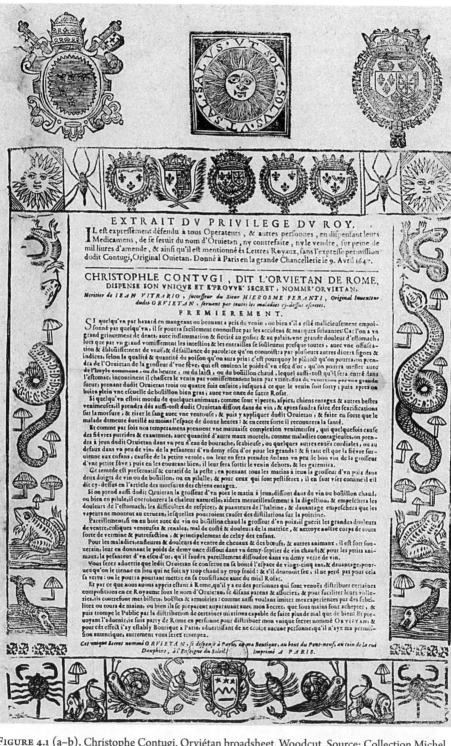

Figure 4.1 (a–b). Christophe Contugi. Orviétan broadsheet. Woodcut. Source: Collection Michel Hennin. *Estampes relatives à l'Histoire de France*, vol. 39, p. 28. BnF Richelieu, Reserve QB-201 (39). Courtesy of the Bibliothèque nationale de France.

The central "sign of the sun" is the most important of these images, and also appears prominently on the small, cylindrical lead containers (*boîtes*) that orviétan was sold in. Even in the twenty-first century, these containers continue to be unearthed by French hobbyists with trowels and metal detectors.[55] The crushed lid in figure 4.2 even includes the Contugi family orviétan motto. As John Styles has shown in the case of early modern London, the integration of printed advertising with a distinctive container design offered useful tools for product differentiation, strategies that were typically adopted by proprietary medicine vendors in response to counterfeiting.[56] The prominence of this trademark—with its obvious links to the authority of the Sun King, Louis XIV—on both the broadsheet and the container lids, also helps illustrate the stakes of the debate between the Contugis and their rivals. In Paris at least, they saw "the sign of the sun" as an inalienable mark, inseparable from their family and the hereditary secret to the true orviétan—despite the fact that similar signs were used by other itinerant vendors, particularly in the French provinces.

FIGURE 4.2 (a–c). Container lid with the Contugi orviétan trademark and motto. Source: User terremythe, on "La-Detection.com—Forum de discussion, identification trouvailles, detecteur de metaux," http://www.la-detection.com/dp/message-75142.htm (accessed November 28 2013). Used with permission.

After the settlement with Boulogne, the conflicts surrounding the Contugi orviétan privilege died down and the parties appear to have lived in peace. In the following year, Louis-Anne Contugi had his privilege confirmed and secured a passport to travel and sell orviétan throughout France. The December 27, 1686, letters patent even specify that he held the privilege jointly with his mother, perhaps in response to the legal difficulties that had arisen when Roberte acted to defend her son's privileges in his absence.[57] Antoine Boulogne for his part would go on to be first apothecary to Louis XIV in 1704.[58] The orviétan privilege was repeatedly renewed and confirmed, remaining in the Contugi family until 1741— although by 1700 it did so under the condition that it be inspected by the Paris Medical Faculty, and after 1736, the Parisian apothecaries' guild had inspection rights to the Contugi boutique as well.[59]

CONCLUSION

Pharmacopoeias and secret remedies were in some ways defined by their opposition to one another: therapeutic innovation was defined against established preparations, familial secrets against published recipes. The case of orviétan shows that despite this opposition, the two were, in fact, mutually interdependent. Although drug monopoly privileges stand as a distinct mode of medical organization—one in opposition to the corporate world of the apothecaries—they nonetheless depended upon pharmacopoeias as a standard of "known" recipes against which innovation could be defined. But even before this criterion was formalized as a part of drug regulation in 1728, the 1680s Contugi–Boulogne orviétan dispute showed that pharmacopoeias had already long been invoked by apothecaries in lawsuits against drug monopolists.

The case of orviétan also raises questions about the role of pharmacopoeias in fixing the identity and preparations of drugs. Medical secrecy, understood as an attempt to deliberately restrict access to knowledge in order to monopolize it economically, also tended to propagate variation in recipes. The lineal transmission of recipes across generations could itself produce variation, as could the work of imitators (or counterfeiters) who endeavored to "reverse engineer" a given drug. These processes help account for the growing diversity among the orviétan recipes that could be found in the pharmacopoeias by the 1680s. Paradoxically, while an individual pharmacopoeia might help to codify standardized preparations among apothecaries within a given jurisdiction, the case of orviétan shows that rather than fixing the drug to a standard recipe, the various European pharmacopoeias available in the 1680s in fact manifested—and contributed to— the diversity of recipes then in circulation.

On a social level, the friction between pharmacopoeias and secret remedies in France was also a manifestation of the conflict between corporate and non-corporate medicine, or between familial monopolists and guild apothecaries. This is abundantly clear in Boulogne's assertion of the superiority of the corporative mode of training in ensuring the quality of pharmaceutical preparations. From his perspective, the Contugis were unjustly trying to monopolize a variant on theriac, an ancient, well-established drug that was virtually synonymous with the apothecaries' craft. From the perspective of the Contugis, however, Boulogne and the apothecaries were attempting to rob them of their rightful inheritance, with the recipes for "pharmacopoeia orviétan" merely serving as a pretext for counterfeiting their trademark and deceiving the medicine-consuming public. For the Contugi family, the dispute appears to have triggered a reevaluation of their monopoly around branding and, to a lesser extent, their privileged point of sale. As medicine commercialized, trademarking offered a third way—alongside pharmacopoeia preparations and monopoly privileges—of establishing distinctions between drugs in the world of healing goods. In their own ways, both parties aimed toward a kind of enclosure of the therapeutic commons, an attempt to claim exclusive ownership over a piece of knowledge—the orviétan recipe—and the commercial benefits that came with it. Both attitudes were inherently proprietary. They differed mainly in scale and type of social group to which they sought to restrict access to the knowledge and practices associated with producing orviétan: corporate ownership in the case of the apothecaries, and familial ownership in the case of the Contugis.

Both parties were taking a gamble by settling out of court. The royal apothecaries were content to sell their "pharmacopeia orviétan," even if they would have to do so by competing against the Contugis and their established brand, which, after decades of sales, doubtless had a cachet among the people of Paris. The professional credibility that came to apothecaries through the institutions of corporatism—apprenticeship, licensing, inspection—would guarantee the quality of the product. Likewise, by laying the emphasis on their brand in the settlement, Contugis were gambling that buyers could never really be sure that the generic "pharmacopoeia orviétan" of the apothecaries was, in fact, the same as theirs, and that, in order to guarantee its efficacy, they would continue to buy their orviétan "at the sign of the sun."

5

CROWN AUTHORITIES, COLONIAL PHYSICIANS, AND THE EXIGENCIES OF EMPIRE

The Codification of Indigenous Therapeutic Knowledge in India and Brazil during the Enlightenment Era

TIMOTHY D. WALKER

HISTORIANS OF medicine have largely undervalued, or failed to appreciate entirely, the importance of Portuguese global exploration and colonial settlement, especially relating to botanical prospecting, the treatment of tropical diseases, and the acquisition and dissemination of non-European medical substances. The Portuguese maritime empire came into being long before that of any European rival state, and was far more diverse geographically, culturally, and biologically. Thus, Portuguese exposure to a broad spectrum of Indigenous healing ideas lasted far longer than that of any other seafaring imperial nation and was key to assimilating and circulating valuable regional medical information, not just back to Europe, but to commercial and colonial destinations around the globe.

In the early sixteenth century, at the initiation of the colonizing era, Portuguese royal interest in medical botanizing had been pronounced and active; official Crown voyage orders typically dictated specifically that maritime expeditions should seek novel local medicines at every anchorage or port of call, even sending expert apothecaries along to assess medical commodities in tropical maritime entrepôts.[1] However, over the subsequent two and a half centuries, this initiative had lapsed, as the Jesuits and other religious missionary brotherhoods had largely assumed responsibility for healthcare in Lusophone colonial regions. But, with the expulsion of the Society of Jesus from all Portuguese sovereign territories in 1759–1760,[2] much of the empire's medical expertise and capacity had been lost. When the Jesuits decamped, they took their valuable secret apothecary records and specialized pharmaceutical knowledge with them, leaving a lacuna that secular authorities slowly realized would require assertive steps on the part of the imperial administration to fill.[3]

Thus, toward the end of the eighteenth century, the Portuguese monarchy and colonial administrators in Lisbon—principally the members of the Overseas Counsel, or Conselho Ultramarino, the body that exercised oversight of the maritime empire—began to take an increasingly active interest in gathering and codifying information about medicinal plants from their colonial enclaves. Sustained high losses of human capital in the tropics—not only among European soldiers, administrators, and settlers but also among valuable slaves in the Atlantic and Indian Ocean regions—also helped to prompt this initiative. Crown authorities, anxious to find effective remedies that could reduce personnel losses through illness or casualties, revived their long-dormant interest in discovering new Indigenous remedies, and commissioned selected agents in the colonies (typically authorities trained in medicine or botany) to catalogue native plants that could be of therapeutic and commercial use to support Portuguese imperial endeavors.

One result of this renewed effort, carried out in culturally diverse colonial settings across the Lusophone world, was the compilation, over approximately a quarter century (ca. 1780–1805), of several new pharmacopoeial lists and guidebooks of materia medica. The colonial agents—physicians, surgeons, botanists, and other natural philosophers—who catalogued indigenous plants or healing substances almost invariably codified them in combination with detailed information about their proper identification, preparation, preservation, efficacy, and application. If considered collectively, these works, sometimes presented to the Conselho Ultramarino in the form of a report or catalogue and other times as fully articulated volume-length manuscripts, apparently intended for publication and distribution in Europe, constitute a marked shift in medical administration and information-gathering practices within the Portuguese sphere. They reveal an unprecedented, centrally directed, and comprehensive state program to harvest medical information throughout the empire, the objective being to leverage such knowledge to shape imperial policy, thus supporting Portuguese efforts overseas in the broadest sense toward improving colonial subjects' health and longevity.

Significantly, these efforts were conducted largely in secrect, without any fanfare or publicity. After all, the Portuguese plan to gather and codify colonial medical knowledge, undertaken as it was clearly for imperial and commercial gain, had manifest strategic objectives. So, a strategic desire to keep such valuable information secret shaped Portuguese tactics to quietly collect, organize, and transfer such knowledge back to the metropole. Most medical field reports were never printed; they circulated only in manuscript within limited internal government channels, which effectively limited their broader potential impact. Note, however, that there occasionally arose an important divergence between the intentions of the authors of these medical field reports (who often wished to see them published) and those of the Portuguese Crown (who sought to keep strategic commercial and medical information a guarded secret). It is clear that the personal professional exigencies of medical practitioners did not always coincide with the strategic interests of the state. Hence, by considering these two general themes—secrecy and methods of collecting medical knowledge—the work presented here seeks to connect directly to two main themes throughout this volume.

As the Portuguese regime embarked on a refined and deliberate knowledge-making effort that reflected Enlightenment-era ideals and methods developed largely in Europe, proponents leveraged long Portuguese imperial experience with Indigenous encounters—protracted exposure to and dependency on diverse Native healing practices in far-flung global colonial enclaves, which in turn relied on local familiarity with medicinal plants of multiple regions. Due to this dynamic of cross-cultural medical exchange, by the mid-seventeenth century practical medicine in Portuguese colonial enclaves had become thoroughly hybridized, with applied remedies in colonial health institutions (whether state-sponsored or religious) relying significantly on Indigenous medicinal substances and methods, derived from various disparate Native healing traditions.[4] These preexisting circumstances allowed for the centralized state effort of the late eighteenth century that is under consideration here.

Until the 1780s, the collection and recording of information about materia medica had been, since the mid-sixteenth century, infrequent and distinctly unsystematic, both in the metropole and in the disparate colonies—at least as far as efforts undertaken solely on the part of Crown authorities were concerned. Instead, this service, arguably fundamental for a small nation-state endeavoring to maintain a global colonial system in the tropics, had been left largely in the hands of religious missionary brotherhoods, principally the Jesuits, who had systematically built up the greatest body of codified practical knowledge in Portugal about medical substances that originated outside of Europe for over two centuries.[5] But medical texts and recipes compiled by members of the Society of Jesus, while detailed and thorough, were meant for Jesuit eyes only; to maintain their grasp on this strategic lore, they deliberately refrained from disseminating their proprietary medicinal information outside of Jesuit circles.[6] Thus, following the Jesuits' suppression, and the compulsory modernization of the Coimbra University medical curriculum in 1772[7] (both policy initiatives of the enlightened though autocratic prime minister, the Marquês de Pombal), Portuguese Crown officials found it expedient during the reign of Queen Maria I to initiate their own state-directed program to systematically gather a centrally organized body of information concerning potentially useful healing substances from the colonies overseas. The abrupt loss of essential medical services that the Society of Jesus had provided throughout the Portuguese realm had simultaneously demonstrated the strategic necessity for state expertise over imperial pharmaceutical knowledge, while underscoring the centrality of effective medical care to the general success of colonial endeavors. The Portuguese case under consideration here exemplifies the practical connection between these two broad historical determinants.

Considered in comparison to similar European imperial undertakings, this Portuguese effort is roughly contemporary to the analogous Spanish endeavor to codify systematically materia medica in their Atlantic imperial sphere, which Matthew Crawford describes in his contribution to this volume (chapter 3). Effected during the Bourbon Reform era of the mid-eighteenth century, Spanish authorities sought simultaneously to strengthen centralized Crown administration throughout the realm and to "revitalize Spain's ailing imperial enterprise through economic and political reform"—thus seeking positive practical outcomes comparable to their Lusophone neighbors, through an improved leveraging of their colonial medical resources.

This essay will present and explicate little-known, often arcane information regarding pharmacopoeias and the codification of therapeutic knowledge within the Portuguese empire during the eighteenth century. My focus will fall mainly on data regarding properties of medicinal substances collected from Brazil and India, and recommended techniques for their application. Through the analysis of selected Portuguese colonial medical texts, and translated excerpts thereof, this essay will provide insight into how Enlightenment-era Portuguese practitioners gathered information in response to calls from the metropole for the systematic codification of healing plant knowledge.

A discussion of state-sponsored pharmaceutical knowledge collection within the Portuguese empire is of necessity global in scale, with compiled information found in pharmacopoeias originating in disparate places of exceptional cultural and botanical diversity. Yet this process resulted in a delicate interplay of medical ingredients from around the globe, combined to make new healing compounds that emerged as a result of the unique context created by Portuguese seaborne trade routes. The modern historian must account for the social context of bioprospecting in each colonial location, not only for the complex ways that Indigenous peoples in Africa, India, Brazil, Malaysia, or Southern China conceptualized the medicinal plants they proffered but also for the ways early modern Portuguese healthcare practitioners utilized them, interpreting their applications and efficacy as suited their own subjective exigencies.

Much of the information presented will be excerpted from little-known but highly representative eighteenth-century Portuguese primary sources: pharmacological and botanical compilations (medical and apothecary guides) in manuscript, written in colonial India and Brazil near the end of the eighteenth century. Among several that will receive consideration, two stand out as particularly notable. The first is colonial surgeon Francisco António de Sampaio's text, titled *História dos Reinos Vegital, Animal e Mineral*, a two-volume manuscript compiled at Cachoeira, Bahia, Brazil, between 1782 and 1789. This manuscript is held at the Biblioteca Nacional of Brazil in Rio de Janeiro. The second text is Bento Bandeira de Mello's untitled memorandum about medicinal plants of northeast Brazil, compiled in 1788. Mello's manuscript is held at the Arquivo Nacional da Torre do Tombo in Lisbon, Portugal.

As will become apparent in the works discussed throughout this essay, nearly the entire weight of scholarly Portuguese medical activity during this era of Enlightenment botanical curiosity fell on imperial enclaves in South America and India; the studies produced in response to Crown requests for medicinal plant information barely dwell on the African colonies at all. With few exceptions, Portuguese practice here mirrors, according to findings Benjamin Breen presents in his contribution to this volume, a broad European disinclination toward adopting African healing practices (chapter 7). Though central to European maritime and imperial economies, Africa was inexplicably underrepresented, and nearly invisible, as a source of medicines found in early modern pharmacopoeias or traded through contemporary commercial channels.

A SCIENTIFIC AWAKENING AND STATE-SPONSORED INQUIRIES ABOUT COLONIAL MEDICINE IN THE LATE EIGHTEENTH CENTURY

During the closing decades of the eighteenth century, the Portuguese monarchy and colonial administrators in Lisbon began a concerted effort to gather and systematize information about efficacious medicinal plants from their worldwide colonies. Their motive was to stem the shockingly high attrition rates—"wastage"—experienced among Europeans and Africans due to disease and injury in the tropics. Typically until the early nineteenth century, newly disembarked European soldiers, settlers, and African slaves in Portuguese colonial enclaves suffered terribly high rates of mortality, their ranks shrinking by 25 to 50% during their first year of service due to brutal work, harsh acclimatization (referred to by contemporaries as "seasoning"), wounds, and tropical diseases.[8] Crown authorities, desperate to find effective remedies that could reduce casualties, revived their interest in mobilizing known or new Indigenous remedies that could be of therapeutic and commercial use to imperial endeavors.

No analogous government effort had been operational since the sixteenth century, during the initial phase of overseas exploration and colonization. Moreover, there had been no published didactic work of note on medicine in the Portuguese territories since Garcia da Orta's volume, focused on Asia, of two centuries before (1563). For nearly two centuries, the Portuguese Crown had been content with the tacit delegation of such empire-wide healthcare efforts to the Jesuits and other missionary orders, or to the Santa Casa da Misericórdia.[9] But the departure of the Jesuits, combined with a growing awareness in Portugal of Enlightenment-era medical advances across Northern Europe, underscored the need for a comprehensive imperial health program directed from the metropole.

Soon after Queen Maria I ascended the throne in 1777, the Conselho Ultramarino began to commission medical authorities in Brazil, India, and Africa to write descriptions of potentially useful medicinal native plants and roots in their respective areas.[10] Similarly, Portuguese Crown authorities, sometimes through the agency of members of their diplomatic representatives abroad, dispatched botanists and natural philosophers to assess and acquire medically or commercially useful plants from other geographic areas of interest, notably in North America, Mozambique, Angola, and Brazil.[11] For example, for twenty years the Brazilian-born naturalist Joaquim José da Silva, who studied with Domingos Vandelli at the University of Coimbra and served as secretary general in Angola (1783–1808),[12] participated in and promoted important "philosophical" expeditions to the Angolan wilderness, all at the behest of the Conselho Ultramarino.[13]

What kind of responses did the Conselho Ultramarino's new initiative provoke? One example can be found in a two-volume pharmacopoeial manuscript, composed in Goa during the last quarter of the eighteenth century, titled *Oriental Medicine: Indic Assistance, for the Clamors of those Poor Infirm Patients of the Orient; For total alleviation of its ills, (...)*[14] Though anonymous, this undated text, totaling 1,312 pages, was intended manifestly as a didactic medicinal guide. Further, it appears to have been compiled in response to an official directive from the metropole. Like the prodigious contemporary manuscript concerning remedies from Brazil by Francisco António de Sampaio, discussed below, this project seems to have been intended for publication as a comprehensive compilation of all known medical lore for the region it covers (coastal Southern and Western India).[15] In a highly telling circumstance, once conveyed to Europe, the manuscript from Goa was deposited in the library of the Lisbon Royal Academy of

Sciences, founded by Queen Maria I in 1779 intentionally as an Enlightenment-oriented institution to facilitate the establishment of empirical learning in Portugal. This factor lends credence to the notion that the work was produced in response to a royal commission.

Through some remarkable historical detective work, Brazilian scholar Fabiano Bracht has established that the author of this noteworthy text was almost certainly one Luís Caetano de Meneses, an Indo-Portuguese physician active in Goa prior to 1786. On the title page, the anonymous author, who wrote beautifully in Portuguese, stated that he was a Native-born Goan, and that the lore he gathered was "acquired from various Professors of Medicine" in the Portuguese colonies in India. The work's first 640 pages describe a total of 782 different medicinal plants from India and other colonized regions of Asia (the remaining pages in the volume describe medical substances derived from "fish, birds, animals, minerals and precious stones"). Caetano de Meneses, if he was indeed the author, provided information in minute detail about these plants' healing qualities, applications, and names in Indian and European languages. Though the provenance of this work is unstated, its dedication (to "The Most Holy Trinity of the One True God"), precise intellectual rigor, and detail suggest that it was composed by a professional medical practitioner of long experience in the Portuguese colonial hospitals or infirmaries of Asia—one who may have received medical training from knowledgeable Jesuit missionaries.[16] The second volume, titled *Medicina oriental: Pharmacia Indiana*, is a companion to the first, providing over three hundred pages of various "Indian Pharmaceutic Compositions." As Bracht observes of this volume, "It is, in fact, a pharmacopoeia, with instructions for the preparation or fabrication of diverse medicines, by following the author's 'Chemical and Galenic' compositions."[17]

Efforts to discover useful Indigenous remedies continued in India, as elsewhere. For example, in a royal directive dated April 2, 1798, the *cirurgião-môr* (chief surgeon) of the Military Hospital of Goa, Dr. José Abriz, who had recently arrived from Lisbon, and the other staff *médicos* were given an opportunity to display their knowledge of Indigenous medicine from Portuguese India. Queen Maria I and the Conselho Ultramarino, seeking medicines to treat tropical diseases throughout the Portuguese maritime network, commissioned the Goa Hospital Militar staff physicians and surgeons to write a description of all the useful medicinal plants found along the southwestern Malabar Coast and employed in the remaining Portuguese enclaves in India. The following year, Dr. Abriz, a European-born, University of Coimbra–trained surgeon, and his colleagues produced a report, extending to nearly forty manuscript pages, in which they provided thorough descriptions of eleven important roots and plants then commonly in use in the medical facilities of Goa, Damão, and Diu, as well as in the

Portuguese East African colonial holdings.[18] The medical practitioners in Goa included their report with the official correspondence of the Estado da Índia (the *Livros dos Monções do Reino*), sent to Lisbon aboard the annual government-sponsored vessel; their cover letter is dated April 29, 1799.[19]

The 1799 Abriz report is actually the sequel to a document submitted in 1794 by the then *físico-môr* (chief physician) of the Estado da Índia, Ignácio Caetano Afonso.[20] Afonso's earlier report had clearly attracted particular interest among colonial officials in Lisbon. However, Ignácio Caetano Afonso was a Native Goan—though he was born into a Portuguese-speaking Indo-European family, his medical knowledge consisted primarily of native Indian plants and their medicinal applications. Having never left Goa, he had little access to formal medical training, but had gained a very favorable reputation as a healer. The governor of the Estado da Índia described Afonso as "a Brahmin . . . favored with natural talents" for healing, saying that "notable cures" had been attributed to him, even though he "had not opened any [medical] book for many years,"[21] and that Afonso had "the sense of a *Médico*, and practiced for many years, which compensated for the defects of his [medical] education."[22] Still, it is likely that the members of the Conselho Ultramarino, inspired by the innovative currents of empiricism that had only recently penetrated learned society in Portugal, wanted the benefit of a second opinion from a medical professional trained wholly in scientific Western medicine.

In any event, though the Conselho Ultramarino sought innovative Native medicines sourced from India and Angola, obviously considering them worthy of investigation, by the eighteenth century the main economic focus of the empire had shifted decisively to South America, due primarily to exceptionally profitable gold and diamond extraction in Brazil, but also to competition from European rivals that diminished trade from Asia. Thus, the primary effort of their inquiry was centered on the less well-known, largely unexplored territories in Brazil, where entirely novel forms of flora and fauna seemed to promise better prospects for medically beneficial discoveries.

INCREASINGLY SYSTEMATIC AND COMPREHENSIVE GUIDES ABOUT BRAZILIAN MEDICAL SUBSTANCES IN THE ENLIGHTENMENT-ERA PORTUGUESE EMPIRE

The century that was drawing to a close in contemporary Brazil had been literally a golden age; extraordinary wealth derived from the colony's booming gold and diamond mines drew many Portuguese immigrants, including a surprisingly meager handful of trained physicians and surgeons from the metropole.[23] There they confronted an entirely unfamiliar disease environment of the Amazon and South American tropics, as well as an Indigenous healing culture shaped by the context of the unique flora and fauna at their disposal, blended with influences to healing methodologies resulting from the massive regular annual influx of enslaved African peoples. Those Western *médicos* with inquisitive spirits produced handbooks, papers, or guides to the novel indigenous healing plants they encountered.[24] Some with a more scholarly bent went much further, creating detailed works for publication, apparently in the hope of reaching a wide audience back in Europe among interested researchers in medicine, botany, or natural philosophy.

In Bahia, a Portuguese-born physician, Francisco António de Sampaio, resident of the inland region near the erstwhile colonial capital city, Salvador, undertook an ambitious project of pharmacological botany, which he titled *History of the Vegetable, Animal and Mineral Kingdoms, Pertaining to Medicine*. His venture—a medical field guidebook broad in its scope and objective—was a particularly ambitious undertaking for someone who was essentially a country doctor in Brazil in the late eighteenth century. Sampaio compiled his two-volume work between 1782 and 1789 at Cachoeira, the main agricultural market town, a center of intense Afro-Brazilian plantation slave labor, on the Paraguaçú River in the fertile Bahian hinterland around the Bay of All Saints. Fair copies of each volume, handwritten and bound together with stunning original painted illustrations of Brazil's flora and fauna, are deposited in the special collections division of the Biblioteca Nacional in Rio de Janeiro.[25]

Sampaio embraced the role of an Enlightenment-era médico who clearly wanted to expose his countrymen to a deeper knowledge and understanding of the traditional Brazilian medicinal plants with which he regularly worked. Indeed, the project (because of internal indications relating to its structure, organization, and parameters) shows telltale signs consistent with analogous works known to have been produced by royal government commission, most likely at the behest of colonial authorities in Lisbon, or possibly Bahia. The two extant manuscript tomes each contain highly detailed descriptions of a variety of native South American plants, a summary of their healing virtues, proper doses to administer to patients, and methods for applying each remedy to the sick.[26]

Sampaio's work depended on his prolonged experience in the Bahian hinterland with European settlers, Amerindians, and Afro-Brazilian slaves. He conveys, through descriptions of regional remedies for common illnesses, some sense of the incredibly difficult conditions under which African slaves, who provided most of the plantation labor and constituted the majority of the population in the district, worked. By considering South American Indigenous healing knowledge as it blended with European and Luso-African medicinal practices, Sampaio's incisive treatise encompassed the social spectrum of colonial Brazil in the eighteenth century.

To this extent, Sampaio's unpublished manuscript follows a model comparable to that which William Ryan identifies in the writing of the notable contemporary London-based naturalist and physician Sir Hans Sloane. In his contribution to this volume (chapter 6), Ryan argues that "medicine and its associated fields of botany, natural history, surgery, and materia medica highlight the particular intimacy of the colonial relation: exposure to and fear of unfamiliar diseases and the highly competitive early modern medical marketplace forced Europeans into close interaction with Amerindians and Africans." This perfectly describes, of course, a context—the reliance upon multicultural interdependence—that the surgeon Sampaio experienced firsthand in rural Bahia, just as Sloane had done as a physician in Jamaica.

Francisco António de Sampaio had been born at Vila Real in northern Portugal, but immigrated to Brazil as a young man. Where he completed his medical studies is not clear, but he never enrolled in any formal course at the Coimbra University faculty of medicine.[27] Most likely he trained as an apprentice with a licensed physician, or through a residency at the Todos-os-Santos Hospital in Lisbon, before leaving Portugal. In 1762, King José I granted him a license as a physician, with the right to practice surgery.[28] In Brazil he became an approved surgeon with a license to practice medicine granted by the Bahian colonial senate.[29] Sampaio then held the post of surgeon at the Hospital of São João de Deus in Cachoeira for nearly two decades. In the 1780s and 1790s, Sampaio corresponded with officials of the new Academia Real das Ciências de Lisboa, and was registered among its members in 1798; his letters reveal that he submitted copies of his work for consideration and inclusion in the academy's library, and that he engaged with a network of like-minded medical botanizers within the Portuguese Atlantic.[30]

Volume one of his work, completed in 1782, described medicinal plants in 219 manuscript pages, supported by another 20 pages of color miniature paintings that skillfully rendered many of the plants described in the text. The plants described in Sampaio's first volume are organized into twelve sections according to their contemporary medicinal applications.[31] European usage in the colony generally mirrored practices gleaned from Native peoples through firsthand observation, but also reflected innovative applications pioneered by the colonizers through their own experimentation. Groups of plants evincing astringent, antivenom, anticolic, antispasmodic, purgative, and antivenereal healing qualities are each treated in their own discrete chapters. Sampaio provided practical information about how to identify, gather, and preserve each plant discussed, together with instructions about various therapeutic applications, proper medical preparation, elaborated recipes, and dosages.

It is perhaps worth noting here that, in regard to profits derived from innovative healing substances and recipes, Portuguese medical practice in the colonies followed a dynamic similar to that which Justin Rivest, in his contribution to this volume (chapter 4), describes as current in early modern France. In the francophone world, useful medical innovation was rewarded with grants of exclusive royal privilege for the sale of efficacious new remedies. Though never formalized to the same degree, such was effectively the same benefit that inventive pharmacists (or, more often, ecclesiastical apothecaries) enjoyed in the Portuguese imperial realm in the seventeenth and eighteenth centuries—their secret compounds could be monopolized for personal or institutional profit, so long as the recipe was kept from dissemination through publication in a pharmacopoeia, or by any other means.

To aid in plant identification for neophytes to Brazil, Sampaio included an alphabetical index of each plant name, carefully noting their Tupí and Guaraní names, as well; this is one of the most extraordinary components of the manuscript. Sampaio's intent was not to be culturally sensitive, of course; his objective was practicality. In a colonial context that relied heavily on non-Portuguese-speaking Indigenous Brazilians to help source medicinal plants from the undeveloped interior, a phonetic guide to useful plant names was an essential addition to his pharmacopoeial handbook. He also commented knowingly on the use and efficacy of non-native medicinal plants (like coffee, pepper, and cinnamon) that the Portuguese had introduced from Europe or their overseas territories in Asia.[32]

Although the painstaking, protracted work was obviously intended for publication for a broad readership in the transatlantic medical and scientific community, for unknown reasons the project never went beyond the manuscript stage. Sampaio's work may have been considered too provincial, or dated, or he simply may have been unable to win the support of an influential patron in the metropole. Had the work been published, it would have contributed greatly to knowledge of regional Brazilian healing plants in the Lusophone world. Instead, it languished as an unprinted manuscript. Despite his link to the Academia Real das Ciências, circulation of Sampaio's fair copy text was probably limited only to a handful of knowledgeable colonial and medical officials in Portugal; therefore, its ultimate impact as a broader conduit of information to a popular audience, though uncertain, was likely quite meager.

The Portuguese Crown permitted no printing press in Brazil until 1808, when the royal court and family took up residence in Rio de Janeiro; until then, all printing projects had to be sent to Lisbon for approval and typesetting.[33] Similarly, the Crown banned printing throughout Portuguese colonies in India and Africa. Such restrictions reflect a deliberate policy to curb the dissemination of what imperial authorities perceived as strategic commercial information, but were also intended to halt the spread of dangerous unorthodox intellectual concepts, including the new culture of empirical science then emanating from Northern Europe. The Portuguese Inquisition Censorship Board, which monitored the flow of ideas within the Lusophone world, was not abolished until 1821; government censorship of scientific texts continued until the implementation of a new Liberal constitution the following year.[34] Clearly, such oppressive regulations and secrecy served to impede the circulation of medical information within the Portuguese sphere; for Sampaio and his contemporary natural philosophers, this repressive, reactionary intellectual climate represented a constant challenge or check to their activities.

Elsewhere in Portuguese America, at about the same time, a similar state-directed effort to gather medicinal knowledge was under way—this one explicitly mandated by colonial authorities in the metropole. In 1788, Brazilian physician and natural scientist Bento Bandeira de Mello submitted a lengthy memorandum on frequently used Indigenous medicines in his home region, the coastal northeast of Brazil. De Mello was responding to a direct royal order from Queen Maria, transmitted through the Overseas Council; he had been charged with creating an alphabetical list of medicinal plants, fruits, and roots from the territories of Pernambuco and Paraíba, with commentary concerning their curative effects.[35] His annotated roster, containing fifty-nine different South American medicinal plants, runs to twenty-four manuscript folios, now archived at the Portuguese National Archive (Torre do Tombo) in Lisbon.

Examples of native healing plants de Mello discussed in his compilation include various types of *ipecacuanha* (also called *cipó*), a reliable emetic and diaphoretic; cinchona bark (also called *quina* or *quineira*), arguably the most important remedy found in the New World—a febrafuge essential to treating malaria and other tropical fevers;[36] *jalapa*, an effective purgative; *copaíba*, the bark and plant oil of which was used internally and externally to treat gonorrhea; and *salsaparilha*, administered against syphilis and skin diseases.[37] More than any others, these particular Brazilian remedies had attracted a broader market within the Atlantic medicinal economy, gaining widespread medical usage not only in South America but elsewhere in the Portuguese Empire; demand for these plants grew steadily until they became significant commodities, both medically and commercially. By the end of the eighteenth century, these Indigenous Brazilian healing products, in bulk or in prepared remedies, commonly could be found in ships' seaborne medical chests[38] or stocking Portuguese pharmacy shelves from Lisbon to Mozambique, Goa, Timor, and Macau.[39]

By way of comparison, it is worth noting here that, in his contribution to this volume, Christopher Parsons focuses on how the collection, processing, marketing, and consumption of a particular North American plant, the northern maidenhair fern (*Capillaire du Canada*), mediated and was mediated by debates about the nature of colonial space in the French Atlantic World (chapter 8). In the Lusophone sphere, plants like Brazilian varieties of cinchona, ipecacuanha, and jalapa are roughly analogous to Parson's case, which "became the principal pharmaceutical remedy" transported from New France to Old. Like maidenhair, the "most well known and most important contribution that Canada made to European materia medica," the above mentioned Brazilian plants represented Portuguese South America in the Iberian metropole. As Parsons writes, "Capillaire de Canada [. . .] became a principal vehicle through which the medical marketplace of seventeenth- and eighteenth-century France came to terms with the ontological status of this and other French colonial spaces." Parsons continues, asserting that the case of Canadian capillaire suggests "that the history of pharmacy in the early modern Atlantic World was entangled with larger political debates about the nature of empire, and that considerations of France's colonial project were weighed through the experience of an otherwise unremarkable plant." The same might well be said for paradigmatic Brazilian plants as medical commodities, and the role they played in establishing or representing the nature and character of Brazil in the minds of Portuguese elites in Lisbon.

Per royal instructions, de Mello sent specimens of many of these plants to the royal botanical garden of the Ajuda Palace in Lisbon, where they were assessed for their medical usefulness, as well as for their suitability for transplant to other imperial regions.[40] Hence, the impact of his work carried farther than the palace chambers of the Conselho Ultramarino. The desired end of this official initiative, of course, was to further Portuguese aims by reducing chronic, unacceptably high wastage of human resources through injury and illness.

DESCRIPTIONS OF BRAZIL'S MEDICINAL PLANTS IN BROADER CONTEXT (EIGHTEENTH CENTURY)

In 1780, another licensed practitioner who had served in Brazil, the Portuguese-born médico Manuel Joaquim Henriques de Paiva, issued a new compendium of medicinal plants and remedies found throughout the Lusophone world. His *Farmacopéa Lisbonense*, published with royal approval and licenses in Lisbon,[41] contained conventional European healing methods, but the work drew heavily on Paiva's experience in South America. The author explained the healing properties of dozens of traditional Brazilian remedies and individual plants, assessing their utility and prescribing methods of application.[42] Paiva benefited from his position as personal physician of the prince regent, Dom João, whom he would later accompany, along with the rest of the Portuguese royal family, when the court relocated to Rio de Janeiro in 1808. *Farmacopéa Lisbonense*, with expanded and corrected editions published in 1785 and 1802, was widely distributed in Portugal and Brazil, more than any other published Lusophone work of the eighteenth century, and provided accurate medical information about South American Indigenous *drogas*, according to contemporary understanding.[43]

Naturally, Paiva's work had been approved and patronized by the Crown; it had been sanctioned by the regime's censorship board and fit perfectly with the royal program of promoting colonial medical information. The publication and success of Paiva's work, as well as the author's privileged social position, may help to explain why Francisco António de Sampaio's manuscript volumes, described above, which were composed by an obscure, unconnected, provincial surgeon in Bahia at almost the same time, never garnered royal support or approval to be sent to press in Portugal.

One further point about this publication bears mentioning. The *Farmacopéa Lisbonense* (1780), like its French counterpart, the *Codex Medicamentarius, sive Pharmacopoea Gallica* of 1818, discussed in Antoine Lentacker's contribution to this volume, became the first national medical guide to remedies principally from Brazil to carry the fully sanctioned approval of the Portuguese Crown (chapter 11). Lentacker observes incisively that, with regard to sanctioning and regulating various government functions, including specifically medical practice, "the centralizing and standardizing agenda [. . .] had deep roots in the bureaucratization of the early modern monarchical state"—an argument that, Lentacker notes, Alexis de Tocqueville had also "famously asserted." Lentacker continues, writing that the "story of the *Codex* of 1818 offers a lens onto the nature and logic" of this governmental transformation toward far-reaching absolutism, a transformation as apparent in royal policies in Portugal of the late eighteenth century as it was in postrevolutionary France, encapsulating, as Lentacker says, "in uniquely transparent ways the changes that the Revolution brought about in the governing of people and things."

Of the many exploratory trips through the Brazilian interior during the eighteenth century,[44] only a few were initiated as a matter of Portuguese state policy. The nine-year "philosophical journey" of Dr. Alexandre Rodrigues Ferreira stands out for this reason, and for the significance of its medical and botanical discoveries. Because of his pioneering descriptions of Amazonian flora and fauna, Brazilians today remember Ferreira as their first Native-born naturalist. Born in 1756 to a wealthy merchant in Salvador da Bahia, he was sent to Portugal to complete his studies at the University of Coimbra. In 1779, after earning a doctorate in natural history, Ferreira was employed at the royal botanical gardens of the Ajuda Palace in Lisbon, where he gained favor with Queen Maria I.[45] In 1783, Maria (who, in an attempt to link Portugal to the Enlightenment-era currents and institutions active across Europe, had just founded the Academia Real das Ciências de Lisboa in 1779) selected Ferreira to explore and map the then barely known regions of the Brazilian tropics (Grão-Pará, Rio Negro, Mato Grosso, and Cuiabá), searching for natural resources that could be exploited for the economic benefit of the empire. Ferreira experimented with promising medicinal native plants that he encountered along the way and catalogued them with annotations.[46] The entire collection of documentation from Ferreira's expedition, including his commentaries about medicinal plants, is held in the manuscript division of the Brazilian National Library in Rio de Janeiro. In 1887, a full description of Ferreira's trip, entitled *Diário da Viagem Filosófica*, was published in the journal of the Instituto Histórico e Geográfico Brasileiro.[47]

In another example of this type of official pharmacological undertaking, just before the turn of the nineteenth century, the newly appointed colonial governor of the Maranhão district in Brazil, Dom Diogo de Sousa, commissioned several descriptive works regarding potentially useful indigenous plants for commerce or medicine. Appointed agents began to compile this information in the interior forests of the district in 1798; his instructions to promote this project had come directly from the Conselho Ultramarino. Three separate works, the fruit of this effort, arrived in Lisbon by 1801: two folios of watercolor botanical illustrations and one manuscript describing the pictured plants' uses.[48] The two compilations of plant illustrations, together containing some fifty-five varieties in all, focus mainly on species that had Indigenous medical applications. These include the stimulant *guaraná*, a widely used healing shrub called *pau d'arco*, and various flora to alleviate fevers, asthma, urinary problems, skin disorders, and even to promote hair growth.[49] The manuscript, titled "Botanical Guide to Some Plants from the Interior of Piauí," was compiled by Vicente Jorge Dias Cabral, a Coimbra University philosophy graduate turned amateur botanist. Cabral's thirty-page text describes the natural appearance and application of twenty-three medicinal plants.[50]

Portuguese imperial designs in the Americas aimed at collecting information about useful plants were not focused only on the Southern Hemisphere; the Lisbon government acted upon opportunities to gather data and specimens in North America as well. With the resolution of the United States' war for independence in 1783, the negotiation and signing of the first Luso-American peace and trade treaty in 1786,[51] and an exchange of diplomatic representatives in the 1790s, the final fifteen years of the eighteenth century saw an ever-expanding volume of trade develop between the United States and Portugal.[52] Tables of ship arrivals to Lisbon harbor, systematically reported in the *Gazeta de Lisboa*, reveal why the Portuguese considered North American commerce so essential: in 1788 the United States, entering with sixty-five merchantmen, ranked fifth among Portuguese trading partners. Two years later, the Americans ranked a close third, with just one fewer ship than the Dutch, but still far behind the British. The United States held a steady share of the Lisbon trade through the eighteenth century and into the early nineteenth.[53] Thus, given Portuguese Crown exigencies with regard to sourcing useful and efficacious plants, ample opportunity appeared to exist in this regard to exploit close commercial relationships with North America.

As the young American republic's economy flourished, the Portuguese began to conduct their diplomatic affairs with a view toward gaining North American mercantile knowledge, workers, or indigenous plants that could be put to use to sustain the Portuguese empire. For example, Cipriano Ribeiro Freire, the first Portuguese diplomatic envoy to the United States (1794–1799), was instrumental in sending—sometimes surreptitiously—seeds from North American cash crops to Portuguese colonial territories in the hope that they could be profitably cultivated there. In Lisbon, Dom Rodrigo de Sousa Coutinho, minister and secretary of state for the navy and overseas colonies, directed that Ambassador Freire send samples of Virginia and Maryland tobacco seeds to Portuguese Africa and Brazil.[54] At the time, the tobacco plant was highly esteemed for its medicinal qualities—André João Antonil, a Jesuit priest and rector of the Jesuit College in Bahia, asserted that chewing or "drinking" (smoking) the properly cured native American leaf, known as the *erva santa* or holy herb, could clear respiratory passageways and alleviate asthmatic discomfort in the chest, provide relief from an upset stomach, help digestion, and even ease "unbearable" toothache.[55] Cipriano Ribeiro Freire also contracted with American merchant Peter Tilly to transport the first seeds for cayenne pepper from North America to the Brazilian province of Pará, where it was transplanted and cultivated with commercial success.[56]

It is worth mentioning that, for their part, the Americans simultaneously were doing everything in their power to gain direct access to trade with Brazil's commercially attractive ports and market towns—which Portuguese imperial authorities kept scrupulously closed to all interloping foreign mercantile interests.[57] Not only did the Lisbon regime wish to block potentially damaging illicit commerce and smuggling that would reduce Crown profits but they wanted to discourage the spread of dangerous republican rhetoric that could encourage independence aspirations throughout the Iberian colonies in the Americas.[58] Moreover, the Portuguese also did not want to lose their near monopoly on the sources of certain strategic plant-based drugs and forest products, like Brazilian varieties of cacao or cinchona, which their own experience had shown could be relocated successfully—and profitably—to other climes.[59]

Attempts at purloining desirable flora from the Portuguese Atlantic colonies seems to have been an ongoing problem with North Americans since at least the middle of the eighteenth century. No less a personage than George Washington had once tried to circumvent government controls on shipping commercially valuable plants out of Portuguese territories, presumably for experimental replanting at his Mount Vernon estate. In February 1768 he wrote to his wine merchant agents in the Madeira Islands to secure a selection of grape cuttings. His delicate phrasing appears to indicate that he was fully cognizant of the sensitivity of his request: "And if there is nothing improper, or inconsistant [sic] in the request a few setts or cuttings of the Madeira Grape (that kind I mean of which the Wine is made), but if in requiring this last Article there be any sort of Impropriety I beg that no notice may be taken of it."[60] There is no indication in Washington's subsequent correspondence that his improprietous solicitation was ever honored.

In fact, rumblings of several independence movements in Brazil around the turn of the nineteenth century caused considerable alarm in Portuguese government circles, in no small part because the riches of *América Portugueza* included a seemingly limitless bounty of efficacious medicinal plants.[61] When faced with the serious prospect of losing their prized South American colony later, in the second decade of the nineteenth century, prescient Portuguese imperial agents transplanted useful medicinal plants from Brazil to other locations within the empire. After Brazilian independence, Portuguese colonists continued to cultivate and exploit Brazilian healing plants in Africa, India, and Europe.

For example, at São Tomé, the tiny island that lies on the equator off the West African coast, the Portuguese planted coffee, cacao, and cinchona trees from the Amazon basin and fertile Brazilian interior. Cinchona, of course, had been known in Brazil since the second half of the seventeenth century as the source of quina, or quinine, the most effective treatment for malaria and other fevers.[62] São Tomé's equatorial climate and large plantations, with their fertile volcanic soil, were soon producing hundreds of kilos of cinchona bark annually for export; quinine from São Tomé saved countless settlers' lives,[63] and facilitated Portuguese expansion into the interior of their African territories during the nineteenth century.[64] Similarly, when facing the prospect of losing their prized colony in South America, the Portuguese proactively transplanted Brazilian cacao to São Tomé by royal order in 1819;[65] the plant thrived on the humid equatorial island and, less than a century later, São Tomé was the world's foremost producer of cocoa.[66] Since the sixteenth century, Europeans had considered chocolate to be a medicinal substance (a belief learned from missionaries, who in turn had gleaned it from Native peoples in Brazil).[67] Chocolate became an essential product for the recuperation of sick or disabled military personnel across the western world into the twentieth century.[68]

CONCLUSION

Over three centuries of Portuguese colonial rule, India's and Brazil's Indigenous peoples and unique flora contributed significantly to the eclectic, cosmopolitan, and syncretic medical culture that developed within the colonial Lusophone world. Systematic Portuguese maritime exploration in the tropics began along the west coast of Africa in the 1430s (and in India and Brazil in approximately 1500), predating by far any other European effort; consequently, Portuguese exposure to tropical diseases, as well as various Indigenous cultural methods of treating them, lasted far longer than that of any other European nation. Necessity combined with long familiarity resulted in the marked Portuguese tendency for receptiveness toward the adoption, exploitation, and dissemination of Indigenous medical practices. Through numerous ecclesiastical, commercial, and medical channels, knowledge of South Asian and South American botanicals and healing techniques were codified in pharmacopoeias or similar didactic texts and circulated throughout the Atlantic World and beyond, enriching medical resources in European imperial enclaves around the globe. Some key Brazilian drugs, like the indigenous varieties of cinchona (subsequently transplanted to São Tomé), actually acted as a catalyst, allowing for the marked expansion of Portuguese power over territories in the tropics—especially into the African interior in Angola and Mozambique—during the eighteenth and nineteenth centuries.[69]

European epistemologies of medicine, however, evolved during the early modern period in no small part as a result of the Portuguese colonial experience. The presence of diverse Portuguese imperial agents in disparate colonial spaces across several continents occasioned the transfer and diffusion of medical knowledge, using commissioned medical reports and descriptive pharmacopoeial lists, to Europe and throughout their entire maritime network, where no direct lines of contact or exchange had existed prior to the establishment of seaborne commercial and administrative links. Indigenous peoples of the Portuguese colonies thus made important contributions to "Western medicine" during the early modern period, but did so through European intermediaries, who codified Native healing practices, interpreting through their own subjective cultural lens, and therefore often altered the original application of Native medicines (or the philosophy behind Indigenous healing techniques) to meet their own ends. This transfer of healing information, because it was accomplished mainly by Europeans using tools of transmission peculiar to them (written administrative reports, pharmaceutical guides, and learned correspondence), depended ultimately on European interpretations, ideas, and concepts about medicine. Thus, medical descriptions of newfound drugs and their effects were usually couched in contemporary Galenic or humoral terms, markedly contrary to the concepts of their cultural origins. A European commercial focus on portable, readily saleable medical commodities helped to inspire a changing approach to illness, with a gradually increasing focus on the idea of medical "specifics"—single remedies intended to address discrete maladies—rather than comprehensive, "holistic" approaches to health often found in Indigenous medical practices.

Exchanges of medical knowledge in colonial Africa, India, and Brazil occurred on a variety of levels and through a range of agents; in any given case, much depended on the preexisting knowledge, skills, and requirements of the persons directly involved in the transaction of healing information. In the missionary context, protracted exchanges were often substantially more complex—and intellectually more profound—than those rapid transactions conducted between sick Portuguese soldiers, *bandeirante* explorers deep in the bush, a harried colonial provincial official, or even a ship's or regimental surgeon and the Native shamans or traditional Ayurvedic *vaidyas* with whom they interacted. Like their martial or mercantile coreligionists, Jesuit priests and lay missionaries often relied on Indigenous cures to treat their own tropical maladies contracted in the service of the Church; however, their greater patience and investment of time for evangelizing ends resulted in a more subtle and detailed understanding.[70] The theological implications of their reliance on Indigenous materia medica, though, must have given pause to evangelicals who were constantly at pains to demonstrate the superiority of Old World religion and culture.[71]

Only in the late eighteenth century, near the end of colonial era in Brazil, did an interest in the exploitation of Indigenous remedies awaken within the core government administrators of the Portuguese Empire, leading to the commissioning of systematic surveys of medicinal plants in India, Brazil, and elsewhere by professionals with botanical or medical training. The resulting documentation, though, was little different from what missionary Jesuit priests had produced over two hundred years before: lists of known indigenous plants, compiled meticulously but largely unscientifically, combined with descriptions of their use according to how colonists and Natives applied them. By this time, centrally organized scientific exploration had long been underway in rival Dutch, English, and French colonies; in each respective imperial enterprise, empiricism had become a common component. Even though Jesuit missionaries in Portuguese colonies had conducted the earliest European medical and ethnobotanical explorations in the tropics, the broad historic impact of later Portuguese scientific endeavor was limited by having begun belatedly—and because, despite the aspirations of some of these reports' authors, little of their efforts ever achieved, or were even apparently meant for, circulation in print within Europe's international scientific or intellectual communities.

Instead, evidence is lacking for a coherent overarching plan—a centralized Crown-directed effort to disseminate this hard-won medical information throughout the empire. Though such a coordinated scheme may have come to fruition eventually, in the event, historical circumstances intervened: the Napoleonic invasions of continental Portugal in 1808 and the consequent flight of the royal family to Brazil, the ensuing Peninsular War, a disruptive constitutional movement in Portugal combined with the declaration of Brazilian independence in 1821, and a subsequent civil war in Portugal (the War of the Two Brothers) effectively ended the Lusophone colonial medicine initiative. So, these painstakingly gathered reports, the knowledge within them still regarded as sensitive strategic imperial commercial information, usually remained hidden from the general public during the nineteenth century and therefore languished, awaiting discovery by modern researchers in recent years.

What lesson should readers take away from this review of various activities and manuscripts aimed at extracting and codifying healing knowledge in the Portuguese-speaking world? One key takeaway is that the Portuguese played a fundamental role, of world-historical importance, in extracting healing knowledge about non-European materia medica from non-European peoples. This chapter has demonstrated that they waged this epistemological offensive on a global scale, with particular focus on India and South America, but with little attention directed toward Indigenous medical practices in Africa.

As a parting thought, consider the apparent paradox that has become clear as this essay draws to a close: although the Portuguese played a major role in making heterogeneous materia medica accessible to Europeans, their efforts to engage healing knowledge and disseminate it on a global scale—having done so earlier than the rest of their European rivals—appear to have become *less* effective during the late eighteenth century, when the modernizing, increasingly Enlightenment-oriented Portuguese state got more directly involved in this process. In the end, that is perhaps the most interesting facet of the story told here.

6

IMPERFECT KNOWLEDGE

Medicine, Slavery, and Silence in Hans Sloane's
Philosophical Transactions and the 1721 London *Pharmacopoeia*

WILLIAM J. RYAN

IN 1686, Sir Hans Sloane (1660–1753)—naturalist, physician, and president of both the Royal College of Physicians and the Royal Society—made his first foray into the *res publica medica* with a paper submitted to the *Philosophical Transactions*, the journal of the Royal Society that he edited from 1695 to 1713.[1] Prior to his appointment as secretary of the society, a position that included editorship of the journal, Sloane drew upon his firsthand experience in Jamaica to provide the journal with a pair of botanical descriptions: "The Pimienta, or Jamaica Pepper Tree" and the "Wild Cinnamon Tree, commonly but Falsly called Cortex Winteranus." At the time of Sloane's publication, both trees were familiar to an English audience as the respective sources of the allspice berry and wild cinnamon bark, commonly used *materia medica*. As the title suggests, therefore, the paper does not claim intellectual value in novelty. Sloane instead stresses the importance of discernment via firsthand observation. Brief, if detailed, specimen descriptions of each tree include accounts of the Jamaican landscape, climate, horticulture, and medical practice. He digresses on the harvest and drying of allspice berries, thereby correcting the common misapprehension of European naturalists that they are a *fructo umbilicato sicco*, or vine dried fruit.[2] The privileged creole observer corrects metropolitan knowledge, demonstrating the value of New World matters of fact to early modern knowledge production and justifying their inclusion in the journal.

In order to do so, however, both author and reader must confront their own intimacy with the practices of Atlantic slavery. Unlike an accompanying illustration, which isolates the figure in the flat plane of a natural historical engraving, Sloane's article digresses on the position of the Jamaica pepper tree in the island's landscape and economy, as well as the specialized knowledge and labor of non-Europeans required to produce the materia medica. For, as Sloane reports:

> There is no great difficulty in the curing or preserving of this Fruit for use, 'tis for the most part done by the *Negro's*; they climb the trees, and pull off the Twigs with the unripe green Fruit, and afterwards carefully separates [sic] the Fruit from the Twigs, Leaves, and ripe Berries; which done, they expose them to the Sun from its rising to setting for many days, spreading them thin on cloaths, turning them then and now, and carefully avoiding the dews (which are very great). By this means they become a little wrinkled or rugous, dry, and from a green change to a brown colour, and are then ready for the Market.... The ripe berries are very carefully separated from those to be cured, because their wet and plenteous Pulp renders them unfit for Cure.... The more fragrant and smaller they are, they are counted the better.

Both the therapeutic efficacy and reliable commodification of this medicament rely on the labor and, significantly, the knowledge of the non-European. The passage repeatedly emphasizes the enslaved Africans' practical expertise about the curing process as well as the importance of careful attention in selecting berries. Sloane reveals an evident intimacy with the preparation: he clearly studied the work closely and recognized the history of settler colonialism and African knowledge of New World nature as integral to the understanding of materia medica. The allspice berry—a fruit so named for its "peculiar mixt smell, somewhat akin to" clove, juniper, cinnamon, and pepper—thus emblematizes the mixed nature of materia medica derived from the West Indies. Plantation slavery is depicted in the pages of the *Philosophical Transactions* and manifests in the spices of a London apothecary cabinet.[3]

In this essay, I address how Sloane's interest in the meanings and uses of New World nature influenced his varied labors of collection, collation, and composition across his medical publications, particularly as editor of the *Philosophical Transactions* and of the 1721 edition of the *Pharmacopoeia Londinensis*. During the first decades of the eighteenth century, the virtuosic physician occupied a central position in British Atlantic science, sitting at the nexus of early modern physick and natural philosophy. He was instrumental in bringing medicine into the purview of English natural philosophy, a fact clearly demonstrated by his work both on the *Transactions* and in overseeing the 1721 revision of London's official pharmacopoeia. While scholarship has attended to Sloane's work as a collector, his two-volume *A Voyage to the Islands of Madera, Barbadoes, Nieves, S. Christophers, and Jamaica* ... (1707), and his career as a fashionable London physician, little attention has been paid to his two-decade span editing the first English scientific journal and its coincidence with the revision of the *Pharmacopoeia*.[4]

Sloane is best known today as the Enlightenment's consummate collector: his vast assortment of curiosities, gathered through a lifetime of global correspondence, served as the founding collection for the British Museum. Sloane also served as president of the Royal College of Physicians and as secretary, and president of the Royal Society, and was at the center of a burgeoning culture of medical letters. Sloane's own participation in the res publica medica demonstrates the diverse forms and formats of publication that governed medical networks of exchange in the Atlantic World. Though best known for his natural history of the British West Indies, *A Voyage to ... Jamaica*, Sloane also produced a list of Jamaican plants, *Catalogus Planatarum Quae In Insula Jamaica* (1686), contributed papers to the *Philosophical Transactions* (some as primary author and others as recipient, commenter, or compiler, spanning the years 1686–1749), published a brief pamphlet, *An Account of a Most Efficacious Medicine for Soreness, Weakness, and several other Distempers of the Eyes* (1745), and also circulated in manuscript his *Memoir of Beaumont* (1740), a collection of case studies covering diseases of the mind.

Collection, collation, and editorship of this kind were central to many efforts to gather healing knowledge in the Atlantic World. As Stuart Anderson's essay reminds us, the production of official pharmacopoeias, such as the *Pharmacopoeia Londinensis* that Sloane oversaw as president of the Royal College of Physicians, involved laborious classifying, organizing, and authorizing of materia medica, thereby rendering medicines consistent within a given jurisdiction (chapter 10). Alongside its political and economic significance, however, producing a pharmacopoeia entailed the complex epistemological process of distilling healing knowledge. As the work of various scholars in this volume attests, the production of printed pharmacopoeias or more informal lists of information on materia medica relied upon varied and diffuse forms of intellectual labor spanning continents and bridging epistemological regimes. One of my interests in this chapter is tracing Sloane's multiple sites of engagement with the dissemination of medical knowledge, from papers in the *Transactions* to the *Pharmacopoeia*, thereby illuminating the diversity of genres, formats, and attendant epistemological registers employed by medical authors in the British Atlantic World.[5]

In my reading, Sloane's work on the scientific periodical mediates between different epistemic systems in the Atlantic World (be they European, African, Amerindian, or a hybrid thereof) as well as between the standardizing imperial discourse that characterizes the *Pharmacopoeia Londinensis* and the inductive openness fundamental to the practice of medicine in the colonial space. The 1721 *Pharmacopoeia* aims at both classification and control of medical preparations and of apothecaries themselves, as evidenced via its incorporation of botanical names into its catalogue of simples; its emphasis on regular measurements in sections on compounding medicines and syrups; and its expunging of a number of dubious cures and preparations that had been included in previous editions. The 1721 *Pharmacopoeia* also demonstrates Sloane's influence through the addition of new materia medica. Among others, the Jamaican pepper appears for the first time, albeit without the detailed attention to the drying process. Therefore, we only see the materia medica's full history in the Atlantic World when we trace it across the varied genres and formats of early modern medical publishing, accounting for the diverse epistemological registers made available therein.

What this essay aims to contribute to our history of medicines is not only the intellectual labor of Sloane but also an understanding of the epistemological function of editorship and the periodical format in the process of building a pharmacopoeia. In what follows, I consider how the *Philosophical Transactions* under Sloane function as an intermediary format in which new and experimental materia medica, drawn from what Londa Schiebinger has identified as the "Atlantic World medical complex," can be integrated into European medical practice and knowledge. In contrast to a previous body of scholarship that saw medical exchange as unidirectional, Schiebinger highlights how "disease, knowledge, and medical remedies ... moved promiscuously between continents, masters and slaves, and imperial monopolies." According to Schiebinger, practicing physicians like Sloane occupy a unique position in the medical complex due to their direct experience with the healing exigencies of the colonial environment and their engagement with metropolitan institutions of science in the centers of imperial power. Sloane's brief labor as a physician in Jamaica was highly influential on his career in medical practice, as a natural historian, and as a central figure in the medical republic of letters. Sloane carried this practical engagement with the diverse healing regimes of the colonial space back to London functioning as, in Schiebinger's terms, a "knowledge broker" at the heart of the global exchange networks that constituted the Atlantic World medical complex.[6] Sloane understood firsthand the complex knowledge of non-Europeans in the New World environment and, through his editorial practice and varied forays into the res publica medica, sought to make that knowledge available to a wide-ranging audience of learned physicians, apothecaries, natural historians, botanists, gentlemen virtuosi, and natural philosophers.

In his role as medical author and editor, Sloane was carefully attuned to the different kinds of medical knowledge different genres could make available. Therefore, his publishing and editing career in this period offers a particularly fruitful site to engage what Gianna Pomata has called the "epistemic genres" of medical publishing. In her work on the history of the medical case study, Pomata describes "epistemic genres" as "those kinds of texts that develop in tandem with scientific practices." As such, these "genres . . . are deliberately cognitive in purpose" and "are linked, in the eyes of their authors, to the practice of knowledge-making (however culturally defined)."[7] During his stint of nearly two decades as an editor, Sloane collected and arranged reports of medical and natural historical phenomena for inclusion in the *Philosophical Transactions*, leaving both content and style generally unaltered. Such reports are therefore characterized by their openness to imperfect knowledge: knowledge that is incomplete, relies on dubious sources, or renders questionable judgments or hypotheses. Sloane's editorial work on the *Philosophical Transactions* explicitly embraces this kind of openness as valid in itself and as a goad to further inquiry, particularly in the pursuit of materia medica. Such openness makes Sloane and the journal vulnerable to critiques, both stylistic and philosophical. The thrust of these critiques, however, casts Sloane's epistemic regime into relief, thereby enabling us to better see how his work on the *Transactions* informs the incorporation of Atlantic medicine into the official *Pharmacopoeia*.

Scholarship on Sloane engages his position at the intersection of commerce, empire, and knowledge production in the eighteenth century.[8] Rather than reading Sloane as the quintessential demonstration of a "view from nowhere," recent work on Atlantic science restores the contested, local conditions to our understanding of early modern natural philosophy.[9] My argument extends this line of analysis to Sloane's editorial practice in the *Philosophical Transactions* and the 1721 edition of the *Pharmacopoeia*. Attending closely to Sloane's editorship opens a key aperture through which to glimpse what Elizabeth Maddock Dillon has usefully described as the "colonial relation." According to Dillon, "the colonial relation names the sustaining structure of economic dependence by the metropole on the colony at the core of capitalist modernity and the bourgeois ascendancy in Europe."[10] Medicine and its associated fields of botany, natural history, surgery, and materia medica highlight the particular intimacy of the colonial relation: exposure to and fear of unfamiliar diseases and the highly competitive early modern medical marketplace forced Europeans into close interaction with Amerindians and Africans, something Sloane had experienced firsthand as a physician in Jamaica. Recognition of and reliance upon such interdependence therefore makes its way into Sloane's own writings, and those he selected for inclusion in the journal, a fact that later editors and satirists would use to critique Sloane and the era of the *Transactions* that he oversaw. Therefore, by attending closely to his contributions to and editing of the medical content in the *Transactions*, we see how the colonial relation was both presented and effaced in the pages of the journal and in the recipes and catalogues of the *Pharmacopoeia*. Attention to Sloane's editorial practice helps us recover the ways that English medicine, literature, and historiography have both written in and written out the colonial.

THE PUBLICATION OF ATLANTIC MEDICAL KNOWLEDGE

Sloane's tenure as secretary of the Royal Society corresponded to a shift in the focus of scientific inquiry during the late seventeenth and early eighteenth centuries. Sloane's 1693 appointment, along with Richard Waller, and his editorship of the *Transactions* helped raise the stature of medicine and its related fields. The beginning of Sloane's secretaryship found the society somewhat in disarray. The *Transactions* had ceased regular publication, one commenter noting that the society did not maintain a single foreign correspondent. In restoring the journal, which at the time was not published by the society but as the personal project of the secretary, Sloane drew upon his private correspondence network and those of his close associates. As such, anatomy, medicine, botany, and natural history, had displaced somewhat the mathematics, chemistry, and physics that had dominated English natural philosophy to that point. While the sciences based on laboratory experiments—what Robert Boyle called "controlled experience"— undoubtedly maintained importance, the inclusion of more papers addressing the observational sciences allowed for increased participation by people of varied stature and from across the expanding British Empire, from New England to the West Indies, West Africa, the East Indies, and China. Medical and natural historical correspondence privileged contributions from firsthand observers, particularly if they were capable of correcting understandings of New World nature. These complex knowledge networks enabled creole elites to participate in a burgeoning republic of letters, thus easing their sense of provinciality. However, doing so required engagement with and acknowledgement of African and Amerindian knowledge and practice, particularly in terms of materia medica. Full understanding of Atlantic world medicines would thus require engagement with European correspondents of nonelite status as well as the non-Europeans who had practical understanding of New World flora and its healing properties.[11]

Despite his elevated stature in Atlantic scientific culture, Sloane was born to middling status in Northern Ireland and attained his social, intellectual, and actual capital through his investments in and knowledge about Jamaica. As his correspondents and literary critics would frequently remind him, Sloane was always, in a way, becoming colonial. In his own contributions to the journal, therefore, Sloane leveraged his access to specimens and natural historical knowledge from the wider Atlantic world. Between 1693 and 1749, he contributed (as an author or translator) twenty-four separate papers to the *Transactions*.[12] Of those twenty-four papers, eighteen carry direct associations, either in title or content, with colonial spaces. As named author, Sloane presents papers recounting natural historical specimens, climatological phenomena, and materia medica from North America (1), South America (3), Africa (2), and the West Indies (5). Other papers translate or recount matters of natural philosophical or medical interest from China, France, Italy, and within the British Isles.

Due in part to this reliance on rarities gathered from throughout the Atlantic World, editors of the *Philosophical Transactions* who preceded Sloane hedged about the nature of the knowledge produced in its pages. Henry Oldenburg, the first secretary and founding editor of the *Transactions*, drew a distinction between the more enduring and assured forms of knowledge produced by society members, and the speculative observations published in the pages of the journal. Oldenburg underscores this in his first "Epistle Dedicatory" and "Introduction," both published in 1666. In the "Epistle" he juxtaposes the distinction between the pieces that will appear in the *Transactions* and the tomes that society members publish elsewhere. Although Oldenburg acknowledges that "weighty Productions require both Time and Assistance, for their due Maturity," he nevertheless finds value in the production of smaller, less polished pieces. "These Glimpses of Light," though no indication of the serious work done by society members, still warrant publication so that "every man may perhaps receive some benefit from these Parcels."[13] Oldenburg's metaphors establish key epistemological expectations for the journal in its early years. The play on heft (e.g., "weighty productions"/"these Parcels") juxtaposes permanence with ephemerality in scientific publishing. The *Transactions* will capture and share scraps of knowledge that are perhaps underdeveloped or may appear inconsequential, both in size and in significance. However, despite (or perhaps because of) this apparent triviality, such papers are nevertheless valuable "glimpses of light."

Oldenburg's interest in these "glimpses of light" recalls Francis Bacon's attention to the "deviating instance" as a catalyst for the reform of early modern knowledge production. For Bacon, anomalous, insignificant, or otherwise unresolved phenomena cultivated curiosity while disrupting received wisdom. Since "the sun, which passeth through pollutions . . . itself remains as pure as before," reform-oriented natural philosophers should pursue sites of possible, if imperfect, knowledge.[14] Early modern scientific journals, including the *Philosophical Transactions* and its French counterpart *Le Journal des Sçavans*, manifested such interest by printing and reprinting reports of, among others, medical wonders well into the eighteenth century.[15] Once made public in a periodical, such papers were expected to goad further researches, an idea Oldenburg emphasizes in his brief "Introduction" appended to the first volume. The best way to "promote the improvement" of natural knowledge, Oldenburg writes, is to "employ the *Press*." For "such Productions being clearly and truly communicated, desires after solid and useful knowledge may be further entertained ... and those ... conversant in such matters may be invited and encouraged to ... contribute what they can to the Grand design of improving Natural knowledge."[16] Oldenburg thus sets the tone for what the *Transactions* will be: a record of the advancement of knowledge in the world that will facilitate speedier and less assured contributions. James Jurin, the secretary just preceding Sloane and under whose editorship Sloane published his first papers in the *Transactions*, emphasizes the generative nature of making public imperfect knowledge. For Jurin, "the Discovery, of one Phenomenon, entices the search after another."[17] Accordingly, Jurin underscores his intention to publish both controlled experiments, akin to those presented before the members of the society residing in London, as well as observations of natural phenomena from among the ranks of society members and their correspondents residing in the far reaches of the Atlantic World. A journal that publishes short, even apparently slight, scientific reports can therefore model a disciplined mode of natural philosophical inquiry, provide a venue for participation in learned culture, and inspire interest in and pursuit of natural knowledge. In these short, periodical pieces, both reader and writer will find the content of and the inspiration for their continued researches. And such researches necessarily imply the New World, a contested, yet undeniably fruitful space of healing knowledge.

Sloane's own publications, both within the *Transactions* and elsewhere, demonstrate an awareness of the epistemological possibility inherent in New World medical literatures. In his *Voyage to ... Jamaica*, for instance, he directly endorsed the importance of pursuing knowledge about New World medicines, both familiar and unfamiliar. "It may be objected," he writes in the preface to the first volume, "that 'tis of no purpose to any in these Parts of the World to look after such Herbs, &c. because we never see them. I answer that many of them and their several parts have been brought over, and are used in Medicines every day, and more may, to the great advantage of physicians and patients, were people inquisitive enough to look after them."[18] And Sloane, of course, had already begun this work with his 1689 *Catalogus ... Jamaica*, a list of the nearly eight hundred plant specimens he brought back from Jamaica.

Sloane's inquisitiveness extended beyond New World botanicals and materia medica and into medical preparation and practice. Among the case studies in his *Voyage to ... Jamaica*, Sloane included "Of *Chegos*, and the Consequences of them." Chegos, or small mites that burrow and lay eggs under one's skin, were an omnipresent threat in unkempt spaces and, while not dangerous, produced significant pain, as Sloane knew firsthand. "I found an uneasiness, soreness, or pain in one of my Toes," Sloane's case begins. "I had a *Negro*, famous for such cases, to look upon it, who told me it was a *Chego*. She (who had been a Queen in her own Country) open'd the Skin with a Pin above the swelling, and carefully, separated the Tumor from the Skin, and then pulled it out, putting into the Cavity ... some Tobacco Ashes which were burnt in a Pipe she was smoking."[19] This instance stands out among Sloane's 128 cases as the only one that treats an illness that he himself suffered from. Furthermore, while other cases reference "*Negro*" physicians, usually as competitors whose misdiagnoses Sloane was brought in to correct, this is the only one to address in detail the practice of a non-European physician. The case goes on to elaborate Sloane's own understanding of the mechanism whereby the chego causes such pain, and to offer another instance of the failed removal of the chego and the consequences thereof for "A very neat lady." With this addition, Sloane highlights the complexity of the operation performed upon his own foot, as well as his limitations as a physician. Despite his defensiveness—Sloane's African healer was, after all, a "Queen in her own Country"—Sloane's case demonstrates the novel threats that exist in the colonial space and the intimate connection with non-Europeans required of patients as well as physicians working there.[20]

Due in part to this association, the print publication of Atlantic medical knowledge in the *Transactions* and elsewhere came under epistemological and professional scrutiny in the late seventeenth and early eighteenth centuries. Despite a shared desire among learned medical practitioners to escape the accusations of secrecy and self-interest that accompanied proprietary medicine as practiced by apothecaries and empiricks, the sharing of materia medica risked exposing trade secrets and, perhaps more damningly, revealing imperfect medical knowledge. A colleague of Sloane's, William Cockburn, responded powerfully to this anxiety in his treatise *The Present Uncertainty in the Knowledge of Med'cines* (1703). Writing in his capacity as physician to the royal fleet, Cockburn attached the advancement of medical understanding to scientific, professional, and imperial motivations. The systematic collection and publication of medical knowledge would advance physic beyond the vagaries of individual practice, thereby elevating medical professionals above "Quacks and Nurses" via the collective pursuit of a higher order of knowledge.[21] Doing so, however, required acknowledging professional ignorance regarding both the cause of disease and the operation of certain medicines.

As an example, Cockburn cites "Jesuits pouder," a derivative of the bark of the Peruvian cinchona tree introduced into European medical practice in the early seventeenth century as a treatment for malarial fevers. "The bark" was frequently cited as an emblem of the medical wonders available in the New World and offered as a goad to their further pursuit.[22] However, in Cockburn's opinion, poor understanding of such an exotic medicine contributes to the belief that it "operates almost like a charm." To combat superstitious quackery, and thereby prevent misuse of other materia medica introduced by imperial expansion, Cockburn encourages "Candid and Honest Physicians to set about a sedulous Inquiry for a true Theory of Disease." Central to this project is admitting stubborn instances of inconsistency in some medicines, as Cockburn does by openly acknowledging the failure of "Jesuits pouder" against certain fevers in Virginia.[23] Participation in the res publica medica—primarily via the attentive observation of materia medica and their use exchanged in networks spanning the Atlantic World—would distinguish the practical English physician from the apothecary, empirick, or quack. However, such distinction required openness to imperfect knowledge, particularly that which was partial or incomplete due to a medicament's origin in unfamiliar landscapes, as part of non-European healing practices, or its association with the racial intimacies of the colonial space.

New World materia medica held far too much promise, both economic and medical, for the European medical observer or practitioner to be dissuaded by potential imperfection. Sloane made this abundantly clear in a brief preface to the *Transactions* of 1699, in which he justified both the style and content of the papers published in the journal by drawing a distinction between "matters of fact" and "Hypothesis." He writes: "There is no doubt but the more discerning will make a great difference between what is related in [the *Transactions*] as Matter of Fact, Experiment, or Observation, and what is Hypothesis. The first sort of Relations ... are, and must always be useful, and the latter may be pass'd over by such as dislike them."[24] This preface is well known in part for Sloane's defense of his laissez-faire editorial hand: Sloane deliberately published articles without significant stylistic alteration, often leaving in salutations or other, at times unrelated, content from the original correspondence in which the article was contained. In addition, speculative, hypothetical, or controversial reasoning from authors was retained. Sometimes the editor offered explicit endorsement, but more often he relied on the silent imprimatur of the work's appearance in the *Transactions*. In contrast to Oldenburg or Jurin, who indemnified the society from the contents of the *Transactions*, Sloane shifted the burden of interpretation from editor to reader, demanding at once a greater degree of openness and discernment from his audience. Furthermore, the inclusion of speculative reports, particularly from the edges of the empire, would contribute to the advancement of healing knowledge.

As an example of this, Sloane invokes cinchona bark in his preface to the 1699 volume of the *Transactions*. Rather than advocating for the bark's therapeutic efficacy, he points to the European adoption of the South American materia medica as an argument in favor of the kind of epistemological stance embodied in his editorial practice. Sloane dramatically reimagines the bark's origin myth:

A poor Indian who first taught the Cure of an Ague, of which the Lady of the Count de Chinchon ... was Sick, overthrew with one simple Medicine, without any preparation, all the Hypotheses, and Theories of Agues, which were supported by some Scores not to say Hundreds of Volumes, and tis plain did mischief by hindering the advantage Men might have received sooner from so innocent and beneficial a Remedy. I say this not to reproach Physicians, who do well to be wary in the use of a new Remedy, till Experience confirms it to be Harmless; but because there are some Specific Medicines mentioned in these Transactions for the Cure of other Diseases, and more are designed for the succeeding Year.[25]

Sloane here not only endorses the openness to new knowledge derived from exotic sources but also identifies the periodical format of the *Transactions* as ideal for counteracting the orthodoxy presumed in the "Scores not to say Hundreds of Volumes" of medical literatures. Unlike the permanence of the medical library, the ongoing and imperfect nature of the periodical allows knowledge to be amended, resituated, and reevaluated.

But access to such knowledge requires the racial intimacy of a "poor Indian" coming to the aid of a "Lady." As Sloane himself well knew, European dependence upon Amerindian, African, or Creole knowledge and expertise in the colonial space was common, if not often represented in imperialist discourse. However, Sloane here makes the open admission of such intimate dependence an object lesson in how Atlantic medicine, and science more generally, will produce greater knowledge. Access to such wondrous medicines would require intimate and potentially dangerous interaction with non-Europeans.[26] And the collection of such knowledge required a broad network of correspondents, often with differing interests, uneven training, and varied epistemological commitments. Such connections not only were necessary but required varied genres and formats to accommodate the possible and potential meanings of non-European healing practices. And rather than exercising a strong editorial hand to bring stylistic or epistemological regularity to the *Transactions*, Sloane preferred to retain the observations and speculations of his correspondents, however imperfect. Attending closely to the epistemic openness that characterizes Sloane's editorial work in the *Transactions* can amplify silences, particularly those around Atlantic World materia medica in the London *Pharmacopoeia*, which he oversaw as president of the College of Physicians.

AFRICAN MATERIA MEDICA IN THE *TRANSACTIONS*

Sloane's direct connections to the West Indies—his firsthand experience in Jamaica, his investment in sugar plantations, his promotion of cinchona bark and cacao—as well as the practical need to gather the raw materials of natural historical inquiry made the cultivation of correspondence from throughout the Atlantic World key to the revival of the *Transactions* as a regular publication. Under Sloane's direction, the journal also featured reports from Dublin and Belfast, a number of Italian correspondents, and work from French physicians. Overall, there is a marked difference in content for the decades of Sloane's editorship, with a rise in the number of explicitly medical papers, be they surgical, anatomical, reports of particularly remarkable cases, or accounts of materia medica.

Sloane's editorial hand is visible (at times through its very invisibility) in a number of ways, but most clearly when he acts as commenter upon a paper included in the *Transactions*. Such comments typically accomplish a variety of epistemological and professional ends. All of the papers on which Sloane offers comment are, first and foremost, medical papers. Sloane lends his own status as a learned physician more directly to these, thereby elevating his fellow medical practitioners above the "neighborhood women" and "Quacks" who serve as foils in the papers themselves. Sloane also dramatizes through his editorial intervention the importance of open consultation and exchange among learned physicians: medicines and medical knowledge should be neither proprietary nor secretive. Additionally, his comments situate each paper within his learned empiricism, thereby demonstrating the broad purview of medical knowledge. Sloane brings to bear his own understanding of botany, natural history, and materia medica in order to situate a correspondent's contribution. In so doing, Sloane models the varied forms of observation required to best understand and evaluate a given materia medica, the importance of imperfect knowledge to the advancement of physic, and how different formats of print publication can accomplish both ends. He does this relatively infrequently (a total of five times), but the situations in which he does so are worth considering for demonstrating his evident willingness to admit imperfect knowledge about materia medica and modeling for readers how to bring empiricist methods to bear on medical practice.

A pair of letters upon which Sloane comments demonstrate the professional and epistemological goals of his editorship for the evaluation of materia medica in the *Philosophical Transactions*. Both letters recount familiar, if not universally recognized, medicaments. A 1693 paper, "Part of a Letter from Mr. T.M. in Salop, to Mr. William Baxter, concerning the Strange Effects from the Eating of Dog-Mercury with Remarks thereon by Hans Sloan [sic] M.D. and S.R.S.," relates the accidental poisoning of a family from cooking and consuming a large quantity of what appears to have been the commonly employed medicinal herb "dog-mercury." Baxter, the original recipient of the letter, forwarded it to Sloane only after local physicians and apothecaries stood divided over the nature of the plant ("[some] say 'tis Dog-mercury, but some say 'tis a sort of Nightshade"). After receiving the letter and a specimen, Sloane confirms the plant via consultation with his herbaria, but remains uncertain of how to assess the case of poisoning. He justifies the publication by urging caution in—but not forbidding—the continued use of dog-mercury: "Whether the quantity or the Quality of this were the Cause of the Effects I know not, but think that every Body will do well to be cautious and wary in the use of it in such Quantities."[27] Sloane appends a comment to correct domestic medical knowledge and publishes the account in order to refine learned understanding of a commonly employed materia medica, and thereby implicitly endorse the regularization of medicinal preparations. Imperfect knowledge—either in erroneously applied therapeutics or in a lack of proper understanding of related bodies of knowledge—can be corrected through publication in the scientific periodical.

Sloane comments on another paper, offering a similar correction and endorsement of "a most Noble and Infallible Medicine" for the treatment of rabies. "Part of a Letter from Mr. George Dampier," which describes how to locate, prepare, and administer (to dogs, cattle, and people) a secret family recipe. As evidence of its effectiveness, Dampier recounts the curing of a herd of cattle, a use that had the side effect of halting the "ill Censures the Neighbours had of a harmless, long Nos'd old Woman or Two."[28] Making known this medicine, first in the letter itself and then via its publication in the *Transactions*, demonstrates the power of proper medical knowledge to dispel superstition. Dampier extends this clarity of vision to the gathering and sharing of the specimen: "I have sent it as it grew, without separating the Earth, Moss, and Grass that grew about it, because you may know it better when you see it."[29] Because of this specificity, Sloane is able to correctly identify the plant—"The simple or Herb mentioned in this Letter is not Jews Ear [as Dampier asserts], but is the *Lichen Cinereus terrestris* . . . and grows commonly in most barren places about London, and all over England"—and offer the exact preparation, without comment, for the readers of the *Transactions*. Sloane's editorial intervention thus underscores the importance of observations that are pointedly discerning about an isolated specimen—able to distinguish between closely related fungi—and acutely aware of the wider vision required to identify that specimen within a landscape of "Earth, Moss, and Grass."[30] Full knowledge of a given materia medica requires a more capacious, yet still discerning vision, one that can identify not only the plant itself but also the landscape and methods of preparation required to properly locate and employ such medicines.

Such varied practices of observation are particularly important in the colonial space due at once to the variety of plants, their evident unfamiliarity to the European observer, and their great potential as sources of materia medica. Sloane emphasizes as much in the 1698 letter he translates and comments upon. "Of the Use of the Root Ipecacuanha, for Loosenesses," amounts to a rousing endorsement of the use of this powerful emetic originally found in Brazil and Peru as a general cure for various forms of intestinal distress. Similar to Dampier, the unnamed author makes bold claims about this "infallible Medicine." The paper provides a detailed regimen—a combination of ipecac with other colonially derived simples, nutmeg, and cinnamon—as well as a defense of this remedy via analogy, once again, to cinchona. "There are a great many who cannot be persuaded that there are any General Remedies," he writes, "but if no Body, at present, disputes the Vertue of *Quinquina*, and if that Medicine is received as Universal by all Physicians, it ought not to appear so extraordinary if there be found other Specificks, that equal this in extent." Despite such boldness, Sloane is willing to endorse, both through publication and in his comment: "Altho' I am of Opinion that the Root mentioned in the foregoing Paper, is not so infallible a Remedy . . . as pretended, yet considering that sometimes those Distempers yield not to ordinary Means . . . I thought it might be beneficial to the Publick."[31] Despite the hedged conclusion (and the fact that ipecac was already well known in European medicine) Sloane publishes the complete recipe, owing in part to the hopeful attachment to Atlantic materia medica: not only new plants, but also new uses and new preparations may be brought to light, particularly if readers and observers are willing to pursue imperfect knowledge.

Such imperfect knowledge necessarily includes unfamiliar and untested materia medica, drawn from across the Atlantic World. In one such example, James Petiver, London-based apothecary, botanist, and frequent correspondent of Sloane's, offers a "Catalogue of Some Guinea-Plants, with their Native Names and Virtues."[32] After identifying his source for the collection of forty plants— "Mr. John Smyth . . . Minister to the Royal African Company in the English Factory at Cabo Corso"—Petiver provides a medical justification for the publication of his catalogue. "It were, Sir," he writes to Sloane, "needless to tell you the many advantages that would accrue to the Art or Mystery of Physick, if the Vertues of all Simples were more nicely inquired into or better known: I shall therefore wave it here, and only present you with an *African Materia Medica*, whose innocent Practice consists of no more Art than Composition, as you may see by the following Method."[33] Petiver imposes a classificatory order on the *"African Materia Medica,"* aiming to make the healing knowledge, as imperfect as it may be, useful for a European epistemic system.

Despite Petiver's diminution of African medical preparations, the content of the catalogue makes clear the potential value of the information communicated. Each entry follows the same basic format: arranged alphabetically according to "Native name" (e.g., *"Aclowa"*), followed by a brief description of the preparation and use credited to Smyth—e.g., "dried and rubb'd on all the Body is good for the Crocoes (or Itsch)." Petiver's remarks are added to Smyth's observations and include a Latinate classification ("*Colutea Scorpioides Guineensis Tragacantha foliis, nobis*") and a specimen description. At times, these descriptions familiarize the exotic plants through analogy, thereby rendering safe their preparation and use. The leaves of the "Ambettaway," for instance, used in Guinea to induce appetite, are compared to the "Common Elm," or the leaves of the "Aguaguin," "made into a plaister and applied to a Cut," resemble the "Common Lilac or Blew Pipe Tree."[34] Petiver aims to regulate West African nature, and its healing uses, through the imposition of European classificatory schemes and the analogy to familiar fauna. Though Sloane does not offer direct comment on Petiver's catalogue, as he does with the above-cited papers, the collection clearly carries Sloane's imprimatur: the secretary is listed in the title as the recipient of Petiver's original letter and Sloane's work on Jamaican plants is frequently referred to in Petiver's own comments on the West African materia medica. Knowledge and understanding of the West Indies and West Africa are considered coeval: understanding of one place can bring control to the natural products of another. Refinement of the raw materials of colonial exchange is accomplished via the intellectual labors of collection, collation, emendation, and publication.

However, Petiver's assertion of such control breaks down as the sources from Smyth either have no available classification or no description, or cannot be rendered familiar via analogy to English or West Indian plants. Nearly one-quarter (9 of the 40) of the entries are presented without comment from Petiver, and another quarter consist only of an added Latin classification, but no specimen description or analogy, as in the following entry: "14. *Assaba*, Warmed in Water, and the Groin rubbed with it, is good for a Buboe. Mr. *J.S.*" Thus, nearly one-half of the entries in Petiver's catalogue consist of the reprinting of African medical knowledge, without comment, although not without European intervention, of course. Each entry in the catalogue is credited to "Mr. *J.S.*," Petiver's correspondent. Smyth's style aims to impose a degree of control over that knowledge even when it escapes or exceeds Petiver's classificatory systems. Each portion of the entry attributed to Smyth follows a similar grammar and syntax: "33. *Pocumma*, Pounded and dried and Bak'd in Bread is good for the Flux." The use of passive voice, an evident attempt at scientific objectivity, transfers agency from the African to the materia medica itself. Nature, albeit African nature, and not colonial exchange, provides access to these varied medicines, and Sloane offers an implicit endorsement via their publication in the *Transactions*.

But such constructions always involve preparations of varying complexity—boiling in wine, warming in water, drying in the sun, or, as mentioned above, "bak'd in Bread"—thereby writing back in the hand of the non-European, particularly when Smyth's brief notes are the only description. For, as we have seen with Sloane's work elsewhere in the *Transactions*, such small fragments of knowledge are provided as prods to further understanding, something that Petiver himself embodies in the production of his entire catalogue. The initial descriptions, limited though they may be, should goad the properly curious reader and observer to the pursuit of more knowledge, greater understanding, particularly as relates to materia medica. Thus, the African is at once present and absent in the preparations: a necessary absence that acts as a prompt to further understanding, thereby unavoidably returning to view through a wider field of observation. Despite European claims to the contrary, Sloane's own writing about the Jamaica pepper tree underscores both the complexity of non-European knowledge about and preparation of materia medica and the importance of attentive European observation of such practices.

And that wider field of vision necessarily glimpses the violence of colonial exchange. In imitation of Sloane, Petiver adds an extra-medical detail to Smyth's observations about one plant. "*Metacoe*," Smyth writes, "Pounded, and so applied, is good for a cut." To which Petiver adds, "With these Leaves, the Blacks also make Match for their Muskets, as my Friend, Mr. Edward Barter informs me."[35] Petiver's telling detail corresponds to a common European assumption that non-European, particularly African, healing knowledge had the potential both to cure and to harm.[36] The possibility of this dual meaning is abetted by the publication of Petiver and Smyth's "Guinean" pharmacopoeia within the pages of Sloane's *Transactions*. While the aims of the former may be to foreclose knowledge through accurate description and classification, the latter encourages the pursuit of potential and possible meanings. The detail about the dual uses of *Metacoe*, pendulous and ominous as it is, requires further understanding. It is a form of imperfect knowledge, not about materia medica but about the colonial conditions that make such information available to European observers and such medicines usable by English apothecaries, physicians, and their patients.

COLONIAL MATERIA MEDICA IN THE *LONDON PHARMACOPOEIA*

Sloane's two-decades-long tenure at the helm of the *Philosophical Transactions* resulted in a mixed legacy for the journal, and for its editor. Sloane's position at the center of an active and global network of correspondents combined with his indefatigable curiosity and zeal for collecting to assure sufficient and diverse content for the publication. In fact, by the end of his secretaryship in 1713 the *Transactions* had resumed regular publication and significantly extended its reach. In addition, Sloane's virtuosic interests not only expanded the content and dissemination of the journal but also extended the empiricist methods of observation and experimentation to which the journal and the society were dedicated.[37] As the allied fields of medicine, anatomy, botany, natural history, and materia medica were brought into the publication, so the generic expectations of papers submitted under such heads came to reflect the editor's empiricist impulses. Nevertheless, Sloane's editorship inspired withering critiques from satirists, who blamed him for eroding the style of the scientific journal and contributing to a decline in the esteem of English natural philosophy.

Most prominent among these satirists was William King, who produced multiple attacks on Sloane, lampooning his collecting, his medical practice, and his cultivation of sociability as central to his scientific research.[38] King's earliest satire presents a pair of mock philosophical dialogues between a "Gentleman" and first a "Virtuoso," and second a "Transactioneer." *The Transactioneer* (1700) takes aim at Sloane's authorship and his editorial practice, as well as Sloane's association with the West. In his preface, King laments the damage wrought upon the journal, the Royal Society, and the reputation of English natural philosophy by Sloane's interest in imperfect knowledge. A more general consequence of Sloane's editorship, however, is the form of social leveling embodied by Sloane himself. According to King, Sloane's varied labors and subsequent fame suggest "that by industry alone a man may get too much reputation, almost in any profession, as shall be sufficient to amuse the world, though he has neither part nor learning to support it."[39] King takes particular aim at many of Sloane's own papers addressing materia medica, including the above mentioned paper on the "Jamaica Pepper"—dismissed by King's "Gentleman" as uninteresting since "every kitchen-girl about town knows Jamaica pepper"—and Sloane's endorsement of ipecac as "harmless and helpful," an insight deemed amusing in its redundancy.[40] Sloane's interest in "imperfect knowledge" of New World materia medica is below the esteem of true natural philosophy. In the second dialogue King takes aim at Sloane's judgment "in the choice of his friends."[41] As Sloane's inclusion of epistolary salutations and valedictions makes clear, his intellectual production relies on an intimate sociability. King's critique identifies a self-indulgent vanity in Sloane's editorial persona—more so than philosophical rigor or utility, praise or imitation of the secretary will assure a place for one's paper in the *Transactions*. And such a knowledge network risks destabilizing the social order, an anxiety all the more urgent when the colonial space is introduced.

Therefore, King's most revealing critique is of Sloane's praise for James Petiver's "African *materia medica*." King's Transactioneer quotes at length from Petiver's catalogue, whose preparations and prescriptions are absurd in their simplicity in the sound of their foreign words. "Hear this African Doctor," the Transactioneer emphasizes: "He has *Aclowa*, good for crocoes or itch; *Bumbunny*, boiled and drunk, causeth to vomit; *Assuena*, boiled and drunk, causeth a stool; *Ambetuway*, causeth an appetite to any sick person; *Attrumaphio*, boiled and drunk, causeth the great sort of pox to skin and dry, and is good against the phrenzy; *Mening* is good for the stoppage of the head; *Apputasy* is good for the scurvy in the mouth. Of the last two he and I have taken abundance, but without effect."[42] By reducing Petiver's catalogue to the terms and preparations observed by Smyth (who receives no mention in King's critique), King presents African medical knowledge as at once laughable and impenetrable, a collection of simples barely recognizable as such due to their unintelligibility. Sloane and Petiver, in their self-aggrandizing sociability, are linked socially, culturally, and bodily to Africa and Africans. In King's retelling, Petiver, labeled an "African Doctor . . . whose physic [is not] beyond his breeding," and Sloane have both taken "*Mening*" and "*Apputasy*" for "stoppage of the head" and "scurvy of the mouth," albeit to no avail. African language itself is presented as ridiculous, doubly so when Sloane and Petiver rely on their presumed knowledge of it to present themselves as worthy of learned esteem. In a later satire, King levies a similar criticism of Sloane's lack of systematic organization in physic. Sloane's medical therapies constitute, according to King, little more than "a House-Wife's Receipt Book, or as Physick was said to be in its first Age."[43] In the satirist's estimation, the creolized Sloane is not a respectable man of science. While both authors—Sloane and King— exhibit anxiety about the instability of colonial spaces, Sloane's work relies fundamentally on the epistemic possibility accompanying such anxiety. King's satires, on the other hand, aim to write Sloane, his medicines, and the incursion of colonial leveling that both represent, out of respectable English society.

Considering King's satires and Sloane's editorship can help us to better understand some aspects of the revision of the *Pharmacopoeia Londinensis*, which Sloane oversaw as the president of the College of Physicians. According to the preface, the 1721 edition of the *Pharmacopoeia* responded to a worrying inconsistency in the preparation of medicines owing to the "inaccurate and imperfect description" of materia medica in previous editions. For this reason, "the President and College" made significant revisions, particularly to the catalogue of simples: a number of substances were eliminated, botanical names were introduced, and regulated measurements were included in the sections on compounding. Notably, some of the materia medica Sloane discussed in the *Transactions* made their way into the *Pharmacopoeia* for the first time, albeit without the full vision of their context as presented in the pages of the journal. *Lichen terrestris* was included for the first time, as was the Jamaica pepper. Both materia medica are reduced, however, by the classificatory scheme of the catalogue of simples. Each simple is presented on pages with three columns: a common name, rendered in Latin (e.g., "Lichen *Hepatica* cinereus"), a botanical name (e.g., "Lichen terrestris cinereus Raii"), and the relevant part or parts of the plant to be employed as a medicament (e.g., "*Herba*").[44] As the preface asserts, "The Catalogue of Simples is entirely new as to its Method; every Plant being distinguished, not only by those names known in the shops, but also by such as are sometimes used by the more accurate writers in Botany."[45] The entire catalogue thus aims to wrest epistemological authority over materia medica from the apothecary to the learned physician.

Significantly, however, a number of entries lack a botanical name. Such is the case with Jamaica pepper, labeled simply as "Pimenta *Piper Jamaicense*" and "*Fructus*." The central column on the page stands empty, the space above and below filled by lengthy botanical classifications of the previous and following entries.[46] The natural historical knowledge gathered by Sloane and his correspondents frustrates the taxonomical impulse of the official pharmacopoeia. And yet, the colonial relation persists as that blank on the page. Similar blank spaces haunt other entries, many (though not all) of them describing New World or African medicaments. Other simples without botanical names include the entries for jalap, cinchona, and one of the few African-derived simples, grains of paradise.[47] The kind of imperfect knowledge that Sloane championed in the pages of the *Transactions* and King mocked in his satires registers in the official pharmacopoeia, albeit as an emptiness in the imperial discourse.

The epistemological and social instability that Sloane allows in the pages of the *Transactions*, and which made him an object of ridicule for King and his editorial philosophy a liability for subsequent secretaries, endures as residue of imperfect knowledge in the 1721 *Pharmacopiea*. Therefore, Edmund Halley, Sloane's successor as secretary of the Royal Society, changed editorial practice at the journal. Halley's first preface clearly announces a new regime. The journal will continue to "present the Publick with such short Tracts, as might otherwise be lost to Posterity." But it will only print "extracts . . . of the material, omitting the Preambles and Conclusions, and the useless parts of such Letters." Additionally, Halley asserts, the *Transactions* will resume their focus on "Natural Phil, Mathematicks, and Mechanics," notably omitting natural history and medicine.[48] Halley's strong editorial hand aims to formalize the *Transactions* while simultaneously reducing the fields of study that would necessarily produce the most colonial papers.

Just as Halley regulates the style, content, and epistemic aims of the *Transactions*, so Sloane's own attachment to the revision of the *Pharmacopoeia Londinensis* serves a similar function. The thrust of the volume is not additive but purgative: of unfamiliar or unreliable medicine, of inaccurate measurements, and of common names. Akin to the occlusion of New World and African materia medica from King's critique, Halley's editorial changes and the subsequent devaluation of Sloane's editorial labor render invisible the colonial relation that had, however imperfectly, been made visible. Sloane's work on the *Philosophical Transactions* can thus be understood as occupying an epistemic middle ground between the messiness of colonial medical practice and the orderliness of the official pharmacopoeia. Looking at Sloane's work in this way, I think, leads to a set of questions about the production of pharmacopoeia and the circulation of healing knowledge in the Atlantic world. How can we track the promiscuous movement of practical knowledge about healing among the various peoples who constituted the Atlantic medical complex? How can we best restore the centrality of colonial nature as well as embodied experience to our histories of health, healing, and medicines? And what further sites of generic possibility will enable us to track the movement of practical healing knowledge among individuals, continents, and epistemic systems? Reading across the fruitful generic landscape of the eighteenth-century medical public sphere, as I have done here with Sloane, can, I offer, better enable us to see how the colonial has been written into and out of our histories of science and medicine.

PART III

Pharmacopoeias and the Construction of New Worlds

7

THE FLIP SIDE OF THE PHARMACOPOEIA

Sub-Saharan African Medicines and
Poisons in the Atlantic World

BENJAMIN BREEN

"EXCEPT ON the coasts, Africa is hardly traded with," Denis Diderot wrote in the *Encyclopédie*. "The interior of this part of the world is not yet sufficiently known."[1] John Jacob Berlu's *The Treasury of Drugs Unlock'd*, a handbook for London's drug merchants and apothecaries written some sixty years earlier, would seem to back up Diderot's pronouncement. Berlu's book described the geographic origins and virtues of 154 drugs in all, including medicinal human skulls (the most potent were said to be found in Ireland), cannabis ("of an Infatuating Quality, and pernicious Use"), unicorn horns, and opium.[2] However, Berlu identified only three drugs as originating in sub-Saharan Africa: grains of paradise, gum senica, and gum animi.[3] By comparison, according to Berlu, thirty-one of the drugs listed in his book were imported from Spanish America, twenty-one from Turkey, and no fewer than forty-six from the East Indies. Likewise, in a manuscript commonplace book recording drugs shipped from the Lisbon apothecary shop of Manuel Ferreira de Castro to Rio de Janeiro in 1738, forty-two medicines hailed from Portugal, twelve from elsewhere in Europe, eleven from the East Indies, seven from the West Indies, and only one (grains of paradise) from sub-Saharan Africa.[4]

Recent scholarship has greatly enriched our understanding of the medical and pharmaceutical cultures of Africans and African-descended peoples in settings like the sugar *engenhos* of Brazil, the cities and ports of the Caribbean, or the tobacco plantations of the Chesapeake.[5] Attention has turned to the question of how enslaved peoples contributed not only to the labor regimes that produced and employed early modern drugs but to the epistemologies governing their use. Not just African hands but also African minds contributed to new articulations of pharmacy in the early modern Atlantic and beyond.[6] As Robert Voeks has put it, the botanical knowledge and medical practices of Africans and African-descended peoples helped create "disturbance pharmacopoeias" that permanently reshaped both medical practices and ecosystems from Brazil to Mexico and beyond.[7] The great majority of scholarship on these topics concerns Africans in diaspora.[8] The present chapter poses a question provoked by this work, and by the absences visible in the lists recorded by early modern participants in the drug trade like Berlu or Ferreira de Castro. If *diasporic* African healing was so important in the pharmacy and medical practices of the early modern Atlantic, then why are cures from sub-Saharan Africa itself relatively invisible in early modern European pharmacopoeias?

I will consider three potential answers to this question. One involves the commercial infrastructure that undergirded the early modern globalization of medicinal drugs. Regions like the Guinea coast and the Portuguese colonial outposts in Angola were closely linked to the global economy, but these were links forged in bondage and overwhelmingly centered on the trade in enslaved human beings, not medicines. Europeans in Portuguese Angola and the entrepôts of West Africa did, however, remark on the exceptional efficacy of some African cures. Describing the Gold Coast of West Africa circa 1700, for instance, the Dutch merchant Willem Bosman praised "green Herbs, the principal Remedy in use amongst the Negroes," which were "of such wonderful Efficacy, that 'tis much to be deplored that no European Physicians has yet applied himself to the discovery of their Nature and Virtue." Bosman continued by saying that he "firmly believe[d]" that these herbs "would prove more successful in the practice of Physick than the European Preparations."[9] But such claims rarely translated into actual efforts to export sub-Saharan cures at scale. It was not until the nineteenth century, as documented in Abena Dove Osseo-Asare's *Bitter Roots*, that African botanicals became extensively incorporated into global pharmacopoeias.[10] When we venture into earlier eras, we find many European accounts of the virtues of sub-Saharan healing barks, herbs, roots, and animal products, but very little evidence of successful incorporation of these substances into the global networks of the early modern drug trade.

A second potential answer to the absence of sub-Saharan knowledge and materials in early modern pharmacopoeias is that many Europeans associated tropical African spaces with insalubrious climates, poisons, and malign spiritual forces. At first glance, this answer would seem to have solid backing in primary source documents (such as those produced by Capuchin missionaries in seventeenth-century Angola) that remark upon the expertise of African healers in creating both poisons and antidotes, and affirm an apparently widespread belief that tropical Africa was climatically and spiritually threatening to European "constitutions."[11] However, this was by no means unique to the region. Many South Asian and American drugs initially similar faced opposition from Europeans who associated their use with demonic influence or feared that they were poisons unfit for European bodies.[12] When compared to the African drugs that we will look at in this chapter, however, New World substances like tobacco, brazilwood, cacao, and guaiacum passed more readily into European pharmacopoeias following an initial period of mistrust.[13] Thus it is not a sufficient explanation simply to say that association with non-Christian spirituality or with poisoned landscapes made drugs impossible to integrate into pharmacopoeias. European apothecaries, physicians, and philosophers actually proved to be quite willing to separate what they perceived to be the "bad" (associations with non-Christian religious practices) from the "good" (beneficial medical virtues), if prompted by commercial imperatives.

A third and final potential answer specifically addresses this divergence between the fate of New World medicaments (both those associated with African-descended peoples in diaspora and with Indigenous Americans) and those of sub-Saharan Africa. Perhaps the early modern Americas attracted a greater number of natural philosophers, apothecaries, and physicians than sub-Saharan Africa, and therefore it was proportionally easier for Indigenous American and diasporic African cures to be integrated into European pharmaceutical practice. Due to the elisions surrounding European involvement in the slave trade, apothecaries and physicians stationed in the slaving ports and entrepôts of Africa did not have the same social visibility, nor the same degree of participation in global print culture. No printing presses were available to them, whereas in both South Asia and the Americas, printing presses had been producing works directly relating to "hybrid" pharmacy from the late sixteenth century, starting with the publication of Garcia da Orta's *Colóquios dos simples e drogas da India* in Goa (1563) and Francisco Bravo's *Opera Medicinalia* in Mexico City (1570).[14]

Taken together, these three answers help explain how sub-Saharan African cultures of healing—despite their on-the-ground importance and the rich diversity of cures available in regions like Angola or West Africa—did not fully participate in the transformations of early modern European pharmacy. African medicines of the sixteenth, seventeenth, and eighteenth centuries figure as a flip side to the more famous commodified medicines of the Columbian Exchange, an alternative pharmacopoeia that, though never forgotten, also failed to become established in early modern apothecary shops or printed texts.

UNAVAILABLE AT ANY PRICE?

As noted above, perhaps there simply weren't sufficiently developed "commodity chains" to allow African medicaments to appear in substantial numbers on apothecaries' shelves or in pharmacopoeias.[15] It is true that natural philosophers and apothecaries in Europe seem to have had a more difficult time obtaining botanical samples directly from sub-Saharan Africa than from other regions in the Black Atlantic, such as Jamaica. Robert Boyle, for instance, took an active interest in what he called "specifick medicines" from Africa, yet seems to have had difficulty moving beyond secondhand accounts to obtain physical samples. Boyle reported his conversation with an English merchant who had experience "with the Negro's of the Inland Countrey" of West Africa (presumably, the experience of Boyle's informant derived from serving as a slave trader). Boyle was told of an extraordinary poison that was "very mortal." A "famous Knight" who served in the African interior was said to have been "poysoned at a parting Treat" using this substance "by a young *Negro* Woman of Quality, whom he had enjoy'd and declin'd to take with him, according to his promise."[16] But Boyle's reliance on hearsay and his lack of access to specific materials posed problems for any effort to experimentally test or commercialize intriguing *materia medica* from the region, as Boyle's colleague Robert Hooke had proposed doing for "bangue" (cannabis) that he obtained via the East India Company pilot Robert Knox during the same period.[17] Although Boyle was invested in the slave-trading operations of the Royal Africa Company (and was clearly also in contact with individuals who had firsthand experience of West and West-Central African slave ports) he appears not to have had reliable access to physical samples from the region.

This contrasts strongly with natural philosophers' access to materia medica from the East and West Indies. For instance, in the same period that Boyle was recording thirdhand accounts of female African poisoners, Royal Society founding member Christopher Merret obtained physical samples of naturalia from a group he called "the Portugal Negroes." Merret was apparently referring to the inhabitants of Malaysia and other Indian Ocean ports, not to Africa. His samples consisted almost entirely of medicines, like *Cibotium barometz* ("poco sempie, accounted a great Cordial") and sagu gum ("from the Islands of Malacca"), that were native to maritime Southeast Asia.[18] The *Musaeum Regalis Societatis*, Nehemiah Grew's 1681 survey of the botanical, medical, and mineral samples owned by the Royal Society, paints a similar picture. The society possessed several vaguely described African mineral samples (like lapis lazuli "found in Gold Mines in Africa") and animal trophies, like the "Hornes of a Wild Bull... brought from Africa."[19] But Grew made no reference to medical or botanical samples from sub-Saharan Africa, with the exception of two Angolan tree barks labelled as *tacusa* and *chicengo* and not mentioned elsewhere.[20] By contrast, Grew recorded dozens of materia medica from regions like South Asia, the Caribbean, and the Spanish American mainland.

Yet blaming these material absences on a lack of commercial infrastructure may be mistaking a result for a cause. It seems clear that it was at least *possible*, given the infrastructure that already existed via the slave trade, for Portuguese, French, Dutch, and British traders and mercantile companies to conduct a sustained trade in sub-Saharan medicinal drugs. The maritime routes of Indies-bound ships (which very frequently watered in West and East African ports) guaranteed a substantial contact between Europe and Africa throughout the sixteenth, seventeenth, and eighteenth centuries. More importantly, however, the Atlantic slave trade led to the formation of European merchant communities and colonial entrepôts from the Senegambia to Cape Town, and up the Swahili coast. By the end of the seventeenth century, economic links associated with the Atlantic slave trade were moving into interior zones as well, with commerce being carried out by go-between figures like the traveling *pombeiro* merchants of the Angolan highlands.[21] The seventeenth-century Parisian drug merchant and apothecary Pierre Pomet described one sub-Saharan African medicament sourced from French slave-trading entrepôts along the River Senegal and harvested in the interior. "Gomme du Senega" (i.e., Berlu's senica gum), was a substance "which is today sold in our Boutiques" and which "is brought to us from Senegal by the blacks and whites who come from the mountains." However, aside from his access to senica gum and two traditional medicaments known to Greco-Roman pharmacy (grains of paradise and euphorbium), even the well-connected Pomet appears to have had scanty connections to sub-Saharan pharmaceutical networks. For instance, he also described the medical virtues of "tamarinds of Senega," but noted that "these are very rare in France."[22] Thus even in regions, like Senegal, where European trading companies and slavers had established powerful footholds, African drugs appear to have been relatively scarce and inconsistent elements of commerce.

The Portuguese archives document the importation of European and South American medicines *into* Angola throughout the seventeenth and eighteenth centuries, although rarely in quantities that satisfied the handful of European medical professionals (usually barber-surgeons and military doctors) who were active in Portuguese Africa. For instance, in 1738, a Lisbon apothecary sent fifty *garrafas* of a drug called Agua de Inglaterra (an antimalarial containing Peruvian cinchona bark) to Luanda via the merchant vessel Nossa Senhora de Nazaré.[23] The records of this apothecary make no mention of any return shipments: a drug with an active ingredient originating in South America was being prepared in Europe and shipped to Africa, with no reciprocal trade in African-derived medicaments. Quite possibly, the ship *Nossa Senhora de Nazaré* was returning from Luanda laden with a cargo not of drugs, but of slaves.

Indeed, it may well be the case that the most important pharmaceutical transfer between sub-Saharan Africa and the territories of the early modern European empires was not one of materials at all, but of knowledge. Although the material deprivation of the Middle Passage made transfers of botanical samples or seeds extremely rare, information contained within the minds of people who had become slaves was able to take root across the Atlantic. One fascinating example of such a transfer can be found in the biography of Domingos Álvares, a Dahomey-born healer and spiritual leader who became popular in Brazil as a *feiticeiro*—a powerful wielder of both poisons and cures—and who was eventually imprisoned by the Inquisition in Portugal.[24] A common thread running through accounts of go-betweens like Álvares is that these healers' powers derived not from their mastery of a preparation of specific medicinal plants, minerals, or animal products but from their ability to fashion magical amulets or pouches (*bolsas da mandinga*) that protected against poison, disease, or injury. One of the most famous Black Atlantic amulet makers was Makandal (whose name appears to have derived from *makunda*, a Kikongo word for magical amulet or talisman). Makandal was accused of using his pharmaceutical and magical knowledge to spearhead a massive poisoning plot against French colonists in 1750s Saint-Domingue.[25] We find mentions of similar amulets throughout the Atlantic world, from the British West Indies where they were components of obeah magic, to West Africa itself, where Willem Bosman complained that even Europeans had "grown very fond of wearing some Trifles about their Bodies, which are consecrated or conjured by the Priest." A law banning their creation in 1720s Jamaica gives a good sense of the general ingredients list: it specifically forbade enslaved healers from "making use of any blood, feathers, parrots beaks, dog's teeth, alligator's teeth, broken bottles, grave dirt, rum, egg shels or any other material relative to the practice of obeah or witchcraft."[26]

The materials contained within these amulets appear to have been less important than the hard-won knowledge required to create them—knowledge that readily passed from Africa to the Americas and Europe, even if the specific botanicals contained within these bolsas changed as those who created them moved between ecosystems. The next section will consider a second potential answer to the question of Africa's absence in pharmacopoeias that hinges on a common thread running between diasporic healers like Álvares or Makandal, and what we know of seventeenth-century medical practice within Africa. Of Álvares, it was said that "today he cures, tomorrow he kills."[27] As Álvares' biographer James Sweet notes, his perceived power as a healer and wielder of powerful plant and animal products hinged on the flip side to his pharmacy: a mastery of poisons. This was a common dichotomy throughout the pharmacy traditions of the Black Atlantic. Was it then the case that African medicaments failed to globalize because of these close associations with poison?

AFRICA AS POISONED LANDSCAPE

"The difference between a poison and a medicament," the Lisbon apothecary Caetano de Santo Antonio wrote in 1711, "is that the poison destroys nature, while the medicament (even when it has inside of it a portion that is poisonous), after having been prepared in conformity with the rules of art, becomes salutary, and a helper of nature."[28] The debates that played out when tropical medicaments entered European pharmacopoeias hinged on this dichotomy. The work of distinguishing between poison and cure often verged into the realm of the spiritual: was this cure the work of God or of Satan? Did it further divine providence or work against it? And what "rules of art" provided spiritually and socially acceptable ways of purifying potential poisons into salubrious medicines?

Long before the influx of Indies drugs in the sixteenth and seventeenth centuries, medieval European apothecaries had established careful criteria for assessing whether a poison's "venomous" or "corruptive" properties could be removed. The fifteenth-century Venetian physician Sante Arduino, for instance, wrote that "the difference between poison and medicine is that poison corrupts the complexion and the substance of the body only through its properties or specific form, such as viper venom," whereas "medicine that corrupts the body" does so via what Arduino called its "elementary complexion," its potentially improvable, unprepared state.[29] In the sixteenth and seventeenth centuries, the boundary between poison and medicine shifted: new chemical techniques promised new methods for testing and altering intrinsic properties. As Alisha Rankin has noted, although poison trials were sometimes mentioned as "theoretical possibilities" in classical and medieval medical texts, it was in the sixteenth century that the experimental testing of poisons became a widespread practice.[30] So, too, did experimental attempts to transform the intrinsic properties of poisons.

At the same time, reports of mariners, ship surgeons, and itinerant natural philosophers raised a new question that destabilized the reliability of trials conducted within Europe: could poisons also be transformed by *where* one encountered them? The Lisbon-based apothecary João Curvo Semedo believed they could. "The meat of pigs," he wrote in his *Polyanthea Medicinal*, "which in the lands of Angola is given to the sick at all times of the year, is positively harmful in Portugal when it is eaten outside of cold and wintry periods."[31] For apothecaries working in imperial metropoles and hubs of the slave trade like Semedo, the challenge of learning "how to separate the good from the bad, the pure from the impure" was compounded by the idea that poisons could become medicines and vice versa depending on where (and by whom) they were consumed.[32] In such a scenario, case histories of colonists, soldiers, or slaves who had traveled long distances could potentially have as much epistemological significance for discerning the value of a novel drug as an experimental trial conducted in Europe.

European accounts of medicine in regions like Guinea and Angola tended to conflate the supposedly poisonous nature of African cures with the sins of the healers who prepared them. As with the healers of Mesoamerica and other Pre-Columbian societies, these figures were almost automatically assumed by early European commentators to be in thrall to witchcraft and Satan. A characteristic (if unusually vivid) opinion was that of the English traveler Thomas Herbert, who visited Luanda in the 1620s and gathered information from Portuguese colonists and slave traders in the region. The Angolans, he wrote, were worshippers of "deformed Idolls" who performed "what the witches urge them to ... soyling their hellish carkasses with juyce of herbs, ryce, roots, fruits, or what the old impostor infatuates them with; the female sex each new Moone defying pale fac't Cynthia by turning up their bummes, imagining her the cause of their distempers."[33] For Herbert, the entire coastline extending from the Gold Coast to the Congo was a place of corrupted pharmacy, "rich in earth, but miserable in demonomy," and the cures produced by these "witches" (a term that probably corresponded to the Portuguese *feiticeira*) had been corrupted into little more than poisons.[34]

Throughout the seventeenth century, apothecaries, surgeons, and barbers (*barbeiros*) working in colonial outposts or onboard slave ships emerged as important sources for navigating the uncertain terrain that connected the realm of the pharmacological with the realm of the poisonous. Surgeons involved in the Luso-Brazilian slave trade, like Luís Gomes Ferreira, were also among the figures who were best positioned to integrate African cures into the mainstream of European pharmacy. Ferreira was a surgeon and slaveowner who treated both enslaved and free laborers in the gold mines of Minas Gerais in Brazil throughout the 1710s and 1720s. In his *Erario Mineral* (Mineral Treasury) Ferreira covered a range of practical topics of geared toward healers working in slave societies. He devoted substantial space to discussing cures for various forms of poisons and *feitiços* (curses), and his writings reflected a considerable amount of time spent speaking with patients from Angola and the "Costa da Mina" (roughly corresponding to present-day Ghana).

On only one occasion, however, did Ferreira recommend a medicine *from*

Africa. In his discussion of the "asthmatic fluxes" that troubled both slaves and Portuguese colonists working in the Brazilian mines, Ferreira described a cure that "neither Doctors nor Surgeons yet have any knowledge of." Much of the ingredient list that followed, like basil and bitter orange, was prosaic. Others ("the urine of a three or four-year-old child") were oddly specific, but far from uncommon. One ingredient, however, stood out: "A fruit by the name of tépe, which comes from Angola, and which there is no lack of in Bahia among those who are curious."[35] Ferreira explained that "os Angolistas" brought tépe "and other things of interest" from Angola to the markets of Bahia.[36] It is tempting to speculate that Ferreira may have learned of the use of this medicine from one of his own Angolan slaves. Regardless of how Ferreira learned of it, this integration of an Angolan medicament into a medical receipt was an exception that proved the rule: Ferreira admitted that "these [tépes] are difficult" because they required personal connections to "some fellow who has business in the Kingdom of Angola" and who was capable of shipping back the "many materia medica which, it is certain, are only to be found in Angola."[37] Ferreira was far more preoccupied with the Mal de Loanda, an ailment that he associated with the noxious climates of Angola and the Mina Coast and which he blamed for the deaths of "two thousand slaves, and many white men."[38] This was a typical pattern in Portuguese pharmacopoeias and medical texts of the period: African medicaments, if mentioned at all, were invariably overshadowed by harrowing accounts of dangerous African diseases, poisons, and climates.[39]

This association of Africa and Africans with poisoned landscapes seems to have played a role in the self-fashioning of diasporic African healers as well. "The Negro Caesar's cure for poison," which appeared not only in dozens of newspapers and almanacs throughout the British Atlantic world but in at least one surviving manuscript recipe book, offers one example.[40] Caesar first entered colonial records as a slave owned by a resident of Charleston, South Carolina. Caesar's cures for both rattlesnake bites and general poisons were, by 1749, famous enough to merit mention in the South Carolina Colonial Assembly.[41] A few months later, the *South Carolina Gazette* published "the Negro Caesar's cure for poison . . . for discovering of which the General Assembly hath thought fit to purchase his freedom."[42] The Commons' decision had resulted from an internal debate over the efficacy of Caesar's cure begun on November 24, 1749. Three slave owners testified to the Assembly that Caesar had cured several whites "who had been poisoned by slaves" and performed cures that had eluded "some of the most skillful physicians in the country." They added that Caesar "was willing to make a Discovery of the Remedy which he makes Use of in such Cases for a reasonable reward."[43] In late April of 1750, the Assembly voted to pay 500 pounds to purchase Caesar's freedom and allotted another 100 per annum to be paid for the rest of his life.[44] This remarkably high annual stipend (roughly double the purchase price of a healthy male slave) and the rapid dissemination of Caesar's cure throughout printed works and manuscripts testify to the high value attached to his knowledge. It is potentially significant that Caesar's claim to fame was expertise in the *cure* of poisons—as with Domingos Álvares, it may have been the case that widespread associations of African pharmacy with poisoning created a parallel belief that African healers were also skilled at creating antidotes.

Caesar's cure, while unusually well documented, was not unique either in its social importance or its characteristic mingling of practices and materials from both hemispheres. As Robert Voeks has noted in the context of colonial Brazil, medical recipes used by African healers, even those in maroon communities, relied substantially on materials from European and Amerindian pharmacopoeias.[45] In Caesar's case, both of the herbs constituting his cure were European transplants. The "plantain" that played a prominent role in his recipe (likely a shrub of the genus *Plantago*) had an ancient pedigree stretching back to its use as a cure for poison in the tenth-century Anglo-Saxon "Nine Herbs Charm."[46] Horehound (*Morrabium vulgare*), another ingredient in Caesar's cure, was originally native to the Mediterranean and enjoyed a long history of use as a curative plant in Southern Europe and the Middle East. It was not the ingredients in Caesar's cure that were novel, but his *practice* and his resulting social positioning as a master of poisons and anecdotes that promised to protect slaveowners against the dangerous botanical knowledge of their own slaves.

A celebrated enslaved doctor in mid-eighteenth-century Guiana known as Kwasi followed a somewhat similar trajectory. Kwasi (or Quashe in some accounts) appears to have been born somewhere in West Africa ca. 1690 and transported to Guiana as a slave. By 1730, he had achieved recognition in the colony as a skilled healer. Like Caesar, he ultimately gained his freedom by popularizing a botanical preparation that was purported to cure diseases and poisons. Known as the "Quaciae bitter," Kwasi's cure may have originated via his interactions with a maroon community in inland Guiana.[47] Kwasi became so celebrated for this and other plant cures that he reached notice of Carl Linnaeus, who wrote of him that "an unknown Negro slave named Qvassi discovered a medicine that he began using for his fellow slaves' severe fevers, and that with such success, that even the masters sought his help."[48] However, Kwasi's earlier self-fashioning as a healer in the maroon communities of Guiana differed significantly from his status as a minor medical celebrity among European medical writers later in his life. Gabriel Stedman attributed Kwasi's original rise to prominence as deriving not from his botanical expertise per se but from his "having got the name of a *lockoman*, or sorceror, among the lower slaves."[49] Stedman noted that Kwasi made a good living selling "his *obias* or *amulets*" to freed black soldiers who purchased them "in order to make them invulnerable." Kwasi was said to have made a handsome profit due to the fact that his amulets contained "neither more nor less than a collection of small pebbles, sea-shells, cut hair, fish-bones, feathers, &c."[50]

In other words, Kwasi first gained prominence among African-descended communities in Guiana as a fashioner of amulets that offered protection against curses, poisons, and other dangers (which had a clear correlate with other West African healing charms known variously as obeah pouches, *minkisi*, or bolsas de mandinga).[51] However, this did not lead to European recognition until he began to sell his "bitters," which slotted far more neatly into the epistemological space of what constituted a valuable tropical drug, occupying a similar position to the highly popular antifever cinchona bark of Peru.

Although the knowledge of obeah men didn't fully integrate with European medicine, it gained increasing attention in the plantation economies of the Caribbean. In 1729, an Anglican rector named Arthur Holt wrote from Barbados, "the Oby Negroes, or conjurers, are the leaders to whom the others are in slavery for fear of being bewitched." It was from these "conjurers," Holt speculated, that ordinary slaves "receive charms . . . [and] get deadly doses to dispatch out of the world such masters or other persons as they have conceived a dislike of."[52] By the middle decades of the eighteenth century, plantation elites believed African and African-descended healers to be such a threat that they required legal persecution. Just two years before the publication of Caesar's cure, for instance, in 1748, the Virginia Assembly banned slaves from participating with whites in making medicines, and in 1751 a South Carolina statute mandated "that in case any slave shall teach or instruct another slave in the knowledge of any poisonous root, plant, herb, or other poison whatever, he or she, so offending, shall, upon conviction thereof, suffer death as a felon." To lessen the risk of slaves attaining and sharing "the knowledge of any mineral or vegetable poison," the statute continued, "it shall not be lawful for any [white] physician, apothecary or druggist, at any time hereafter, to employ any slave." Indeed, slaves were banned even from entering "shops or places where they keep their medicines."[53]

BUYTRAGO'S FAILURE: THE ROLE OF MEDICAL ELITES IN ESTABLISHING GLOBALIZED PHARMACOPOEIAS

In the first decade of the eighteenth century, a Portuguese cavalry officer named Francisco de Buytrago stationed in Angola learned from "the people of this land" that there existed in the forests a medicine that he came to call the *casca da vida* (bark of life). Buytrago's account of the bark of life offers an emblematic case of how the virtues of African materia medica and pharmaceutical practice circulated (or rather, failed to circulate) in print. Buytrago's manuscript text consists of two sections; the first a survey of some of the materia medica he encountered in 1710s and 1720s Angola and Brazil, and the second a short treatise specifically relating the "miraculous virtues" of the casca da vida. The unpublished account, which Buytrago seems to have written in the expectation that it would see print, offers a glimpse into how Catholic Iberian cultures of healing began to merge with African *feitiçaria* and pharmacy in the eighteenth century. However, it also exemplifies the difficulties of transmitting sub-Saharan African pharmaceutical knowledge into European print culture.

Buytrago's failed attempt to publicize casca da vida exemplified a larger pattern: whereas elite physicians and natural philosophers like Hans Sloane made well-publicized trips to centers of diasporic African medical practice like Jamaica or Cartagena, far fewer licensed physicians or natural philosophers were active in the ports of coastal West or West Central Africa.[54] Buytrago, for instance, was a soldier, not a doctor. It seems likely that his failure to publicize both his Angolan bark and his own purported feats of healing stemmed in part from Buytrago's lack of connections to the Portuguese medical and natural philosophical elite. Perhaps, then, sub-Saharan African pharmacopoeias failed to globalize in part because of a lack of "go-between" intermediaries (like Hans Sloane, Francisco Hernández, or Garcia da Orta) who combined a sincere interest in non-European epistemologies with prestigious training and connections that allowed them to publicize their work.

The title of Buytrago's manuscript, "Arvore da Vida, e Thesouro descuberto" (Tree of Life, and Discovered Treasure), evoked both the Garden of Eden and the obsession with making profitable "discoveries" in the *sertão*.[55] The structure of his text reflected this dual concern with "treasure" and spirituality, with the first half narrating the "miraculous" cures of demon-possessed people (*demoninhados*) in Angola, Brazil, and Portugal that Buytrago attributes to the medicinal bark of the "Arvore da Vida," and the second half transitioning into a more workmanlike descriptive list of "the names of the most unique things which are found in the Kingdom of Angola, and their virtues." From the opening pages of his text, Buytrago argued emphatically for the medical value of African nature: "In the Kingdom of Angola and in the provinces and neighboring lands of that region, there exists such a profusion of barks, and herbs, and other things of such singular virtue and efficacy... that they exceed those of all of the world in power, and in the greatness and variety of the plants and herbs that these folk employ."[56] Unlike his compatriots, Buytrago also unreservedly praised the skills of African healers, lauding the transparency of their "surgeons and physicians, who do everything in the sight of the Sick, in order to free them from any ill suspicion." For Buytrago, the *botica* (pharmacy) used by these healers was indeed nothing short of "perfect." But Buytrago's manuscript never found a publisher. Indeed, it has barely even been described in contemporary historiography. For the past three centuries, the manuscript appears to have rested unread and unremembered in Lisbon archives.

Buytrago proudly describes himself as a cavalry officer, but his real function was as an enforcer for the Portuguese and Brazilian slave trade. And perhaps this role can help explain the relative lack of status among European advocates for African pharmacy. The disdain was palpable in an account by the Dutch West India Company factor Willem Bosman, writing about the Gold Coast region of West Africa circa 1703, who noted that "he who here acts the part of a Doctor, is also a *Feticheer* or Priest." Bosman lamented that these physician-priests had gained adherents "even [among] some Europeans who not only think favourably of, and believe this Idolatrous Worship effectual, but ... are likewise grown very fond of wearing some Trifles about their Bodies, which are consecrated or conjured by the Priest." These healers, Bosman claimed, used "contradictory Ingredients" in such a way that "the Remedies used here frequently seem pernicious in the Case wherein they are given."[57] Yet despite "how contradictory and improper soever these Med'cines may seem," Bosman nevertheless admitted that in practice "[they] are found very successful." This was thanks not only to the power of local medicines—"about thirty several sorts of green Herbs, which are impregnated with an extraordinary Sanitive Virtue"—but to the expertise of the African physicians, whom Bosman found himself praising effusively despite his own misgivings. "I have seen several of our Country Men cured by them," he recollected, "when our own Physicians were at a loss what to do ... I have several times observed the Negroes cure such great and dangerous Wounds with them, that I have stood amazed thereat."[58] However, although Bosman and other Europeans on the ground in sub-Saharan African ports may have marveled at the ability of *African* medical professionals, they also disdained efforts by European physicians and apothecaries who sought to emulate them.

Chemical medicine promised a set of techniques that, in theory, were able to turn suspect substances into commercially viable medicaments, just as distillation allowed the waste products of the sugar produced in the vast Brazilian sugar *engenhos* to be repurposed as intoxicating rums and cachaças. Yet although African medical practices also relied upon material transformations, such as amulets purported to turn bones or dirt into powerful technologies of healing, these objects could not be chemically manipulated in the same manner as the pharmaceutical simples mastered by iatrochemists or apothecaries dedicated to "modern" methods. Such objects were not so much medicaments as they were the physical legacies of spiritual practice, imbued with a healing property that was as much supernatural as it was physical. An African amulet was a unique item and could not be refined, distilled, or purified. While a bark or mineral that could be recombined with other materia medica or altered using chemical methods, a fetish object was indivisible.

Medical texts written *in* seventeenth- and eighteenth-century Africa are sparse, but collectively paint a picture of a failure to bridge this epistemological divide. Reports from surgeons and apothecaries sent back to the Portuguese Overseas Council are notable for failing to follow the path of Buytrago: these medical professionals evinced virtually no interest in local healing traditions or botanicals, and instead bemoaned the lack of funds available for shipping European medicines to Africa, alongside complaints about the "venomous" climate.[59] A typical letter, sent by the *chirurgião-mor* (chief surgeon) Luiz Goncalves de Andrade in 1666, explained that a soldier needed to be sent home to Brazil because the "conjunctions of the moon ... and climate are contrary to his nature."[60] One manuscript from 1760s Mozambique, however, offers a tantalizing hint of an attempt to blend African cures with chemical medicine of the sort that enjoyed high status in eighteenth-century European pharmacy. The author, Antonio Pinto de Miranda, described the Kingdom of Monomotapa (present-day Mozambique) as abounding in both poisons and antidotes. Foremost among the latter was a cure by a Portuguese colonial officer in the inland province of Tete: "Manoel Gomes de Nobre has a little gourd with some roots against poison, which mixed with oil are an effective and likely remedy. ... Take a large quantity of *solimão*, some brains of the sea horse, and gall from alligators of the type that are of the most poisonous in the world, the most poisonous that can be found. Place this inside the gourd with roots and oil and it will take out the poisons within."[61] This description of a poison trial involved the use of a mercury-based chemical medicine named *solimão* alongside local materia medica like the gall of poisonous African alligators.[62] This remedy was then stored, African-style, in a gourd. The anecdote hints that European and African knowledge surrounding poison was beginning to hybridize and enter into everyday practice. However, the fact that it remained in manuscript is significant: this was decidedly *not* the type of cure that could become popular in Europe, not least because it was not the product of a medical professional at all, but, like Buytrago, the work of a moonlighting soldier.

CONCLUSION

In seeking to answer the question posed at the outset of this chapter, we must be wary of setting up a binary between two monoliths: "African healing" and "European medicine." After all, the amulets or pouches associated with West and West-Central African cures had clear correlates with the longstanding usage of medicinal amulets in Europe, as did the use of animal and human parts believed to contain supernatural powers. The body parts of various animals believed to have "sympathetic" virtues, for instance, appear quite regularly in lists of the wares of apothecaries and drug merchants. The "grave dirt" mentioned in a Jamaican law banning obeah amulets had a direct correlate in the London pharmacopoeia's entry for *mumia*, which purported to be actual remains of an Egyptian mummy, but frequently consisted of an ad hoc mixture of human remains, burial wrappings, and bituminous soil surrounding a grave.

It was a spiritually charged mixture that would not have been unfamiliar to the makers of obeah pouches and amulets sold by diasporic healers like Kwasi or Makandal. Indeed, these African-made objects had clear correlates with the long-standing usage of medicinal amulets in Europe. As late as 1747, we find the Anglo-Scottish physician Thomas Short describing a variety of magico-medical amulets that were apparently still in contemporary usage in England. For instance, Short noted that "Common People" wear "a Piece of the [Piony] Root hung about the neck in an amulet . . . to prevent Convulsions," noting that "they are also made of Human Scull, Poppy Roots, &c."[63] Short recommended amulets for several other ailments: an amulet of the root of hounds-tongue for "the lousy Disease," an amulet of mistletoe for the falling sickness, or an amulet of a certain type of iris root for "Bloody Fluxes of the Belly [and] Uterus."[64]

It is not the case, therefore, that West and West-Central African cures would have necessarily appeared unacceptably foreign or unusual to medical consumers in early modern Europe. In some ways, they were all too familiar. They simply didn't fit into either the commercial model of the European trading companies or the self-presentation of European apothecaries and physicians, who increasingly sought to distinguish themselves via their chemical manipulations, and who almost invariably avoided mention of overt connections to the African slave trade. In other words, at precisely the time that European medical authors were coming into sustained contact with African medicinal drugs, the receptiveness of these authors to cures from Africa was on the decline. They were, after all, cures carried over by slaves, slave traders, and soldiers, and were increasingly associated with the tensions between foreign and local cures that, as Stuart Anderson notes in this volume, increasingly characterized the emergence of national pharmacopoeias (chapter 10).

The disjuncture was compounded by the fact that African-descended healers in the Americas (who were better positioned than healers within Africa to gain attention from elite medical authorities) tended to blend European or Indigenous American botanicals, not African. It may well be argued that this is an unavoidable result of the violence and disruption of the Middle Passage. Yet as Judith Carney has shown, a number of food crops and animal species did in fact cross the Atlantic between Africa and the New World, effectively as stowaways on slave ships.[65] Thus there was at least the potential for Native African medicines to make the crossing as well. However, although many sub-Saharan

African medicinal plants would go on to gain international prominence in the nineteenth and twentieth centuries, only grains of paradise entered into widespread use by apothecaries either in North and South America or in Europe prior to the nineteenth century.[66] And this special role of grains of paradise (*Aframomum melegueta*) seems to have been a product of the widespread traditionalism of European pharmacy rather than any effort to incorporate new materials: the substance, after all, had been mentioned by Pliny, and was therefore not counted as a "modern" drug.

It might be tempting, at this point, to argue that African cures stood as a kind of hidden obverse to the pharmacopoeia of seventeenth- and eighteenth-century Europe. Because African medicaments were filtered through the medium of the slave trade and the association of Africa with poisons, and because they typically reached European authorities at several levels of remove, the pharmacopoeias of sub-Saharan Africa proved virtually impossible to integrate with the everyday wares of apothecaries in Europe. They did, however, demonstrate the types of suspect practices that apothecaries might want to avoid, serving as a kind of epistemological negative space. What wasn't fit for the apothecary shop was fit for the plantation poisoner, and vice versa.

Yet what we have here isn't a simple binary between European medicine and African healing. On the level of popular medicine—the sort of healing that took place among barber-surgeons and in city squares, in ordinary households, and onboard ships—there was far more commonality between early modern European and African practices than historians have allowed for. Practitioners like Caesar, Kwasi, or Domingos Álvares, who could move fluidly between African and European epistemologies, were more common than we think. And we see this not only in Inquisition trials but in a shadowy form in books by authors like Semedo, who included a number of medicines with Bantu and Kikongo names in his *Polyanthea Medicinal*, and in the republication of Caesar's cure in domestic English household remedy books and journals.[67]

Throughout the eighteenth century, however, European medical elites increasingly condemned those cures or practices that, in their view, could not be refined and improved using the arts of chemical medicine—or experimentally tested using the new natural philosophy. At issue here was the question of professional self-fashioning: the period of most sustained contact with African healing traditions and African nature coincided with one in which it became increasingly important for medical authors to demonstrate that they, too, were devoted to empiricism and "modern" chemical methods.

Africa remained deeply associated with poisons and antidotes in the minds of Europeans, even if substances *from* Africa rarely appeared on store shelves. African healers in diaspora worked within this perception, carving out their own spaces as experts who could tip the balance between life and death. Vincent Brown has written about African healers in eighteenth-century Jamaica who participated in what he calls a "mortuary politics." In other words, the enormous death toll suffered by both whites and blacks made death into a political event, one that gave enslaved Africans a small window of opportunity for asserting power and epistemological authority.[68] But what was an opportunity in the colonies—the figure of the African poisoner turned healer—became a roadblock when it came to establishing medical authority in late seventeenth- and eighteenth-century Europe. African pharmacy, like African bodies, resisted integration with the mainline of an Enlightenment science of pharmacy that made claims to universalism, but remained rooted in older prejudices and elisions.

If we take an expansive view of what early modern pharmacy encompasses, however, it is virtually impossible to draw a clear line between materia medica and spiritual rituals, amulets, and charms, including the use of supposedly supernatural substances like animal teeth or bones, human body parts, and grave dirt. The distance between a cure like cinchona bark and something like an obeah pouch was narrower than we might first assume. It was arguably not until the rise of scientific pharmacy in the nineteenth century, and specifically the isolation of alkaloids like morphine and quinine, that European pharmacopoeias became notably distinguished from this larger world of charms, amulets, rituals, bleeding, vomiting, smoking, eating, and drinking. Even before this, however, the seeds of a divergence were visible.[69] Those substances that could be standardized, repackaged, remixed, and consumed in isolation from their original cultural context had the opportunity to become commodities. Those that depended upon unique and proprietary preparations, those that were implicated with "dark" networks like the slave trade, and whose perceived powers were activated not by ingestion but by ritual acts, were often pushed outside the bounds of the medical.

A few further questions can be drawn from the story told here. First, did the Iberian legacy of the feitiço and of medieval cross-pollinations with "Moorish" medico-spiritual practices make consumers in Iberia and colonial Latin America more amenable to African cures than those in the other European nations and colonies? Second, did efforts to police the boundary between "poison" and "medicine" influence the later division between illicit and licit drugs that would emerge in the nineteenth century? And finally, how should we, today, draw the lines around the history of pharmacy? How far does it continue to shade into the realm of spirituality and magic? If we are committed to reconstructing nonelite and non-European voices in the history of early modern medicine, then perhaps we need to find a place for poisons and charms on our shelves, even if early modern apothecaries couldn't.

8

CONSUMING CANADA

Capillaire du Canada in the French Atlantic World

CHRISTOPHER PARSONS

IN THE forests of eastern North America, northern maidenhair fern (*Adiantum pedatum*) is simultaneously remarkable and easy to miss. On the one hand, as a fern it is "graceful" and "slender" and a beautiful member of "rich, deciduous woodlands."[1] On the other, it is ubiquitous, growing across half the continent in great numbers and only one of over two hundred species of the genus *Adiantum* that grow the world over. It is these two contradictory features of the plant— global and familiar—that make it a particularly interesting plant to study in the seventeenth- and eighteenth-century Atlantic World. When European explorers, colonists, botanists, and medical practitioners began to travel beyond the Old World, they discovered populations of plants such as *Adiantum* that presented both commercial opportunity and an epistemological challenge. They offered a new source of familiar and valuable botanical commodities and an opportunity to meditate on the nature of the New Worlds to which Europeans had come.

In 1749, Jean-François Gaultier briefly described northern maidenhair fern in New France, the colony in which he had lived and worked for seven years as a royal physician and a corresponding member of the Paris-based Académie Royale des Sciences.[2] In his manuscript *Description de plusieurs plantes du Can-*

ada, he wrote "Adiantum American Cornuti. This is the *Capillaire du Canada* that grows in the forests and in lightly humid places: much is collected, care is taken to have it dried in the shade, it is sent to France where it sells well and where it is more valued even than in Canada."[3] There is little wonder that Gaultier did not feel the need to detail the morphological or medicinal character of the plant to the same extent as other plants, such as *Angelica canadensis tenni*, *Larix canadensis*, or *Sarratina Canadensis, foliis*.[4] Unlike these other "Canadensises," Canadian maidenhair had been described in the preceding decades in great detail by celebrated botanists (such as Joseph Pitton de Tournefort and Etienne-Louis Geoffroy) and by major pharmaceutical writers (such as Pierre-Jean-Baptiste Chomel and Nicolas Lemery).[5] It was highly valued in France, where the name *capillaire du Canada* signified a stronger version of a plant that was also collected in and around Montpellier and used in the treatment of chest ailments.[6]

We might ask how naming the plant "du Canada" created economic value and transformed the plant into a legible pharmaceutical in this period. Gaultier wrote his description of capillaire at a time when Canada itself was a shifting and uncertain label. As the principal French colony of the continent, the terms *Canada* and *New France* were often conflated and confused in everything from travel writings to missionary accounts and even in official correspondence.[7] As historians such as Catherine Desbarats, Allan Greer, Gilles Havard, and Paul Mapp have amply demonstrated, French colonial spaces in North America were both geographically and socially unstable.[8] Most eighteenth-century authors understood Canada to be but one colony in a larger New France that could include the extent of the American continent and that stretched from New Orleans to what is now Nova Scotia, principally along the Mississippi and Saint Lawrence Rivers and throughout the Great Lakes watershed.[9] Canada was therefore only one among many geographical spaces then being produced through maps, *mémoires*, and administrative efforts to delineate and control colonial space. Administrators expended considerable effort to give abstract divisions of the continent—Louisiana, Acadia, New France, and Canada—legal purchase, and explorers, merchants, and colonial promoters invested these categorizations with affective, ecological, and economic value as France's colonial project became continental in the late seventeenth and early eighteenth centuries. These boundaries concretely shifted with the creation of new colonies, such as Louisiana in 1717, and the loss of others, such as Acadia with the treaty of Utrecht in 1713, but they shifted discursively in response to far more subtle changes in the French Atlantic World as well.[10]

This essay joins these studies by focusing on how the collection, processing, marketing, and consumption of a North American plant mediated and was mediated by debates about the nature of colonial space in the French Atlantic World. Maidenhair is a common fern, with little to make it stand out of the forest understory. It is a congener of French capillaire; it is undeniably a different species as we understand the term today, but it shares a common botanical ancestor with the French variety, and both are readily and visibly recognizable as related plants.[11] If northern maidenhair has often featured prominently in historians' efforts to reconstruct the medical worlds of French colonial North America, it has attracted little sustained attention, even from those medical and colonial historians who highlight its economic importance in the seventeenth- and eighteenth-century French Atlantic World.[12] Yet during the eighteenth century, it became the principal pharmaceutical remedy transported from New France to the old; capillaire, aside from American ginseng (*Panax quinquefolius*), experienced a brief boom in the 1740s and 1750s, and is the most well known and most important contribution that Canada made to European *materia medica*.[13] Capillaire du Canada therefore became a principal vehicle through which the medical marketplace of seventeenth- and eighteenth-century France came to terms with the ontological status of this and other French colonial spaces.

The pharmacopoeia became an important genre in which claims about this colonial space were produced and consumed. Historians who have studied the delineation of colonial space in North America have most often done so through the careful study of textual materials that debated the appropriateness—or even the reality—of labels that often seem self-evident to us today (Canada or Louisiana, for example). Evidence of the practices through which these and other colonial spaces were created, however, survive in multiple registers, many of them material. Allan Greer's study of colonial surveying throughout the Americas, for example, highlights how legal spaces were produced on the ground.[14] Colin Coates has deftly demonstrated how legally and culturally defined spaces were made real through the active intervention in Laurentian landscapes.[15] More broadly, scholars who have studied the articulation and practice of Frenchness in this period have also directed us toward the careful consideration of objects, such as food, clothing, drink, and furniture.[16]

The abstract colonial spaces of metropolitan maps were also mobilized and contested in the pages of medical and pharmaceutical texts. As I ask what was added by calling capillaire Canadian, I am following the lead of scholars who have examined how the collection and consumption of "exotic" goods in seventeenth- and eighteenth-century Europe mediated knowledge about the extra-European world. Capillaire du Canada was neither exactly foreign nor entirely familiar. It was not, for example, as simply "exotic" as coffee, tea, or other consumables that began to be imported during this time.[17] Yet it was a plant that could not easily be identified as "indigenous" either, as authors noted observable differences between French and American capillaire regularly, and the foreign origin of the plant seems to have added some commercial value.[18] Instead, capillaire evidenced the "geographical instability" that was common to many of the goods brought to Europe from the Americas and the extra-European world; goods that were "fundamentally multivalent."[19] Tracing the history of this plant's use and reception reveals the complicated path to the "hybrid" medicine described elsewhere in this collection, such as in Joseph Gabriel's account of the *United States Pharmacopoeia* (chapter 12).

How did those apothecaries who prescribed and sold capillaire du Canada imagine Canada, and how did their customers and patients understand the effect of a Canadian influence on their bodies? We can look to capillaire less for a coherent statement about what Canadianness meant in the seventeenth- and eighteenth-century Atlantic World than for evidence that the means through which this question was asked extended well beyond elite administrative or scientific worlds. The investigation of what made capillaire du Canada different from just capillaire—or the debate about if any difference existed at all—demonstrates how questions about the nature of colonial spaces were assayed through the study of its medicinal effects, and were mediated by considerations of commercial value that came from the plants' extra-European origins. The case of Canadian capillaire suggests then that the history of healing knowledge in the early modern Atlantic World was entangled with larger political debates about the nature of empire, and that considerations of France's colonial project were weighed through the experience of an otherwise unremarkable plant.

WAYS OF KNOWING A NEW WORLD

In 1703, Guillaume Delisle's map of *Canada ou de la Nouvelle France* visualized the entirety of France's North American empire with an unmatched precision (figure 8.1).[20] Nonetheless, it had difficulty precisely locating Canada. Across a stretch of land that begins with Lake Superior and ends at the mouth of the Saint Lawrence, we can read "CANADA ou NOUVELLE FRANCE," a common figuration of the name of France's northernmost American colony. Beneath this was, in red, another large "CANADA." To the east of this, however, we can locate another, smaller, "CANADA" written across the Saint Lawrence River. Where the large red "CANADA" seems to tie the name to a large region colored in green, the smaller "CANADA" cut across demarcated regions as it transcended the Saint Lawrence. The effect is an overall ambiguity about what or where Canada is. It was both among the largest names on the map and smaller than other colonies within North America that were often included in catalogues of New France such as "ACADIE" or "TERRE NEUVE" and Indigenous territories such as the "IROQUOIS" or the "OUTAOUACS."

As numerous scholars have demonstrated, maps such as Delisle's extended a cartographic project that had begun "as one part of Louis [XIV] and Jean-

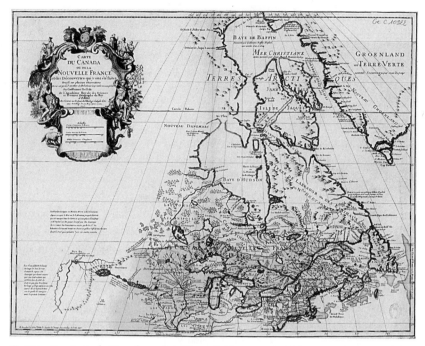

FIGURE 8.1. Carte du Canada ou de la Nouvelle France et des Découvertes qui y ont été faites / Dressées sur plusieurs observations et sur un grand nombre de Relations imprimées ou manuscrites. Par Guillaume De l'Isle de l'Académie des Sciences et Premier Géographe du Roy. Paris, 1700. Courtesy of the Bibliothèque nationale de France.

Baptiste Colbert's larger effort to rationalize French administration and more efficiently exploit French resources" into the Atlantic World and that pushed the gaze of modern cartographic techniques beyond France's shores.[21] Any collection of colonial-era maps can attest to the influence of new geographical technologies and the sophistication of cartographic practices on the visual representation of French North America in the seventeenth and eighteenth centuries; put simply, it became possible to represent American geography and topography with a far greater precision over the course of a period that witnessed the arrival of experts in new practices and institutional support for the establishment of colonial surveying. Maps produced by cartographers working on both sides of the Atlantic—such as Jean-Baptiste-Louis Franquelin, who worked in Québec between 1674 and 1693, and Vincenzo Coronelli, who produced two globes for Louis XIV from Paris in 1683—visualized information gathered from explorers, merchants, and missionaries.[22] The Saint Lawrence River was charted with new precision by hydrographers (such as Franquelin and Jean Deshayes) who arrived in the colony in 1685 and began to chart the river in 1686 before returning to France soon after.[23] As a historian of cartography has written recently, all of these were efforts taken by the state to "better administer its colonial space."²⁴ This was part of a larger project aimed at making both France and the early modern French world legible from Paris and Versailles.²⁵

Delisle's map was then indisputably a tool of empire. Yet his difficulty in placing Canada—and the confused relationship between Canada and New France—points as well to debates about the status of these colonial spaces that had been in play since the earliest moments of French colonialism in the Americas.²⁶ Over the course of the seventeenth century, these terms were mobilized within a variety of discourses and by a variety of actors. New France and Canada became the object of legal disputes to demarcate royal claims and the exercise of commercial rights.²⁷ As an adjective, "Canadian" gradually shifted over this period from a qualification of Indigenous peoples and their territories to describe the characteristics of the American-born French colonists; it became a North American synonym for the Caribbean creole.²⁸ By the end of the century, explorers such as René-Robert de La Salle rhetorically placed Canada in opposition to the Illinois and Great Lakes regions that they proposed were the future of France's American empires.²⁹ These terms remained fluid, therefore, and were useful because they pointed to the possibilities of empire rather than its essential characteristics. Delisle, who explained that his map was based in part upon "the most recent Memoires," thus channeled the contested spatial imaginary of French colonialism in North America, rather than producing it by himself.

Maps are among the clearest visualizations of the changing spatial imaginary of French colonialism in North America, but they were just one register among many of increasingly complex delineations of colonial, imperial, and Indigenous space on the continent. Studies of the natural environments of North America had long proved fertile ground for imagining colonial futures. Giovanni Verrazzano had seen echoes of the utopian Arcadia described by the classical poet Virgil in the American woods, and he was the first to use the name New France to lay claim to the continent on behalf of the French crown in the early sixteenth century.³⁰ In 1558, André Thevet (cosmographer to Henri II, François II, Charles IX, and Henri III) described his encounter with a botanical blessing of France's imperial ambitions in North America in his *Singularitez de la France antarctique*. At the gardens of Fontainebleau, he explained, stood a transplanted American tree that was considered

> for a long time useless and without profit until one of our people wanted to cut one. As soon as it was cut to the quick a liquor came out of it in quantity. This being tasted, was found to be of such good taste that some thought it to be equal to the goodness of wine—so that some people collected an abundance of this liquor and helped refresh our people. And to see and experiment on the

source of this drink, the said tree was sawed down and its trunk being on the ground a miraculous thing was discovered in the heart of the tree: a Fleur-de-lys well pictured what was admired by all. About this some said that it was a very good presage for the French nation, which in the passage of time through the diligence and zeal of our kings could conquer and some day bring to Christianity this poor barbarous people.[31]

In these ways, encounters with American flora were treated as opportunities to gain insight into the nature of the continent and avenues to imagine colonial futures there. While the discovery of a fleur-de-lis was spectacular, even more subtle botanical discoveries could facilitate European study of these new places and lay the foundations for French claims to sovereignty there. If maps projected a concept of the colony from above, plants such as these encouraged understanding colonial spaces from the ground up.

Nonetheless, many studies of American places remained essentially confused about the geographical spaces that they evoked.[32] An excellent example of this comes to us through the work of the onetime Jesuit Louis Nicolas, who was the author of a natural history and illustrator of a *Codex Canadensis* that represented human and natural facets of the Great Lakes region, where he proselytized in the later seventeenth century before ultimately leaving the Society of Jesus and returning to France. He, like Thévet, sought to explain the character of France's North American colony through the representation of indigenous flora. Even as he made it clear that he sought to write a faithful natural history of French North America, what was certain was that he remained as concerned with the physical features of a plant as he was with what might be called its occult or emblematic characteristics.[33] One of the plants that he foregrounded was the passionflower, which, writes Jorge Cañizares-Esguerra, "clearly represented the whip and the pillar, Christ's red wounds and blood, the three nails of the Crucifixion, the crown of thorns, and the lance that pierced Christ's side."[34] The image, as it appeared in Nicolas's *Codex Canadensis*, was presented without comment or explanation on a page otherwise depicting the white and red cedars of North America. The casual placement and lack of comment might invite the reader to understand that this, like the cedars around it, was a common and otherwise unremarkable feature of North American environments.[35]

As Nicolas represented the flora and fauna in images in his *Codex Canadensis*, he consciously sought to engage with the work of Juan Eusebio Nieremberg, a Jesuit who had described the plant in his 1635 *Historae naturae*, and who had suggested that the plant grew in the Spanish Americas. It is impossible to know precisely where Nicolas could have come into contact with this image, but the exactness of his sketch suggests that he had Nieremberg's text in front of him as he drew. Nicolas's representation of the passionflower preemptively abandoned a dialogue with contemporary naturalists and scientific practice as it recorded and accentuated specific features of the plant he claimed to have seen in the interior of North America. His introduction of the passionflower into the environments of French North America conflated the natural theology and natural historical study of American flora and positioned North American environments as a source of new insights into the relationship between God and both the human and natural worlds. Building upon work such as that of Jakob Gretser, a German theologian who in the early seventeenth century had suggested that "the flower itself symbolized Christ's reign in Nature" and that "Providence had selected America as the place for the passion flower to be discovered," Nicolas subtly suggested an ecological and floral continuity across the Americas rather than specific botanical or ecological regions.[36]

It was not uncommon, then, to blur the origins of American or extra-European plants and animals in French texts even as these objects were presented as evidence about the nature of the places from which they had come. While specimens of American flora were transported to France as textual descriptions, images, and physical objects throughout the seventeenth century, the message that they carried about France's North American colonies remained complicated and frequently confused. Many of those who sent them seemed uncertain about the nature or extent of the places to which they were indigenous and therefore about what parts of the world they could speak to.

NEW FRANCE AND ATLANTIC PHARMACEUTICAL NETWORKS

Those who have studied Nicolas's natural history have often commented upon the fact that it blended elements of botanical study that seemed reminiscent of earlier modes of thought, such as emblematic natural history, but that it was undeniably empirical and offers some of the most naturalistic descriptions of American flora and fauna that we have today.[37] It should not be surprising, then, that if Nicolas's work included discussions of the passionflower it was also among the first to include an allusion to the popularity of northern maidenhair in France and concrete reference to its sale as materia medica. As he explained, "This simple is one of the rarest and the most sought-after in the whole country. Not that there is not many in all the woods; but it is precious for its virtue of refreshing the chest."[38] Nicolas did not provide a drawing of the plant, but he signaled the importance of capillaire as the last in his list of plants in his natural history. It may not have been the most remarkable plant from New France, but capillaire became a principal means through which Canadian ecologies were experienced and understood in France.

Nicolas's reference to French interest in northern maidenhair is the first hint of a transatlantic trade in *A. pedatum*, but he was not alone in drawing attention to the plant. Capillaire was first named as such in the work of Pierre Boucher, a landowner and colonial promoter whose 1664 *Histoire veritable et naturelle* was intended to induce Louis XIV to invest in the colony. He also hoped that his king would support a military expedition against the Haudenosaunee (Iroquois) communities to the south who, he contended, claimed the best territories for the agricultural colonies that were France's destiny in North America.[39] Many of the plants that Boucher celebrated therefore were among the most prosaic of the Canadian landscape. He promised his readers that the seasons were regular and could support French ecological regimes and that wine could be made with colonial grapes and bread with Canadian wheat; this was a celebration of predictability and familiarity. He included capillaire in a list of the commodities that awaited settlers alongside other familiar plants and in a chapter titled "Names of wheats and other grains brought from Europe, that grow in this country."[40] Where an exotic plant such as the passionflower might augur well for French colonialism and the success of Catholic evangelization, plants such as capillaire emphasized the intelligibility of this New France and its essential familiarity.

In the writings of early colonial promoters, such as Samuel de Champlain, it was the widespread availability of commodities such as an indigenous hemp and a wide variety of familiar genera of trees that would make Canada a profitable facet of France's Atlantic empire.[41] In the years after New France was made a royal colony in 1663, imperial planners and administrators looked to find new sources of familiar commodities. Jean-Baptiste Colbert, the architect of Louis XIV's colonial policy, encouraged an assessment of these and other familiar botanical commodities. The first royal administrator of the colony, Jean Talon, wrote of forests that could become ships, fields that could become fodder, and French crops that could support a robust colonial population that would contribute to the crowns revenue and resources.[42] Those indigenous plants that drew his attention were thus those that were familiar to French merchants and consumers, but he was just as interested in introducing such profitable commodities as hemp and flax.[43]

Colonists in seventeenth-century New France had few exotic remedies to offer European markets.[44] In the middle of the century, a laborer for Jesuit missionaries named François Gendron reported a mineral cure for cancers but to little effect.[45] Far more eventful was a trade in *rognons de castors* (or castoreum) that grew out of the fur trade that drove much early settlement and colonial expansion into the heart of North America.[46] Early modern apothecaries saw the Canadian rognons as an inferior replacement for that produced by eastern European beavers but still thought that it could help in the treatment for a variety of conditions, ranging from epilepsy to deafness and paralysis.[47] Similarly, some colonial authors suggested that moose hooves might also treat epilepsy, but there is little evidence that there was much interest in Europe or that the moose were obtained by Indigenous hunters for this trade.[48]

For Canadian medical historians, the paucity of these contributions amply demonstrates the relatively light footprint of France's North American colony on pharmaceutical practices on either side of the Atlantic and the "dependence" of colonial apothecaries upon the remedies of their metropolitan counterparts.[49] Renald Lessard, for example, has argued convincingly that "indigenous and Canadian contributions to pharmacy and medical practice are in general relatively limited and hardly changed official medical practice."[50] Stéphanie Tésio has similarly compared after-death inventories of apothecaries in the Saint Lawrence Valley and Normandy between 1669 and 1800 to demonstrate that only 3.96% of identifiable remedies used by apothecaries were Canadian in origin.[51]

Surviving lists of medicines requested by medical practitioners demonstrate the debt of Canadian pharmacy to European remedies. In 1665, the printer Sébastien Cramoisy published a list of the items requested by the "Hospital Nuns of Kebec."[52] The author was Marie Forestier, dite de Saint-Bonaventure-de-Jésus, superior of the sisters who provided medical care to the capital of New France and the surrounding areas.[53] The items were requested for the maintenance of the Hôtel-Dieu at Québec that had been founded in 1637 and that, by the time that Saint-Bonaventure-de-Jésus wrote her request, had become the principal site for the practice of Old World medicine in the New World colony. The hospital, she explained, had become a victim of its own success, with the need to lay out patients in the chapel and "a scarcity in every department."[54] The list included Old World remedies (such as senna and rhubarb), Asian botanicals (such as cinnamon and nutmeg), and even American remedies from central and South America (such as jalap).[55] Her list presents a sort of pharmacopoeia and suggests the Atlantic character of medicine in colonial New France, with drugs that would have been familiar to medical practitioners throughout France and the Atlantic world.[56] Later lists of medicines requested by the Hôtel-Dieu identified similar elements of their pharmacy in need of replenishment (including chamomile, jalap, and almonds) alongside "Kina" and other remedies that would not have been out of place in a contemporary European pharmacy.[57] Analysis of after-death inventories by Renald Lessard has demonstrated the ubiquity of "exotic" drugs such as jalap among New France's medical practitioners throughout the eighteenth century.[58]

It was only in the latter part of the seventeenth century that northern maidenhair became a staple in the descriptions of American flora and French pharmacopoeias. Among the first to confirm the trade that Nicolas hinted at was Antoine Laumet de Lamothe Cadillac, the founder of Detroit and eventual governor of Louisiana. In 1694, when he wrote to colleagues in France, he was the commandant of the French post of Michilimackinac, in the northern reaches of what is now Michigan in the straits between Lake Michigan and Lake Huron.[59] He explained that he had sent "three barrels of *capillaire* syrup that I had made here by a good hand, one for you, one for Mr de Latouche, and the other for the marquis de Chevry; I had asked the captain of the *Bretonne* to take care of it, but as I have not received any news of anything I am persuaded that they have been lost, or perhaps my letters have been intercepted, I have no news from you this year."[60] Details on the production of a syrup from the fern anticipated the regular production of a botanical commodity that would become standard in eighteenth-century Atlantic commerce, and the intended recipients of the package were promoters of France's American colonies overseas, such as the Marquis de Chevry, whose family was involved in the commercial exploitation of Acadian fisheries.[61]

By the end of the seventeenth century, capillaire du Canada was regularly sent to French Atlantic ports and distributed throughout the country. The Baron de Lahontan reported, for example, that it was sent from Québec to "Paris, to Nantes, to Roüen & to several other cities of the kingdom."[62] The statistics gathered by Renald Lessard trace out a relatively stable trade that remained constant throughout the last decades of the French regime in North America, even if the quantity and price varied year by year (figure 8.2).[63] Northern maidenhair seems to have been at least partially collected by Indigenous communities who came to rely upon the plant as a supplement to a fur trade that was declining in value.[64] In 1710, for example, the Jesuit Father Davaugour reported collecting the plant as a regular seasonal occupation of the Wendat community at Lorette, explaining that "they gather, toward the end of August, quantities of a plant useful in pharmacy and of no mean value in Europe, which druggists call 'Capillaire.'"[65] In 1712, the surveyor Gédéon de Catalogne reported that the Iroquoian peoples of Kahnawake similarly collected the plant for sale in nearby Montreal.[66]

The growing value of northern maidenhair made it the principal means through which the medical practitioners of New France participated in Atlantic networks of pharmaceutical trade. The records of the sisters of the Hôtel-Dieu de Québec provide some insight into how these exchanges were affected. In 1728, for example, the sisters sent capillaire to the Rochellais merchant M. Dupas in exchange for two pounds of coffee.[67] In 1740, the two Duplessis sisters—Marie-Andrée de Sainte Hélène and Geneviève de l'Enfant Jésus—sent four different shipments of the plant to different French Atlantic ports. Their relationship with the Dieppois apothecary Jacques-Tranquillain Féret is particularly well documented and demonstrates that "*capillaire* was the drug that they sent the apothecary in the greatest quantities."[68]

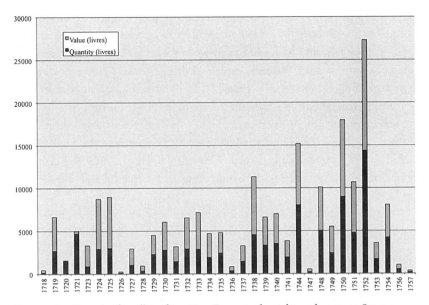

FIGURE 8.2. Exports of capillaire from New France in the eighteenth century. Source: Renald Lessard, *Pratique et praticiens en contexte colonial* (PhD dissertation, Université Laval, 1994), 174.

THE VALUE OF BEING CANADIAN

From the first arrival of northern maidenhair in France, authors drew attention to its larger size as the primary difference from the more familiar French capillaire that was most often collected in the south of France. French audiences received their first impression of the plant in Jacques-Philippe Cornut's *Canadensium Plantarum*. This was the first "regional" flora produced in France, although numerous researchers since have demonstrated that many of the plants that the book described were in fact neither from Canada nor even North America.[69] What united those described was often their presence in the garden of Jean Robin, onetime gardener to the king and maintainer of the Paris faculty of medicine's small garden near the cathedral of Notre Dame. It was here that Cornut observed what he called *Adiantum Americanum*, although he also referred to descriptions of the plant gathered in its native habitat. There was little sense here that the plant was more important than any other that had been brought back from French explorations or acquired in trades with other well-connected gardeners, even if it is one of the first plants featured in the book.

Later authors in Canada emphasized both the morphological difference between Canadian and French maidenhair and its pharmaceutical interchangeability (or even superiority). While noting that the leaves of capillaire du Canada were bigger than the more commonly known variety collected near Montpellier, for example, the Jesuit Pierre-François-Xavier de Charlevoix also promised that "its quality is well above that of other *Capillaires*."[70] He wrote that "nowhere else is this plant so tall and lively as in Canada."[71] In the early eighteenth century, the colonial administrator Charlevoix explained that "*Capillaire* grows much taller here, & is infinitely better than in France."[72] More often still, usage and morphology alike were implied in the simplified name of the plant. The capillaire du Canada was frequently simply listed in catalogues, denied even a cursory discussion of the morphological and medicinal differences that warranted the qualifier "du Canada."[73] The name the plant carried therefore simultaneously laid claim to familiar medicinal properties, economic value, and ecological and morphological features.

The value of the plant was confirmed by contemporary pharmaceutical texts that foregrounded an embodied experience of American difference. In his 1698 *Traité universel des drogues simples*, Nicolas Lémery described a transatlantic commerce in the plant that already seemed to be fairly well developed. He provided insights into the nature of the plant provided by recently invented tools such as the microscope and by celebrated botanists such as Joseph Pitton de Tournefort, but the geography that he traced out seemed far from certain. "We brought from the Canadas, Brazil & and several other places, a type of dry *Capillaire*, much larger than ours," he explained, before continuing to relate that "it is called by C. Bauhin, *Adiantum fruticosum Brasilianum* & in French, *Capillaire de Canadas*: its stem is narrow, hard, of a reddish brown or [purpurine], leaning towards black, is divided into several branches, that carry small leaves almost like ordinary *Adiante*, oblong, toothed on one side, whole on the other, soft, tender, odiferous. This *Capillaire* is the most esteemed of all, because it has the most odor."[74] This passage revealed the geographic uncertainty surrounding the fern. Its French name highlighted a Canadian origin but its Latin name pointed instead to Brazil. The maidenhair fern indigenous to France was explored through microscopy and botany but the equivalence (and even superiority) of the foreign analog was highlighted by its scent.

A contemporary of Lémery, Pierre-Jean-Baptiste Chomel, reiterated the importance of the "subjective" knowledge of scent and taste in assaying northern maidenhair in his pharmacopoeia. Chomel's 1712 *Abregé d'histoire des plantes usuelles* described a marketplace in which many of the five known types of *capillaire* were rare in Paris. While some "ignorant herborists" made use of entirely different plants in *capillaire*'s place, it was more common to "substitute that of Canada, that is not rare in Paris," which was, in fact, "more agreeable to taste."[75] In France, the plant's value and efficacy was registered through these ephemeral impressions. Canada was something that could be tasted, smelled, and experienced in apothecary shops throughout Paris.

The pharmacopoeias such as these seemed ultimately uncertain about the distinction between Canadian and French maidenhair, but their structure repeatedly emphasized a rough equivalence. Lémery's pharmacopoeia proceeded from a more general description of the familiar capillaire as the basis for the genus and the *"Capillaire de Canadas"* was presented as a species of this more familiar type. After a brief digression, a general discussion of pharmaceutical qualities returned and Lémery explained that "the *Capillaires* contain little phlegm, much oil, a modicum of salt. They are good for the chest [*pectoraux*], open the stomach [*aperitif*], they excite sputum, they soften the accretions of the blood, they provoke menstruation in women."[76] This strategy was perhaps best demonstrated by Joseph Pitton de Tournefort's *Traité de la matiere medicale*, where, after listing *Adiantum Americanum* as the first of several known types of the plant, the author wrote simply, "Here are the *capillaires* which are commonly used."[77]

In later texts, the *capillaire du Canada* and the familiar maidenhair collected in the south of France were equated more directly. By 1726, the *Dictionnaire universel de commerce* of Jacques Savary des Bruslons concluded that "there is truthfully only the *Capillaires* of Canada, & of Montpellier."[78] Editions of Lémery's *Pharmacopée universelle* suggested that it was "in Languedoc, in Provence, [and] in Canada" where the plant had "much odor & virtue."[79] Texts continued to point to differences in the size of the plant that could be observed at firsthand in gardens such as the Jardin du Roi, but assured readers of the pharmaceutical equivalence of Canadian and French maidenhair.

Authors on both sides of the Atlantic therefore highlighted the size of the plant and observable differences in taste and quantity. What made the plant Canadian was, alongside an increasingly specific geographic origin that dropped reference to Brazil, its virtue, a quality recognizable to the trained sense of eighteenth-century chemists and apothecaries. Differences in the shape and size of the plants were easily overcome by authors who focused on the common registers of the plant's potency such as smell and taste and by guiding readers in the production of preparations that rendered morphological differences invisible.

The importance of these sensory cues was underscored when capillaire du Canada was sent to France as a ready-to-use remedy. In effect, whatever the morphological difference, the plant could be transmuted into a recognizable remedy through the power of French pharmaceutical practice. The proper preparation of these commodities was vital. The medical marketplace of seventeenth- and eighteenth-century France demanded that valuable American drugs be packaged as such and feared the loss of desired medicinal properties when they were not. Nicolas Lémery warned his readers to be attentive to the means by which the capillaire that they sought to purchase had been transported from Canada. He cautioned that "it is so common in several places in America, and principally in

Canada, that the merchants package their merchandise in it in place of hay, when they want to send them to distant countries. It is by the means that we receive much of it. But it is better when it comes packed in paper bags, or closed in boxes, because the odor is better conserved. One must choose it new, green, odiferous, whole [and] moist to touch."[80] On occasion we have a little sense of how capillaire was prepared before being sent to France. The Duplessis sisters, for example, sent maidenhair in sacks and barrels, but we otherwise have little sense of how they prepared the drug.[81]

Often, however, the plant was sent as a syrup to France. When he wrote from Michilimackinac in 1694, for example, the sieur de Cadillac complained about the loss of just such a syrup.[82] Administrative and personal correspondence demonstrates that some capillaire was sent from North American ports already prepared as a syrup. This was a syrup that, Lahontan explained, was "very appreciated in France to soothe pulmonary affections."[83] Eighteenth-century authors advised their readers on how to produce a syrup from the plant that would preserve the plant's medicinal properties both in France and, presumably, as it crossed the Atlantic Ocean. The medical author Philbert Guybert wrote that to produce a syrup of capillaire, "the herbs cleaned and washed will be put in three liters of hot water to infuse for twenty four hours, in a glazed terrine . . . at the end of which time you will pour it all in a basin . . . the decoction will be clarified with egg white and yolk, & cooked with three parts sugar into the consistency of syrup."[84] Chomel added that "it is infused . . . like tea; a good pinch in . . . boiling water, to which one adds then a little bit of sugar."[85] In New France, the syrup was on occasion sweetened with maple syrup or sugar, or was made with the addition of other sweeteners such as raisins.[86]

The ambiguity of pharmaceutical knowledge about capillaire du Canada therefore mirrored that of what it meant to be Canadian in an era in which France's colonial project faced a multitude of cultural and economic challenges. It still hinted at the transformative powers of French expertise to draw the virtue out of North American environments and to translate uncultivated wilds into valuable commodities that improved the financial and physical health of the kingdom. Yet the "du Canada" remained as a reminder of a marker of the morphological difference and the physical distance that separated New France from old. Increased experience with the plant made it less familiar rather than more.

Although capillaire became a valuable commodity because of its close similarity to a familiar French plant, in the eighteenth century it became increasingly exotic. Instead of a new source of old remedies, Canada seemed to provide entirely new opportunities because of the product of encounters with Indigenous peoples and cultures. Linnaeus's student Pehr Kalm, for example, asserted that knowledge of maidenhair was properly Indigenous. He explained that the plant which throughout *Canada* bears the name of *Herba capillaris* is likewise one of those with which a great trade is carried on in *Canada*. The *English* in their plantations call it *Maiden-hair;* it grows in all their *North-American* colonies, which I travelled through, and likewise in the southern parts of *Quebec*. It grows in the woods in shady places and in a good soil. Several people in *Albany* and *Canada,* assured me that its leaves were very much used instead of tea, in consumptions, coughs, and all kinds of pectoral diseases. This they have learnt from the *Indians*.[87]

Likewise, the military officer Louis-Antoine de Bougainville who was otherwise as dismissive of Indigenous cultures as he was of American environments seemed to offer capillaire as evidence of what could be learned from closer attention to the medical knowledge of aboriginal peoples. In his account of his experiences in Canada during the Seven Years' War, he wrote that "there are many rare plants of which the *Sauvages* know the properties well, it is to be wished that we had some capable botanists to study with them; the *capillaire* is stronger than that which is collected in Europe."[88] Even the most familiar parts of New France now seemed foreign and common remedies became exotic.

CONCLUSION

The transition of capillaire du Canada from familiar to foreign hints at the larger transformation of what Canada—and a colony in Canada—meant to French audiences. For a brief moment, it seemed that whatever foreignness existed could be transformed as the Canadian raw materials were transmuted into recognizable French commodities that were prized for the slight hint of the exotic that they possessed. This drove the large-scale transportation of the plant from New France to old and ensured that the capillaire du Canada became a principal vehicle through which French consumers experienced the French colony. In the end, however, capillaire responded to changes in how the colony was perceived. Even as this increasing exoticism contributed to its commercial value, it could do little to improve the image of a colony that would be given up for Guadeloupe within a decade as part of peace negotiations with Britain.[89]

Pharmaceutical texts were participants in broader conversations about the nature of empire in the extra-European world. The story of capillaire might seem inconsequential when considered alongside the larger histories of interimperial and interethnic conflict in this period. Yet it demonstrates how debates about the nature of French colonialism in the Americas carried beyond the halls of Versailles and into nonelite spaces in the French capital. The medical marketplace of eighteenth-century France tested both the experience of colonial places on French bodies and their value to French tastes. The significance of the "du Canada" of capillaire du Canada was weighed through careful assessment of individual experiences and the economic interests of apothecaries and their patients. In the process, following the marketing and consumption of capillaire in France reveals the fundamental instability of geographical knowledge within pharmaceutical networks that traded in and theorized about the exotic flora beyond the shores of France. Capillaire, even as it remains an utterly unremarkable plant in backyards and understories today, can therefore point us toward understanding the ways in which imperial and colonial histories were entangled with the daily economic, social, and medical lives of metropolitan Europe.

9

RETHINKING PHARMACOPOEIC FORMS

🙡

Samson Occom and Mohegan Medicine

KELLY WISECUP

BETWEEN 1754 and 1756, Mohegan Samson Occom compiled an herbal, a list of fifty-two "Herbs & Roots," as he titled the document, with information about each plant's healing virtues and the malady that it cured.

The herbal contains recipes for medicinal concoctions that cured common disorders, from a sore throat and broken legs to the French pox, or syphilis. Occom wrote most of the herbal in English, but he also used Algonquian words to identify six herbs. In 1754, Occom was a young man who had just agreed to work as a schoolteacher for the Montaukett community on Long Island, having had to take leave from his studies with minister Eleazar Wheelock because of ill health. He had not yet given the execution sermon for the Wampanoag man Moses Paul that would become a bestseller in 1774, nor was he the minister famous in England and North America alike. When he wrote the herbal, Occom was beginning his work as teacher; he soon married Mary Fowler, a Montaukett woman, and cultivated a farm while raising his family at Montauk.[1]

"Herbs & Roots" exists today in two separate pieces, booklets that were probably hand-sewn by Occom himself.[2] The manuscripts were not printed in the eighteenth century, but they appeared several times in twentieth-century collections of Occom's work: both booklets are reproduced in Joanna Brooks's collection of Occom's writing, and excerpts appear in a collection on *The History and Archaeology of the Montauk* and in Edward Connery Lathem's *Ten Indian Remedies: From Manuscript Notes on Herbs and Roots*.[3] However, all printed versions of the herbal omit a number of notes, textual fragments, and noninscriptive marks that are present in Occom's manuscripts. For example, the first and last pages of the first booklet contain lists of farming equipment and notations that indicate how much Occom paid for the equipment. Near the end of the first booklet, Occom wrote that he had paid a Montaukett man, Ocus, for teaching him the virtues and uses of the fifty-two plants listed in "Herbs & Roots."[4] The first booklet includes as well a journal-like entry in which Occom records traveling from Montauk to Mohegan and back in August of 1756 with "Henry," possibly Henry Quaquaquid, who later preached with Occom and who, along with Occom's mother, had joined minister David Jewett's congregation around 1739.[5] In addition to these travels and economic transactions, the herbal records debts that are spiritual and material: in both parts of the herbal, Occom included fragments of a letter to an unnamed Christian and probably white woman, in which he expressed his gratitude to God for showing favor to him. Finally, Occom employed dashes of varying lengths to separate the entries throughout "Herbs & Roots." The various elements composing the herbal provide insight not only into Occom's medical and rhetorical practices but also into the strategies with which he established relationships among Native, Christian, and colonial forms of knowledge. At the same time, Occom's rhetorical strategies in "Herbs & Roots" call for a reconsideration of the form of the list, especially its relation to so-called oral and literate communication practices and cultures.

While several studies of Occom's writing have mentioned the herbal in passing, "Herbs & Roots" and Occom's medical knowledge, more generally, have received far less attention than his religious practices and writing.[6] Yet as Occom's journals and sermons make clear, medical knowledge remained important to him throughout his life and his career. Occom continued to practice medicine after he was ordained as a Presbyterian minister in 1759; indeed, his medical and religious practices often intersected. For example, Occom noted in several journals that he provided medical care for people whom he encountered on his preaching tours.[7] Moreover, in 1766, Occom wrote a medicinal recipe on the back page of one of his sermons, a detail that further suggests that he provided both medical and spiritual care for his congregations, perhaps simultaneously.[8] As these examples suggest, Occom's interest in medical knowledge was early and ongoing, and it accompanied his Christian education under the tutelage of Eleazar Wheelock and his ministerial work.

This chapter examines the manuscripts of "Herbs & Roots," and it analyzes both the lists of plants as well as the seemingly extraneous letter fragments and financial records. By considering the herbal in its entirety, I show that the lists of herbs and roots not only record Occom's knowledge of fifty-two plants but also make visible the environmental, pathological, and spiritual contexts that gave the plants power and meaning. The herbal represents relationships among plants, diseases, and bodies; between plants and medical practitioners; and among various people in New England. Thus, while studies of Occom's writing and work have emphasized his Christianity and his participation in Wheelock's missions and schools, his herbal provides an opportunity to consider how Native medical practices shaped Occom's knowledge and rhetorical strategies.[9]

At the same time, "Herbs & Roots" asks us to adopt new perspectives on pharmacological knowledge, its circulation, and its textual representation in the Northeast. Around the same time that Occom created a record of his medical knowledge, British colonists in North America were attempting to adapt English pharmacopeias for the North American climate and plants. This work occurred in gardens, fields, and forests as well as on the page, both in printed books and in manuscript recipe books. Colonists could access American editions of English medical texts and copies of English editions transported across the Atlantic, such as Nicholas Culpeper's *Pharmacopoeia Londinensis, or, the London Dispensary* and Thomas Short's *Medicina Britannica: Or a Treatise on such Physical Plants as are Generally to be found in the Fields or Gardens in Great-Britain: Containing a Particular Account of their Nature, Virtues, and Uses*.[10] These authors' careful attention to specifically English plants, complexions, and bodies meant that colonists had to carefully adapt them to the different botanical and geographic contexts of North America. While this adaptation often occurred unofficially, through trial and error, it also sometimes took place in more formal, textual contexts, such as when Philadelphia botanist John Bartram annotated the 1751 Philadelphia edition of Short's *Medicina Britannica* by indicating which plants he had observed in the colonies and providing their common names. Bartram's notations map classical and English medicine onto American contexts by using preexisting English categories to classify and describe local plants and bodies. This adaptation indicates how medical practitioners and botanists recontextualized pharmacopoeias by rewriting them for new places, plants, and bodies. As Bartram points out, America has many "good medicinal Plants, which may be administered to the People with great Advantage, if properly adapted to the Season, Age and Constitution of the Patient; the Nature, Time and Progress of the Disease: Without which Caution, it is not likely that the Practice should succeed generally."[11]

Bartram's notes aimed to help colonists adapt plants "to the Season, Age and Constitution of the Patient; the Nature, Time and Progress of the Disease."[12] His annotations provide American names for plants, indicate whether some plants grow wild or only in botanic gardens, and mark if a plant listed in Short's text is absent from North America. When placed alongside Short's instructions for preparing specifics and for observing the "Time, Constitution, or Nature of the Disease," Bartram's annotations allowed British colonists to select, prepare, and use plants successfully.

As Bartram's attention to patients' season, age, and constitution and to the nature, time, and progress of the disease suggests, eighteenth-century pharmacopoeias insisted on the importance of local contexts, even as they circulated throughout the Atlantic World. While some herbals and pharmacopoeias did remove plants from the local, geographic, and environmental contexts in which they originated to standardize their descriptions, this process of decontextualization and recontextualization was often incomplete, and often purposefully so. Medical writers and botanists like Bartram compiled plants with information about their local origins, pertinent textual information, and knowledge about the bodies for which they were best suited. Such texts made local contexts crucial to representing and using medicinal plants appropriately. Bartram's adaptations of English pharmacopoeia mapped North American plants as part of the British Atlantic World, acknowledging differences but also emphasizing correspondences among English and colonial texts and bodies. (For an analysis of similar representations in a French imperial context, see Christopher Parsons's contribution to this volume, chapter 8) In this way, these texts complicate characterizations of eighteenth-century natural history and forms of classification as systematically cataloguing objects within a field or table that made those objects legible in relation to one another but decontextualized them from their local contexts.[13]

Occom's "Herbs & Roots" further disrupts emphases on pharmacological decontextualization, because Northeastern Indigenous medical practices rested on particular relations between herbs, roots, and their environmental contexts, and because Occom's lists of medicines and cures obtain meaning from their ongoing relations to and immersion in Indigenous environmental and intellectual contexts. For many Native medical practitioners, herbs and roots derived their medical virtues and use from their position within environmental, interpersonal, and nonhuman contexts. As Gladys Tantaquidgeon, a descendant of Occom's sister Lucy and a Mohegan medical practitioner and anthropologist, wrote in 1977, medicinal plants and herbs possessed "spiritual natures equally as sensitive as our own," and practitioners had to understand and acknowledge these natures before they could use the plant appropriately.[14] As I explain in more detail below, plants were intimately connected to certain illnesses, as indicated by the shape of their leaves or their taste, and these correspondences designated some plants as ideal cures for particular maladies. Finally, both plants and medical practitioners were indebted to nonhuman forces for their power over disease, and practitioners appealed to these nonhuman entities in order to use herbs and cure diseases effectively.[15]

As a consequence, medical practitioners did not so much discover, extract, and classify the virtues of herbs as they served as mediators between plants and nonhuman forces, on the one hand, and patients, on the other. As Tantaquidgeon has explained of Native medical practitioners, they are the "media through whom the Creator sends his healing power to alleviate distress caused by the physical and mental ills which attack the frail bodies of mankind."[16] Plants and medical practitioners alike were located in a series of networks outside of which their virtues and knowledge, respectively, lacked power or efficacy. Occom's "Herbs & Roots" is one node in Northeastern medical networks, which endow it with meaning. In these ways, the lists I analyze in this chapter pose a contrast to the nineteenth-century US pharmacopoeia emphasizing Enlightenment rationality that Joseph Gabriel discusses in this volume, pointing to alternate ways of defining and transmitting medical knowledge that existed alongside—and that colonists often posited as a threat to—colonial and US medical practices. To trace some of these Indigenous networks and their narratives, I first examine Occom's various methods of identifying herbs. I consider the ways in which "Herbs & Roots" established Occom's relationships to other people, in particular, his medical teacher Ocus, New London merchant Nathaniel Shaw, and the unnamed woman to whom he drafted a letter in the final pages of both booklets. Finally, I suggest that "Herbs & Roots" suggests ways in which scholars might reevaluate the form of the list.

IDENTIFYING HERBS AND ROOTS: APPEARANCE AND PLACE

At first glance, Occom's lists of plants and roots appear to provide scarcely enough information to locate and identify plants, much less to complete recipes. The entries often begin with a name for the root or plant, and they sometimes include information regarding how to prepare concoctions, usually instructing the practitioner to take part of the herb or root and to boil it in water for a certain length of time. Finally, most listings identify the malady that the herb cures, and some explain how to administer a concoction to a patient or how to apply a poultice. However, most entries contain even less information; for example, some do not name the herb or root, instead offering only the name of the illness against which the herb is effective. Yet unlike the loss of epistemological control manifested in a lack of familiar classifications in James Petiver's "Catalogue of Some Guinea-Plants, with their Native Names and Virtues" that William Ryan traces elsewhere in this volume (chapter 6), the apparent sparseness of Occom's herbal is not a loss or lack. Occom's descriptions of herbs and roots not only present important information regarding the plants' virtues and uses but they also locate the plants in environmental and pathological contexts that endow the herbs and roots with their medicinal virtues. Several patterns of identification are present in the herbal: Some roots and herbs are identified on the basis of a key characteristic, while others are described in terms of their ability to heal a particular illness or disorder.

In many entries, Occom identified roots and herbs by naming one of their primary or distinguishing features, and he followed this identification with a statement of the malady against which the plant was efficacious. Four examples display this structure explicitly:

> wehsuck or Bitter Root good to kill Lise 12
>
> A Long Notched Leaf good Boil 17 / Master over Witch Rt 18
>
> Long Fever herb take it the Leaves & throw them into hot water and Put them upon the wrists hollow of the feet and upon the forehead 20
>
> Prickly Leav'd and Thorn Rts most of the thorn boild in about 3 Quarts of water till Consumed to a Quart—good for Heart burn 36[17]

In these entries, each of the names—"wehsuck or Bitter Root," "Long Notched Leaf," "Long Fever herb," and "Prickly Leav'd and Thorn Rts"—highlight a defining element of the plant's appearance, taste, or texture. As they identify these qualities, the names also establish the healing virtues of the respective plants. In Native medical systems, one mode of classification identifies plants according to their appearance or qualities, which correspond to and thus reflect their medicinal properties. This classificatory practice is founded upon the belief that plants visibly signal their virtues and, by extension, their relationship to diseases. As anthropologist Frank Speck explained of Northeastern Algonquian medical practices, "relationships exist between the disease and either the appearance, the imaginary quality, smell, taste, or even the name of the herb used in the treatment."[18]

Mohegan medical practitioners gave plants names that signaled their virtues and, consequently, the maladies they cured. For example, Tantaquidgeon records that "'Canker lettuce,' shin leaf (*Pyrola elliptica*), is steeped and the liquid used as a gargle for sores or cankers in the mouth," while "'Peppergrass' (*Bursa bursa-pastoris*) seed pods are made into a tea for the general benefit of the stomach. Its pungency is thought to kill internal worms."[19] In these records, as in Occom's herbal, the names of plants act not only as labels for identification but also as keys or clues to the virtues and uses for the plants. In "Herbs & Roots," "Prickly Leav'd and Thorn Rts" possessed traits similar to the inward stings brought on by heartburn; they consequently provided an ideal antidote with which to counteract that malady. The name "Long Fever herb" directly links the herb to the disorder it cures; the herb's name indicates that it possessed attributes that suited it to healing fevers. In some instances, a quality or feature of the plant signaled its medicinal qualities: for example, "Bitter Root" could counteract lice, perhaps because its bitterness drove the lice away. Similarly, as Occom wrote in another entry, "Indian Elm good for sore mouth 7."[20] Speck commented that New England Natives employed elm for bleeding lungs because the "slippery quality of the medicine will 'smooth down' the irritated throat and lungs."[21] In this case, the elm possessed qualities that counteracted the nature of the malady. The herbal's descriptive names for plants thus allowed Occom to represent correspondences among plants, their medical qualities, and the illnesses they counteracted. Far from being incomplete or partial descriptions, Occom's entries provided information about herbs' and roots' characteristics and accompanying uses.

At the same time, Occom's practice of identifying herbs according to their properties represented his own knowledge of and relationship to specific environmental locales. As Daniel E. Moerman has pointed out in his study of Native American ethnobotany, common names for plants—such as "Long Notched Leaf" or "Long Fever herb"—were not universal but varied by location. Montauk would have constituted one such location: according to Lloyd G. Carr and Carlos Westey, Long Island is composed of several different "biotic areas," each of which differ "floristically."[22] Moreover, Natives' names for plants varied in each region throughout New England, so that, as Frank Speck concludes, "special herbalisms are distinctly matters of local tribal knowledge, rather than group knowledge."[23] Descriptive names such as "Bitter Root" or "Prickly Leav'd and Thorn Rts" would thus have been place-specific, probably unique to Montauk. Moreover, Occom employed Algonquian words—such as "wehsuck" for "Bitter Root"—to name several herbs and roots, and these words further display his knowledge of specific places and plants as they were known to Natives. The names for plants in terms of their appearance and virtues provided a system whereby Occom could represent both the identity of the herb and its use; they also represent Occom's relationship to Montauk: his knowledge of the land, its indigenous plants, and their local names.

LOCATING HERBS AND ROOTS: DISEASES AND INVISIBLE CURES

Occom sometimes provided no information about the appearance or qualities of the herb or root; in these cases, he identified plants not with a descriptive title but by naming only the disease or malady that they cured. In these instances, the lists of herbs, roots, and maladies served as a device that allowed Occom to bring to mind the body of knowledge associated with each illness and its herbal cures. These listings not only represent relationships among plants and diseases but also include in these relationships the medical expert who would interpret such descriptions. For example, Occom included entries on:

A Rt. For fits. Pound the Rt and Soke in water about half an hour. 4 Rts will doe. 19

an herb good for worms 24

An herbe good for Rattle Snakes bite 34

Herb good ^{to} heal brocken bones ^{about the} fingers and feet 4724

In each entry, the name of the malady stands in for the herb's or root's proper name; no further instruction for identifying the "Rt" or herb is provided. The herbs and roots listed here seem to be so intimately related to particular maladies that an account of their appearance or qualities is unnecessary. Indeed, each record represented correspondences between a particular disorder and a plant's medicinal properties. The plants listed in this manner are connected to certain diseases by virtue of their properties, which would have been represented by taste or texture or by a sign on the plant's or root's surface, perhaps in markings that pointed to some feature of the malady. Yet rather than identifying these markings, Occom named the disease to which the plants' appearance and virtues corresponded. In these cases, a practitioner could work back from the malady to the appearance and identity of the plant. Such entries consequently call into play a practitioner who, like Occom, was knowledgeable of such connections and who would thus be capable of locating and applying appropriate herbs and roots to the named maladies. The lists of herbs in terms of the disorders they cured thus represent the medical knowledge that Ocus relayed to Occom and that the two men would have drawn upon to heal various illnesses.

Occom's herbal requires that plants and their virtues be considered in multiple, extratextual contexts, for isolating plants from their environmental and pathological frames of reference would render the herbal incomprehensible. In this way, "Herbs & Roots" must complicate our understanding of eighteenth-century European taxonomies, which created a system in which all known plants could be listed and their similarities and differences made visible by comparisons that the taxonomy made possible. In such natural histories, plants' identities were derived not from their environment but from their location in the list and their relationship to all other entities in the taxonomy.[25] As a result, European men of science could study exotic herbs without considering the contexts in which those herbs originated. But Occom's herbal does not decontextualize plants. Instead, it communicates information about roots and herbs by presenting plants in terms of their connections with maladies and with specific places. Rather than obtaining their identity and virtues from Occom's descriptions or their presence in the herbal, the herbs and roots remain connected to contexts outside the herbal, such as the botanical environment on Montauk or their relationships with diseases, and the plants derive their healing power and their identities from their location in these contexts. In this way, "Herbs & Roots" referenced a preexisting and external body of environmental and botanical knowledge upon which medical practitioners drew in order to identify and use herbs, and the herbal derived its meaning from material contexts beyond the text and its lists of herbs and remedies. Finally, the lists of herbs and roots required a knowledgeable interpreter to bring full meaning to the entries, and as such, they represent not only the relationship between a plant and disease but also between the practitioner and plants.

The words in Occom's lists do not correspond to one object or referent alone; instead, they display a series of correspondences and meanings associated with the word and the plant it represents. The rhetorical strategies at work in these descriptions of herbs and maladies parallel the modes of representation that Mohegans employed to create and ornament baskets. Designs on wood-split baskets drew upon a "Mohegan symbolic 'vocabulary'" that basket makers and informed observers could "read" in order to interpret their significance.[26] The messages communicated by these designs required a knowledgeable interpreter to remember the events or information associated with the designs. In the same way, the entry "A Rt. For fits 19" allowed Occom to recall the relation between fits and a certain root, as well as the root's specific appearance, how to prepare it, and how to apply it to the body.[27] The entry had meaning for Occom because the root and its identification were part of a "symbolic vocabulary" that he had learned to read and apply in order to cure illnesses.[28] Finally, designs on baskets communicate more than "simple representations of nature"; they present familial and tribal histories, information about medicinal plants and herbs, and the spiritual forces in things.[29] Again, a knowledgeable reader had to interpret these representations of spiritual forces, just as Occom's entries in the herbal required a medical expert to interpret their meaning and make visible the connections among plants, diseases, and spiritual forces.

The listings in "Herbs & Roots" suggest that Occom possessed knowledge of the nonhuman forces at work in roots and herbs, forces that endowed plants with power over bodies and minds. In addition to listing herbs whose appearance displayed their ability to counteract illness, Occom also recorded his knowledge of herbs whose power to influence bodies and minds was not attributable to visible qualities. Occom included an entry for "wores Rt good to Draw young mens to young women 31."[30] Native medical practitioners employed such herbal "love medicine" in order to determine whom individuals would marry, as well as to aid couples experiencing marital difficulties or an individual who had failed to obtain the favor of a lover.³¹ Love medicine was applied in several different ways: the practitioner might perform a ceremony with parts of a root, with the goal of influencing the mind of the young man. Alternatively, the applicant might carry the root about herself, with the expectation that the young man would be drawn to it. In both cases, the root acted at a distance; that is, it influenced the man's mind without necessarily being applied to his head. "Wores Rt" worked invisibly, and its powers thus could not be traced to the root's apparent qualities.³²

Occom's knowledge of invisible herbal qualities is fascinating in itself, but this entry is all the more interesting when we consider that Occom had received both a Mohegan and a colonial education by the time he wrote "Herbs & Roots." He thus would probably have been familiar both with colonial, Christian attitudes regarding the investigation of plants with invisible virtues and with Native views of such plants. Throughout the seventeenth century and into the early eighteenth century, Europeans in the colonies and back in Europe had sought to explore the hidden realms of nature by investigating phenomena with invisible, or occult, causes. As scholars such as David D. Hall and Karen Ordahl Kupperman have shown, British colonists acknowledged the influence of invisible forces upon their everyday lives, and they made use of a variety of practices to understand and control those forces.³³ And as Benjamin Breen rightly emphasizes in this volume, these parallels between European medicine and medical practices of African (and, I'd add, Native American) healers should make scholars cautious of assuming vast differences between these medical traditions. By the mid-eighteenth century, however, Euro-Americans increasingly sought to distance themselves from the idea that the occult virtues of herbs—those that did not have a clearly identifiable natural cause—could be discovered and employed without delving into inappropriate, diabolic forms of knowledge. Moreover, by the late eighteenth century, medicine in the northeastern colonies had been professionalized, and licensed physicians insisted upon employing herbal practices that had been authorized with observation and experimentation.³⁴ While colonists did not completely repudiate their belief in "the power of 'the hidden' in nature," they did privilege the study of observable characteristics and causes, and they sought to resist metropolitan biases against colonial knowledge by identifying nature's secrets and inappropriate forms of knowledge with Native Americans and Africans.³⁵ As a consequence, as Susan Scott Parrish argues, "African and Indian ways of knowing emerged from and were irremediably connected to both literal poison and a poisonous episteme, poisonous because magical and non-Christian."³⁶ Some colonists did continue to acknowledge the efficacy of Native herbal knowledge, but they also insisted upon scrutinizing such knowledge to ensure that it was not connected to allegedly inappropriate investigations of a plant's invisible powers. Differentiating among practices of investigating herbal virtues thus became a way of marking differences among the peoples who lived in the Americas.

As a Mohegan Indian, Occom may have had a different relation to natural knowledge that colonists considered dangerous and inappropriate—such as information regarding invisible herbal virtues—and he approached this knowledge from a different perspective than colonists did. Tantaquidgeon's description of plants as possessing "spiritual natures" points to Natives' interest in plants' other-than-human natures, which endowed plants such as the "wores Rt" with invisible virtues that could be employed for a variety of purposes.[37] As A. Irving Hallowell has pointed out, in Algonquian and Ojibwa languages, it is possible to characterize some objects as well as persons as "animate."[38] It would thus be misleading, from a Native perspective, to characterize some cures as having "natural" causes and others as having "supernatural" causes, for herbs and roots all possessed the potential for animation. While colonists insisted upon upholding these categories in order to align themselves with the study of natural causes, Native medical practitioners acknowledged plants' nonhuman natures in order to access and employ a plant's powers. It would thus also be misleading to divide herbal virtues and Native medical practices into natural and supernatural categories. From a Mohegan or a Montaukett point of view, acknowledgments of plants' nonhuman causes or the use of "wores Rt" as love medicine hardly constituted a penetration of poisonous realms of nature; instead, these actions were necessary if practitioners were to use plants in ways that respected their spiritual natures and invisible powers. As a consequence, practitioners who knew the virtues of love medicine and other herbs that worked invisibly adhered to strict protocol, which included "profound faith in the Creator, strict observance of the laws of nature, and a deep regard for the traditions of their tribe."[39] From this standpoint, then, Occom's inclusion of "wores Rt" in the herbal reflected his understanding of plants' invisible virtues and his respect for the nonhuman forces present in herbs and roots, rather than a dangerous journey to investigate secret, poisonous medical knowledge.

However, as a Christian, Occom would likely also have been aware of colonists' views and anxiety regarding occult virtues. He had converted to Christianity and had completed his education with minister Eleazar Wheelock by 1749, meaning that he could have learned of the dangers that colonists perceived in investigating invisible medicinal virtues. Yet Occom also emphasized similarities between Native and British views of nonhuman powers, reminding audiences of their traditions of such knowledge. In his 1761 "Account of the Montauk Indians, on Long Island," Occom described *powaws*, medical practitioners who could both heal people who had been poisoned and poison people. Occom noted that a powaw informed him that "they get their art from the devil, but then party [sic] by dreams or night visions."[40] Occom's statement attributing some of powaws' powers to the devil might suggest that he did associate secret medical knowledge with diabolic revelations, but he went on to note that the powaws' medical practices—even those not directly connected to knowledge of observable, natural virtues—were effective, and were "no imaginary thing, but real."[41] Occom concluded by claiming, "I don't see for my part, why it is not as true, as the English or other nation's witchcraft, but is a great mystery of darkness."[42] Occom reminded colonial readers that they had recently shared with Native Americans a belief in the effects of supernatural powers upon their everyday lives. Moreover, he left open the possibility that some of powaws' knowledge, such as information obtained in dreams, was not acquired from a diabolic source and could consequently complement medical practices that Christians considered appropriate.

Occom's statements suggest that some medical practices with invisible causes had inappropriate origins, but not all did, and communication with the devil did not mean that powaws' medical practices could not heal patients. Both the "Account" and "Herbs & Roots" show that Occom continued to investigate spaces and practices that colonists considered non-Christian, even after his education under Wheelock. Medical knowledge and practice thus offered one context in which Occom established a relation between Native and Christian cosmologies. Occom's listing for "wores Rt" and his description of powaws' powers suggest that he did not find hidden or invisible medicinal virtues to be inherently diabolic. Rather than refusing to investigate herbs that possessed spiritual causes, Occom both integrated these herbs into his medical practice and applied his Christian education by demarcating practices inspired by the devil from appropriate medical knowledge. Occom's medical knowledge was constituted by his application both of Native conceptions of things as animated by nonhuman forces and of Christian views of some of these practices as potentially dangerous.[43] His herbal shows that not all Christians found the investigation of nature's secrets to be dangerous or damaging to their identities. Thus, while Jeffrey Glover has pointed out the difficulty early Americanist scholars face in interpreting the "magical abilities" that Natives claimed for certain objects by noting that scholars "can't argue for magic as a historical cause," Occom's herbal highlights the importance of considering claims to roots' "magic"—or occult virtues—from a Native perspective, in which nonhuman entities and their actions play important roles in both religious and historical contexts.[44]

"Herbs & Roots" shows that Occom did not perceive his investigation of herbs' occult virtues and his Christianity to be mutually exclusive. Indeed, the herbal indicates that alternate, "magical" explanations for healing existed alongside Anglo-European theories privileging natural causes, even if colonists did not always acknowledge these explanations.[45] Occom's attitude regarding occult virtues demonstrates his practice of what Robert Warrior has called "intellectual sovereignty," which Warrior defines "not [as] a struggle to be free from the influence of anything outside ourselves, but a process of asserting the power we possess as communities and individuals to make decisions that affect our lives."[46] Occom did not cut himself off from the Western, Christian education he received from Wheelock, for he employed that education as well as the medical knowledge he obtained from Ocus to accomplish ends beneficial to people on Montauk. Natural and spiritual causes remained interconnected in the herbal, as did the respective epistemologies with which each explanation was associated in the eighteenth century.

LEARNING HERBS AND ROOTS: EXCHANGE AND AUTHORITY

The herbal's extratextual elements demonstrate the ways in which Occom employed his "intellectual sovereignty" to ensure the wellbeing of his family and his fellow Native Americans.[47] These elements develop the medical networks represented in the herbal by allowing Occom to define his relationships to people in both Native and colonial Christian communities. Occom described his relationship to Ocus at the end of the first booklet, writing:

May ye 17th, A.D. 1754

Ocus has now Learnt me 52 Roots—and I have this Day Paid him in all 27 York money—God is the witness and my Name Samson

Samson Occom[48]

Occom's statement of his payment to Ocus showed that he had fully compensated his mentor for the "52 Roots" that Ocus had taught him to identify and use. However, this statement was not simply economic in nature, for it also established Occom's connection to Ocus and, by extension, Occom's medical authority. In his study of cross-cultural exchange, anthropologist Nicholas Thomas explains that, in New Georgia, objects exchanged can be converted into power or prestige, thus creating a complex system of exchanges that does not fit squarely within models either of gift exchange or commoditization. As an example, Thomas points to cases in which shamanic power is purchased, rather than transmitted genealogically or to a chosen successor. Thomas writes that in such exchanges, the transfer of "ritual agency" is accomplished through the exchange of payment and knowledge.[49] The new practitioner's authority is then performed in the course of medical ceremonies: "The practice of ritual may often have involved reference to the source of knowledge, to the empowering authority, but in such contexts the allusion to the name figures as authentication and validation, rather than as the expression of longer-term indebtedness."[50] Native herbalists in New England, similar to New Georgian practitioners, regarded their medical knowledge as private property; however, they usually transmitted their knowledge to a chosen successor, often a family member.[51] Algonquian medical practitioners were often women, and their "cures were regarded to be, to a certain extent, secret property."[52] Their successors were chosen not based upon who paid them for their knowledge but upon practitioners' observation of which of their young relatives expressed interest in their medical knowledge. While it is possible that Ocus may have chosen Occom as his successor, the fact that Occom represents their relationship in financial terms suggests that he acquired his medical knowledge in a less conventional fashion.

Occom may have defined his relation to Ocus so precisely in order to explain his rather unorthodox manner of obtaining knowledge of herbs and roots and in order to reinforce his medical authority. Occom's reference to the precise amount he paid Ocus signified neither his "longer-term indebtedness" to the medicine man nor his status as a member of Ocus's family.[53] Instead, Occom's statement of purchase represented his connection to a "source of knowledge [and] empowering authority"; that is, the fact that his knowledge derived from someone already regarded as a medical expert in the Montaukett community.[54] In this way, the statement validated Occom's medical knowledge and, by extension, his status as a healer. Moreover, the receipt shows that Occom was not only Wheelock's protégé; he was also part of a long line of Montaukett and Mohegan medical practitioners, as manifested more recently by Gladys Tantaquidgeon's collection of herbal remedies, many of which she learned from her aunts and "grandmothers" in the early twentieth century: Fidelia Fielding, Emma Baker (daughter of Rachel Fielding, whose great-grandmother was Occom's sister, Lucy Occom Tantaquidgeon), Lydia Fielding, and Mercy Ann Nonesuch Mathews (a Nehantic woman).[55]

In the same way that Occom's statement of indebtedness to Ocus signaled his medical authority by representing their relationship, so Occom's letters to the unnamed Christian woman represented his spiritual authority by articulating his relationship to Christians and to divine favor. Occom wrote one draft of the letter in each of the booklets, near the end of the first booklet and at the very end of the second. In the first, Occom wrote,

> Madam,
> I look upon my self to be most indebtted [sic] to God above many [of my] Poor Nation, In that he has shewn Such distinguishing Favours to me, in giving of me greater advantages to know his ways.[56]

Occom located himself both in Native communities, with his reference to his "Poor Nation," and in Christian communities, with his note that he enjoyed

God's "distinguishing Favours." Additionally, Occom imitated English missionaries' practice of referring to Indians as "poor" in order to make their readers feel pity for Native Americans. As Laura M. Stevens has pointed out, the word "poor" had a number of meanings, but when used in Puritan missionary writings, "*poor* conveys several meanings at once: it indicates material need, a lack of civilization, spiritual impoverishment, and brutal treatment by Catholic colonists."[57] Such descriptions of Native Americans allowed English readers to characterize themselves as "benevolent" when they supported missionary projects to save the "Poor Indians," and this discourse of pity also, Stevens argued, contributed to the US policy of Indian removal by justifying US Americans' inaction.[58]

As Stevens has noted, Occom revised the trope of the poor Indian in order to expose the hypocrisy of white missionaries in his autobiography and his 1772 execution sermon for Moses Paul, a Wampanoag Indian who was sentenced to death for killing a white man while intoxicated. In these instances, Occom exposed the ways in which calling Indians "poor" justified "racial discrimination," and he revealed the "abysmal treatment that such pity rationalized."[59] In his description of his nation as "poor" in the herbal, however, Occom identified himself with his "Poor Nation" and suggested that he, out of all Natives, was most indebted to God. By claiming both extreme spiritual poverty and great indebtedness, Occom placed himself squarely within Native communities that colonial missionaries described as "poor."

In the second booklet of his herbal, Occom revised the letter to the woman and his statement of his relation both to Christians and to his "Poor Nation." Occom added to the letter while also revising his first draft. He wrote:

> Look upon my self of all Creatures most Indebted to god in that he has Shew [sic] Such Distinguishing Favour to me in giving greator advantages to me, to know him and his ways, Particularly by Shutting up the Hearts of many of his People, to take notice of me with an Eye of Pity and Compassion, in that they have received me in their Favour and have manifested their Pity to me, in Endless Instances amongst which your self is one.

> and I shall always look upon my Self greatly beholden to you, even as Long as I shall have my wright Senses about me, for the Book and Money that you have Sent me by the Hands of the Revd Mr Wheelock of Lebanon, may God Reward you out of his [...] Treasure of Spiritua^l Rewards and gifts, the Preachr says, Blessed is he that Considereth the Poor—and Since you have shown such Favou^r to me, May not I have ease in trust your addresses at the T^{h}rone of grace.[60]

In this revised version of his letter, Occom altered the statement of his relation to the woman and to all Christians, by noting that he considered himself indebted to God out of "all Creatures," white and Native alike. Additionally, Occom emphasized Christians' actions and assistance to him, rather than the state of the Mohegans as a "Poor Nation." His statement of obligation to the woman emphasized their relationship as fellow members of God's creatures, and Occom defined his spiritual authority as a Christian who could "trust" in the woman's "addresses at the time of grace." As he revised his letter, Occom moved from emphasizing that he was a favored member of a "Poor Nation" to claiming an identity that he could share with both whites and Natives—that of a Christian. In the same way that noting his debt to Ocus signified Occom's medical authority, so the letters to the woman displayed Occom's status as a Christian on the basis of his debts, both to God and to "his People." The letter positioned Occom in a series of relationships that included Native Americans, white colonists, and Christians of all nationalities; in this way, it complemented Occom's reliance on both Christian and Native views of herbs with invisible virtues.

The physical space—that is, the paper—shared by the letters and the herbal integrates Occom's Christian religious beliefs and his relationship with the woman into the herbal's medical networks. It is not clear why Occom drafted the letter to the woman in the pages of the herbal: perhaps he needed paper and used a spare page to compose early versions of the letter, or perhaps he wrote it after providing medical care for someone or while waiting for an herbal concoction to steep. Regardless of why the letter fragments appear in the herbal, their presence and proximity to Native modes of representing and using medicinal plants established connections between Native and Christian worlds. Significantly, Occom did not separate these contexts: his letter to the woman precedes his statement of his relation to Ocus, and both are interspersed with medicinal information. Indeed, in the botanical entry following the first draft of the letter to the woman, Occom addressed an unnamed subject, writing: "Take some Weecup and Sweet Fern for the Boy—And for your self the same Weecup and Sweet Fern or some Sage, or Hysop—and take some Bone and Burn it thoroughly and Pound it Fine and it about half a Spoon full at a time with a little water just before or after meal."[61] It is not clear whom Occom addresses in this entry, but the fact that the recipe follows his letter to the Christian woman raises the possibility that he was writing a cure for her. If so, then the herbal represents the ways in which Occom's medical knowledge and practice connected Native and Christian worlds: Occom's education with Ocus made possible his offer of a cure for a Christian woman. Moreover, the letters contribute to Occom's statement of his authority in both Native and Christian contexts, for together they represent his debts to his teachers, benefactors, and to spiritual forces.

Occom's statement of his spiritual status—his "greater advantage" out of all creatures to know God's ways—in the herbal's pages included his medical knowledge in God's "ways."[62] Occom's spiritual and medical knowledge were far from incompatible, as his aforementioned inclusion of an herbal recipe on the back page of a sermon from 1766 indicates. Following a sermon on the doctrine of regeneration, Occom jotted down a recipe for a medical concoction composed of roses and balsam to be taken every morning and night.[63] The spatial proximity of the sermon and recipe points to the ways in which, as a minister, Occom cared for both bodies and souls, sometimes simultaneously. Similarly, the letter fragments in "Herbs & Roots" display Occom's early concern for the bodies and souls of his "Poor Nation," and his use of his medical and spiritual education to provide healing for Natives and whites alike. Taken as a whole, then, the herbal demonstrates that Occom defined his relations to both Native and Christian contexts by including interpersonal connections with white colonists in preexisting medical networks, composed of people, plants, and nonhuman entities. "Herbs & Roots" presents Occom as a medical practitioner who acquainted himself with and drew upon forms of knowledge and power in Native and Christian traditions in order to heal both bodies and souls.

The present locations of the herbal's two parts further manifest Occom's participation in Native, colonial, and Christian worlds. The first, smaller booklet is held at the New London County Historical Society (New London, Connecticut), along with Occom's journal for 1785–1786, and a letter, from Occom to Nathaniel Shaw. A wealthy merchant who built the house in which the historical society is now located, Shaw lent money to Mary Fowler, Occom's wife, while Occom was in England on a fundraising trip for Wheelock's school.[64] Writing from London in 1767, Occom stated that he was "much oblig'd to [Shaw] for your Care and assistance to my Family, in my absence," while assuring Shaw that he would repay the money loaned to Mary Fowler. Occom concluded the letter with a postscript in which he requested that Shaw "supply my wife with Money if she wants 10 or (pounds) let her have it."[65] Occom drew upon his economic relationship with Shaw in order to address personal concerns; namely, his family's wellbeing in his absence. While it is not clear how part of Occom's herbal came to be held at the New London County Historical Society, Occom's connection to Shaw suggests that the herbal could have been located in Shaw's papers at some point and moved with them to the historical society.[66] The herbal's location manifests Occom's connections with prosperous colonists and his willingness to use his knowledge of colonial commercial networks for the benefit of his family. Moreover, the manuscript's position in New London, a town located down the Thames River from Mohegan, where Occom was born and lived for part of his adult life, represents Occom's connection to local, Mohegan geographies as well as the difficulties providing for his family that Occom faced as a consequence of his missionary work and his transatlantic travels.

The second part of Occom's herbal is located at Dartmouth College's Rauner Library (Hanover, New Hampshire), in one of the largest archives of Occom's journals, sermons, and letters. The Dartmouth location of the herbal manifests Occom's long relationship with Wheelock, who founded Dartmouth College and whose papers and correspondence are likewise held at Rauner Library. The herbal's presence at Rauner represents Occom's Christian education and his commitment to missionary work and to Wheelock's school for Indians. Nonetheless, it is somewhat ironic that the herbal is located at Dartmouth, for one of Occom's reasons for cutting off contact with Wheelock in 1771 was that the minister had moved the mission school from Connecticut to New Hampshire, a move that, Occom argued, indicated an investment in educating Anglo-American boys and men, rather than Native Americans.[67] The two archives in which "Herbs & Roots" are located thus point to Occom's professional and familial relationships, his economic transactions, and his ongoing commitment to his family and to education for area Native Americans. In this way, the two sections of the herbal and their present locations not only communicate medical knowledge but also represent Occom's engagement in various networks throughout New England.

EMBODIED LISTS

The networks represented and constructed in "Herbs & Roots" require a reconsideration of the list as a rhetorical and epistemological form. Studies of literacy and of writing have treated lists as playing a key role in differentiating so-called oral and literate cultures. For example, Walter Ong argues that lists hold an important position in what he sees as the progression from orality to literacy, for he suggests that "writing was in a sense invented largely to make something like lists," citing Sumerian cuneiform script used for accounting.[68] As "oral" cultures sought to record information in a form that would allow easy access yet did not require memorization, Ong suggests, they began to use writing to list things. Unlike Indigenous recording devices such as Incan *khipu* and Lakota winter count calendars, Ong argues, lists were not memory aids but represented "an objective tally."[69] Indeed, we might see written lists as an ideal form of what, according to Ong, writing offered: "abstractly sequential, classificatory, explanatory examination of phenomena or of stated truths."[70] For Ong, then, lists made possible the consideration of objects or ideas as abstract truths, removed from their contexts.

In contrast to Ong's analysis, Occom's list of herbs, roots, and their virtues does not separate information from its origin or knower. Instead, the list calls into being an expert reader who can position the records of plants and illnesses in their appropriate contexts and subsequently interpret the herbal. The words in the list do not simply refer to objects or to abstract facts; they delineate objects' connections to interpersonal and other-than-human contexts as well as objects' roles in exchanges of medical knowledge. Occom's list thus does not so much replace memory as it provides a "symbolic vocabulary," written in English and Algonquian languages, that links plants, bodies, illnesses, and nonhuman beings.[71] The herbal thus suggests the futility of dividing communication systems—and, by extension, cultures—into categories such as "oral" and "literate": the list is a written document employing alphabetic signs, but these signs evoke Mohegan and Montaukett medical networks and their accompanying conceptions of power and knowledge.

In this way, the herbal poses an alternative to Enlightenment natural historical and botanical projects to create a list in which all elements would be universally comprehensible. In the Linnaean system of classification, for example, Latin provided a language that was universally accessible to people of all nationalities (at least in theory), while systematic rules of classification identified plants by locating them within classes, orders, families, and so on. As Londa Schiebinger points out, a plant's name had "no essential connection to the plant, but was to be something agreed upon by convention," so that all plants could be catalogued in exactly the same way.[72] As naturalists identified and classified plants within this universal list, they also extracted plants from their environments, precisely so that plants could be consistently catalogued. By contrast, in "Herbs & Roots," plants remain deeply connected to the natural and spiritual contexts that endowed them with power. Their identities are defined by their position in place-specific medical networks, which are represented in but not confined to the herbal. Occom's knowledge of these networks, not his ability to identity herbs via an external system, allowed him to access and employ herbs' and roots' healing properties.[73]

The herbal consequently requires consideration of a range of communication practices in order to interpret and apply its information. As Robert Warrior points out in his study of Native nonfiction, "many actions in Native life are neither primarily oral nor even linguistic, such as ceremonially presenting someone with provisions, taking part in a ritual feast, being part of a societal dance, or cooking a communal meal."[74] The medical knowledge and practice that forms the foundation of Occom's herbal is one such "embodied discourse."[75] Indeed, the written manuscript is only one manifestation of Occom's medical practice, as the herbal itself makes clear by referencing actions of learning, collecting, and using herbs. Occom's medical practice would have been composed of rituals for collecting and preparing plants, applying medicines, and observing bodies, as well as alphabetic script and speech. Finally, Occom employed nonalphabetic lines or dashes of varying lengths to indicate the relationships among various plants.

Rather than placing Occom in one religious community or communication system over another, then, "Herbs & Roots" exemplifies the ways in which Occom drew upon available modes of communication in order to represent the interpersonal, environmental, and spiritual relationships—Native and Christian—that defined his medical authority.

PART IV

Pharmacopoeias and the Emergence of the Nation

10

NATIONAL IDENTITIES, MEDICAL POLITICS, AND LOCAL TRADITIONS

The Origins of the *London, Edinburgh,* and *Dublin* Pharmacopoeias, 1618–1807

STUART ANDERSON

DURING THE early modern period several different kinds of text were published under the title "pharmacopoeia." Only some of these were "official" in the sense of having the authority of a monarch or state, and even fewer applied to an entire country. Yet, in less than two hundred years (between 1618 and 1807), the countries of the British Isles published three separate national pharmacopoeias, covering England, Scotland, and Ireland; these were the *London, Edinburgh,* and *Dublin* Pharmacopoeias, respectively. That they were all given the title "pharmacopoeia" reflects an emerging movement to create texts that would both convey authority and exercise control of one group of medical practitioners over another.

In this chapter, I adopt David Cowen's definition of an official pharmacopoeia as "a compendium of drugs and formulas which is intended to secure uniformity and standardisation of remedies, and which is made legally obligatory for a particular political jurisdiction, especially upon the pharmacists and pharmaceutical manufacturers of that jurisdiction."[1] Cowen also identifies a further criterion—that it must be prepared by an official pharmacopoeia commission, one that has the authority of the monarch, state, or parliament. As Antoine Lentacker points out in this volume (chapter 11), the term "pharmacopoeia" was itself a creation of the Renaissance, when it emerged as a newly coined Latin term derived from Greek roots. Its first known appearance was in the title of a 1548 dispensatory published in France. In the early modern period, not all publications titled "pharmacopoeia" fulfilled all—or even most—of Cowen's criteria.

Although much has been written about the three British pharmacopoeias individually, most of this analysis has focused on content rather than origin or process. There is little comparative historiography beyond content, yet their origins and processes had both similarities and differences. Their origins arose from increasing friction between rival groups of medical practitioners over professional boundaries, from a wish to emulate developments occurring elsewhere in Europe, and from a genuine concern about the quality of drugs in use. The processes by which they were eventually produced reflected the shifting political and religious conflicts of the time. Exploration of these origins and processes is important, as it throws light on the circumstances under which some pharmacopoeias came to be "official" while most did not.

In the United Kingdom, pharmacopoeias were "made legally obligatory" by means of a Royal Charter authorized by the monarch. The "particular political jurisdiction" was frequently contested by those whose activities would be regulated by it; and the "official pharmacopoeia commission" was invariably a college of physicians. Indeed, it is noteworthy that in all three cases, the development of the pharmacopoeia was driven by physicians rather than by those who would be required to use them, but whose cooperation would nevertheless be needed. The circumstances necessary for a pharmacopoeia of this type to come into existence were thus complex and demanding. The physicians needed to be organized into a college authorized by a Royal Charter. Relations with other medical practitioners had to be considered, and occasionally their advice might even be sought. There needed to be people able and willing to undertake the work. And the final text had to be approved by College members acting as the "official pharmacopoeia commission."

The process was subject to widely differing views on a diverse range of issues, from the merits of Galenic versus chemical medicines, to the nature and purpose of the pharmacopoeia itself. Into this mix entered religious and political beliefs, along with views on national identity and England's relations with other countries. The pharmacopoeias appeared during a period of great turmoil in British history; conflict, terror, and violence, arising from religious differences and battles between monarchs and their parliaments, were the background against which other events including the development of pharmacopoeias took place. Yet for the whole of the period, England had some form of political union with both Scotland and Ireland.

A narrative of British political history is thus central to explaining how three pharmacopoeias came to be published, why they were developed when they were, and why there were substantial delays in their development. The history of the origins and development of the pharmacopoeias is not just one of medical debates but is also one of the political and religious history of Britain. Thus, the problems of making pharmacopoeias "official" go far beyond disputes between medical practitioners. While an "official" pharmacopoeia received the endorsement of the state and thus had a greater chance of achieving the goal of standardizing medical practice, the tradeoff was greater entanglement in the political and religious issues of the day. The elite physicians who founded Royal Colleges and prepared pharmacopoeias depended for their livelihoods on the patronage of monarchs, their entourage, and the landed gentry. Political and religious allegiances changed; if physicians found themselves on the wrong side of history, their lives could be imperiled.

Wider political, religious, and scientific narratives are rarely referenced in previous scholarship. Urdang, for example, in his analysis of the first *London Pharmacopoeia* in 1618, notes that the social and political background was "Queen Elizabeth I and King James I" and the cultural background was "Shakespeare."[2] Likewise, Cowen, in his analysis of the origins of the *Edinburgh Pharmacopoeia*, places the emphasis squarely on attempts by the physicians to control pharmacy, and he argues that delays in its publication were the result of internal disputes about the contents of the pharmacopoeia.[3] Yet an examination of the relationship between England and Scotland, of religious turmoil, and of relations between physicians and rulers helps to inform both origins and delays. And Kirkpatrick's account of the origins of the *Dublin Pharmacopoeia* refrains from commenting on Ireland's difficult relationship with England during the years between the first suggestion of a *Dublin Pharmacopoeia* in 1721, and the reprinting of the final edition in 1851.[4]

Whether they should apply to the whole country or just the capital city was an issue with all three pharmacopoeias. Scale of place is a theme that has been considered before; in his 1946 paper Urdang found that, in many countries, "an own pharmacopoeia became gradually a matter of national ambition, a part and a proof of national sovereignty and unity."[5] He notes that in 1618, James I was king of Great Britain, Scotland, and Ireland, and thus in a position to impose a single pharmacopoeia across the entire British Isles.[6] Despite this, James I commanded that the *London Pharmacopoeia* should apply only in England. Why was this so? Why were later editions of the *London Pharmacopoeia* not applied in all countries of the British Isles? Why were separate pharmacopoeias developed in Dublin and Edinburgh?

These are the key questions considered in this chapter, where the focus is on identifying the problems and complexities of developing pharmacopoeias, rather than on successes of the state, as with the genesis of the French Codex located within the French Revolution's project of reforming language, as described by Lentacker in chapter 11. This was also the case in the United States, where pharmaceutical markets were increasingly dependent on the linkages established between the *Pharmacopeia of the United States of America* and the legal apparatus of the emerging regulatory state, as described by Joseph Gabriel in chapter 12. The answers to the questions posed lie in the complex interaction between external factors of political and religious turmoil and internal tensions arising from competing interests of national identity, medical politics, and local traditions.

ENGLAND: PHYSICIANS, COLLEGES, AND PHARMACOPOEIAS TO 1518

Although the first edition of the *London Pharmacopoeia* was published by the Royal College of Physicians of London in 1618, it was the culmination of a long process that stretched back decades. From 1447, spicer-apothecaries had the exclusive right to inspect drugs and spices in apothecaries' shops.[7] Written sources available to them were limited; the oldest text in Old English was *Bald's Leechbook*, or *Medicinale Anglicum*.[8] Versions of ancient herbals were available, and these early pharmacopoeias remained the basis of drug therapy up to the late medieval period.[9] After Caxton set up his printing press in 1476 books could be produced more quickly and in greater numbers; William Turner's herbal was published in three volumes between 1551 and 1568.[10]

In 1511, an act was passed that required that "no person within the city of London or seven miles of it should practise as a physician or surgeon unless they had been examined by the bishop of London."[11] In 1518, efforts were made to found a college of physicians, and on September 23, the king signed the Royal Charter creating the Faculty of Physicians of London.[12] The driving force was Thomas Linacre, who had qualified as a physician in 1484. In 1487, he went on a mission from Henry VII to the pope, staying in Italy until 1499.[13] On returning to London, he promoted the idea of developing a *London Pharmacopoeia* following the publication of the *Nuovo Receptario*, published in Florence in 1498, as described by Emily Beck in chapter 2.[14] The petitioners included Cardinal Wolsey, the Lord Chancellor; Linacre was Wolsey's personal physician. When the next parliament sat in 1523, the College took the opportunity to seek incorporation by statute, giving it additional powers. The act extended the remit of the London College to the whole of England. When its pharmacopoeia was published, it too would apply to the whole of England. As noted by Matthew Crawford in chapter 3, this model was also used in eighteenth-century Spain, where the newly established pharmacopoeia represented the Crown's official policy regarding the preparation of medicaments, and was enforceable throughout Spain's territories—not just in Spain but also in the Americas.

If there were plans for a pharmacopoeia early in the life of the College, they were soon overtaken by events. Henry VIII's demand for a divorce resulted in passage of the Ecclesiastical Appeals Act in 1532, as a result of which the king became the final legal authority and head of the Church of England. All his subjects were required to formally acknowledge this; not to do so was high treason. The religious conflict that followed greatly affected the physicians because of their dependence on patronage. Despite being small, the College could not escape external interference.[15] Several of the leading fellows suffered heavily for their religious convictions. Under Edward VI, John Clement, the College's president in 1532, was exiled to Flanders; he returned after the accession of Mary I in 1553, only to go into exile again with the accession of Elizabeth I in 1558. Some switched religious allegiance as events unfolded. John Fryer, president of the College in 1549–1550, was twice imprisoned, first as a Lutheran, and later, having returned to the Catholic faith under Elizabeth I. Others, such as Sir William Butts, made fortunes by switching sides; he became one of Henry VIII's physicians, and received a grant of confiscated monastic lands.[16]

With the accession of Elizabeth I in 1558 the College discussed the possibility of preparing "one certain, uniform Pharmacopoeia for the use of all the apothecaries of London."[17] Its aim would be to counteract the fraud and deceit of those who sold "filthy concoctions and even mud under the name and title of medicaments for the sake of profit."[18] They limited their ambition to London; Clark suggests that this may have resulted from a loss of interest in the provinces.[19] It was soon realized that producing a pharmacopoeia would be a large undertaking. The matter was adjourned to the next meeting when all the fellows were to attend; each was to state his views clearly, and they hoped "something complete and outstanding" could be produced. As "the matter seemed weighty" and it "seemed a toilsome task," the deliberations were deferred again.

National and international events again help to explain the lack of progress. In 1586, a plot to assassinate Elizabeth I and put Mary, Queen of Scots on the English throne was foiled, ending with the execution of Mary in February of 1587. A bond of association was drawn up to be signed voluntarily by noblemen and others, pledging their support to the queen. The College of Physicians received a copy, which was read out to the fellows. Fourteen took the oath, which confirmed allegiance to the queen and the Church of England, but two did not. The College also experienced difficult relations with other practitioners. In a dispute in July of 1586, the Grocers' Company (which included the apothecaries) urged the physicians to set a fixed composition for Genoa Treacle, but considered publication of a pharmacopoeia unnecessary.[20] They pressed for "the reformation of divers abuses" and the resolution of their "griefes and complaintes."[21] In 1588, they petitioned the queen for a monopoly of compounding and selling medicines. Their parent body, the grocers, supported them, but the physicians opposed the petition, and when presented in Parliament in 1589 it failed.[22]

Initial Attempts to Prepare a *London Pharmacopoeia*, 1589

With a return to more peaceful times, the College resumed its normal business and the pharmacopoeia project was reconsidered. On October 10, 1589, it was "proposed, considered and resolved that there shall be constituted one definite public and uniform dispensatory or formulary of medical prescriptions obligatory for apothecary shops."[23] Over the next five years, a large number of committees were formed. In 1589, ten committees of fellows were appointed to undertake the work of selection and compilation. Each dealt with different groups of medicaments, the products being divided between the president and twenty-two other fellows.[24] Separate committees considered syrups and decoctions, oils and waters, liniments and ointments, pills and electuaries, lozenges and eye salves.[25] The queen's physicians took on powders and confections. All the fellows were to present their written opinions on the tasks they had been given at the Michaelmas meeting.[26]

Two weeks later, on October 23, 1589, a new committee was appointed to edit the whole work. This proved somewhat premature, as they had nothing to edit. The task was deferred to the next meeting, when it seems little progress was made; selecting items for inclusion in the pharmacopoeia proved an onerous task. National events could hardly be blamed; Elizabeth remained on the throne throughout the period, although the assassination of Henry III of France in 1589 caused great concern in England. But five years later, on December 13, 1594, a new editorial committee of eight was appointed, which included five of the former six members. The work again proceeded slowly, for the pharmacopoeia is not mentioned again until 1614, twenty years later. No explanation for the delay is given, but political and religious turmoil undoubtedly played a part.

By the end of the Elizabethan era, religious conflict in England had largely subsided. Elizabeth was succeeded in 1603 by James I, already King James VI of Scotland. Although a Protestant, his dispute with Parliament led to further violent unrest. In 1605, an attempt was made by Catholic sympathizers to blow up king and parliament. Although it failed, it heralded a new wave of anti-Catholicism, from which the College's fellows were not exempt. But difference in scientific beliefs (as well as religious ones) played a part in further delaying preparation of the *London Pharmacopeia*. At the College, the teachings of Galen held sway over those of Paracelsus until its statutes were revised in 1601.[27] But the College found it increasingly difficult to avoid the rise of chemical medicine, as candidates presented themselves as well versed in chemical medicine, but inadequately in Galenic medicine,[28] a tension reflected in the views of the existing fellows.

Final Efforts to Publish a *London Pharmacopoeia*, 1614

The final impetus to complete the pharmacopoeia came from action taken by the apothecaries.[29] Increasing numbers became dissatisfied with being part of the Grocers' Company and wanted to form their own. Their leader was Gideon de Laune, the Huguenot apothecary to Anne of Denmark, queen to James I. With others, he presented a bill to Parliament in 1610 for the establishment of an apothecaries' company with a monopoly of the trade. The Grocers' Company opposed the bill and it failed. In 1614 he tried again, submitting a petition for a Royal Charter to the king. The College of Physicians discussed the proposal at a meeting in April, and a majority of those present supported it.[30] Although it faced considerable opposition, it eventually received the king's seal of approval on December 6, 1617, resulting in the foundation of the Worshipful Society of Apothecaries of London.[31] This time separation was fully supported by the College of Physicians.[32]

The prospect of a strong body of apothecaries prompted the physicians to revive the pharmacopoeia project. In 1614, eight College fellows were appointed to examine foreign pharmacopoeias. The initiative was led by the president, Dr. Moundeford, and three other fellows, Sir William Paddy, Dr. Lister, and Dr. Atkins.[33] The pharmacopoeias of Bologna, Bergamo, and Nuremberg (copies of which were listed in the College library catalogue of 1660)[34] were compared with their own drafts, probably papers that had survived from 1594.[35] Things moved a little more quickly; in 1616, an editing committee was appointed and the collaborators sent their papers to this body. It then appeared that many that had been prepared had been lost, a misfortune blamed on the carelessness of the president, Dr. Forster. Forster's successor, Dr. Atkins, put much more energy into the project.

The king's intention in inaugurating the Society of Apothecaries was to protect his subjects from unskilled and ignorant empirics who compounded "unwholesome, hurtful, deceitful, corrupt and dangerous medicines."[36] The wording of the charter was intended to reassure both the Grocers' Company and the College of Physicians that the new society posed no threat to either. He had anticipated that it and the College of Physicians would enjoy a productive working relationship. However, the College expected that the new society would be subservient to it, and with this in mind the plans for publication of a *London Pharmacopoeia* were accelerated. A pharmacopoeia would "bring science to bear by standardizing and improving the dispensing of medicines."[37] But the wording of the charter did nothing to restrain the apothecaries from both prescribing remedies and attending patients, in other words, from practicing medicine. This led to disputes between the two groups that were to continue for two hundred years.

Publication of the *London Pharmacopoeia*, 1618

There remained fundamental differences in medical philosophy, particularly the place of Galenic versus chemical medicines. A key figure was Sir Theodore Turquet de Mayerne, Swiss by birth and French by adoption, and a member of the Paris Faculty of Medicine. In 1603, following disputes over the use of chemical medicines, the Paris Faculty prohibited licentiates from using them under severe penalties. Mayerne was expelled from France ostensibly for advocating the prescribing of mercurials and antimonials.[38] He settled in London in 1611 on the invitation of King James I, who made him his first physician. He was subsequently appointed physician to Charles I and Queen Henrietta. Mayerne was instrumental in seeing the *London Pharmacopoeia* through to completion.[39] He is thought to have been behind the inclusion of many chemical medicines, including flowers of sulphur, mineral acids, and preparations of iron and antimony into the pharmacopoeia.[40] The manuscript was completed and two of the fellows were tasked with correcting the proofs.

It was only at this stage that the apothecaries were consulted.[41] Six prominent apothecaries were required to "give daily attendance," which they assiduously did. These were the master, the two wardens, and four of the assistants.[42] The last committee meeting was held in the spring of 1618; Mayerne was to write a dedicatory letter to King James I, and several of the fellows wrote the preface. The manuscript was completed and in type by April 9, 1618. The king's proclamation, dated April 26, 1618, required all the apothecaries in the realm to compound their medicines in accordance with the *London Pharmacopoeia*.[43] Although pharmacopoeias that applied to cities or regions had been published previously elsewhere in Europe, the *London Pharmacopoeia* was the first to be applied to a whole nation.[44]

The interpretation of events surrounding its publication, withdrawal, and subsequent reissue has been the subject of extensive debate. The apothecaries claimed that the physicians did not take their advice and, as a result, the pharmacopoeia contained a large number of errors when it appeared in May 1618.[45] The *London Pharmacopoeia* was written entirely in Latin. Munk wrote that the issue of May 7, 1618, was published "surreptitiously and prematurely, by the printer in the absence of the President."[46] On his return the president supposedly "found the book full of errors."[47] So, on December 7, 1618, the College brought out a corrected edition (figure 10.1),[48] with an epilogue blaming the printer for having "snatched away from our hands this little work not yet finished off."[49] The

FIGURE 10.1. Second issue of the first edition of the *London Pharmacopoeia*, December 1618. © Royal College of Physicians of London.

statement blaming the printer for premature publication may have been made to cover up a serious difference of opinion within the College, one school wanting to keep the pharmacopoeia simple, and the other wanting it to be a grander combination of formulary and textbook.[50] The latter view prevailed; the revised edition published in December combined a formulary with detailed text. Urdang suggests that the later version represented the "victory of baroque abundance over renaissance simplicity."[51]

SCOTLAND: PHYSICIANS, COLLEGES, AND PHARMACOPOEIAS TO 1603

Like the London one before its publication, the first edition of the *Edinburgh Pharmacopoeia* in 1699 followed a long sequence of events concerning the evolution of medical practice in Scotland and relations between physicians, surgeons, and apothecaries during a period of extreme political and religious turbulence. By 1505, Edinburgh already had an Incorporation of Surgeons and Barbers. In 1599, Glasgow established a Faculty of Physicians, later to become the Royal College of Physicians and Surgeons of Glasgow. Edinburgh's College of Physicians, which published the pharmacopoeia, followed only in 1681. The context of these developments was the relationship between Scotland and England between the Union of Crowns 1603 and the Act of Union of 1707. Relations between England and Scotland had been hostile for centuries, with the two countries at war for long periods.[52] In the sixteenth century, religious conflict with its roots firmly in the English Reformation erupted in Scotland, further complicating the relationship with England.

In 1503, the daughter of Henry VII of England, Margaret Tudor, married James IV of Scotland, joining the two monarchies through marriage.[53] In Edinburgh, the surgeons and barbers established a guild in 1505.[54] Elsewhere in Scotland, the surgeons joined with the apothecaries in the grocers' guild. During the reign of Mary, Queen of Scots (1542–1587) a widening Protestant–Catholic split occurred. Mary's son, James VI of Scotland, imposed a degree of peace and political stability, and it was in this period that the Glasgow Physicians obtained a Royal Charter and founded their faculty in 1599.[55] James VI succeeded to the English throne in 1603, becoming also James I of England. With this Union of the Crowns, England and Scotland shared a monarch for the first time, a situation that continued until 1707, eight years after the publication of the *Edinburgh Pharmacopoeia*.[56] James reigned over both countries until 1625, although he moved to London and Scotland became a "kingless kingdom." The creation of medical colleges in Scotland became very difficult; petitioners for Royal Charters needed access to the king, and for nearly a century there were few opportunities to do so.[57] The peace following the Union of the Crowns did not last. An attempt made

NATIONAL IDENTITIES, MEDICAL POLITICS, & LOCAL TRADITIONS

in the English Parliament to join England and Scotland by an act of Parliament in 1606 was not successful.

During the seventeenth century, several attempts were made to found a college of physicians in Edinburgh. The first initiative came from the king himself, as part of his plan to rule his kingdoms as one United Kingdom. He invited one of the leading Edinburgh physicians, George Sibbald, to petition him in 1618, the year in which the first edition of the *London Pharmacopoeia* was published. The king issued a royal warrant in 1621, entrusting the establishment of a college of physicians to the Scottish Parliament. It was based on the model of the London College; it would include only physicians, and would supervise only those in Edinburgh itself. However, Sibbald submitted an amendment proposing that the College should have jurisdiction over surgeons and apothecaries as well as physicians, and that this should apply to the whole of Scotland. This was opposed by the Incorporation of Surgeons of Edinburgh and the Faculty of Physicians and Surgeons in Glasgow. No agreement had been reached at the time of the king's death in 1625.[58]

A second attempt was made in 1630. A young but well-connected Edinburgh physician, Dr. John McClure, arranged for a petition to be put to Charles I for the creation of a college.[59] The Edinburgh physicians were asked to submit "heads and articles" for consideration by the Privy Council. They dropped their demand to have authority over the whole of Scotland, but again proposed that they should have jurisdiction over surgeons and apothecaries as well as physicians. Again, the surgeons, as well as the universities, objected and this second attempt to form a college of physicians in Edinburgh failed. Resistance continued for another twenty-three years. But in 1642, Civil War broke out in England, ending with the execution of Charles I in 1649 and the creation of a republic under Oliver Cromwell. Cromwell was anxious to expand, and his army quickly occupied Scotland.

Initial Attempts to Prepare an *Edinburgh Pharmacopoeia*, 1680

Disputes with other practitioners resurfaced and medical politics soon dominated the agenda again. In 1645, the Incorporation of Surgeons and Barbers in Edinburgh began to admit apothecaries, and those who had served dual apprenticeships in surgery and pharmacy then formed a new fraternity, who were able to undertake all forms of medical activity. The physicians felt threatened by them, and in 1653 a third attempt was made to found a college. A group led by Dr. George Purves submitted a list of "public abuses in matters of medicine" in Scotland. A draft charter for such a college was drawn up; it would have jurisdiction over all physicians, surgeons, and apothecaries in Scotland, with the exception of surgeons in Glasgow and Edinburgh. In 1657, Purves presented himself at the College of Physicians in London.[60] The College instructed its registrar to give every assistance to Dr. Purves in looking through their statutes.[61] But the opposition of the Edinburgh surgeons, the physicians and surgeons in Glasgow and the Scottish universities was renewed, and continued until Cromwell's death in 1658, the commission's recommendations dying with him.[62]

The surgeon-apothecaries had become the largest body of incorporated medical men in Edinburgh.[63] The physicians were exposed to competition from them and the activities of those falsely presuming to "profess and practise physic." By 1681, the Edinburgh physicians felt an urgent need to protect their interests and those of a "gullible public" against unqualified practitioners and unsafe medicines. They again attempted to obtain a Royal Charter giving them certain privileges and responsibilities. The leading figure was Robert Sibbald, born in 1641 into a wealthy family. He was a staunch Scottish nationalist; he was nine when Cromwell ordered the invasion of Scotland. As a young man he had "become determined to marshal the talents and resources of Scotland to carry his country forward into the modern world."[64] In 1660, he traveled to Leiden to study medicine, continuing his studies in Paris. He then moved to London, where he met senior figures of the Royal Society. He returned to Scotland in 1662, determined to create in Scotland the learned institutions he had admired in London and Paris. He now had a "grand design for the advancement of Scotland."[65]

A second attempt to join England with Scotland by means of an act of Parliament was made in 1667, during the reign of Charles II, but this too failed. In the same year, Sibbald began to win the confidence of the leading physicians in Edinburgh. In 1680 he invited them to meet regularly at his lodgings to discuss matters of mutual interest. Most of those who attended were from Scotland's landed gentry. From 1680, a number of its members, including Sibbald and Andrew Balfour, began to draw up a pharmacopoeia for general use in Edinburgh.[66] The opportunity to raise the subject of a college of physicians came with a case brought by the Faculty of Surgeon-Apothecaries against an apothecary who had performed a surgical operation. As part of the inquiries into it a meeting was held of all the Edinburgh physicians, at the end of which Sibbald proposed establishing a college. He was confident that this time the attempt would be successful.

In 1681 the king's brother, the Duke of York (later James II), was in residence in Edinburgh.[67] Sibbald and Balfour were granted an audience; they produced the warrant issued by James I in response to the petition presented by his uncle, George Sibbald, in 1618. Although the Scottish Parliament had been entrusted with acting on it, no action had been taken. To avoid the previous objections Balfour insisted that the proposed college would be "metropolitan and not national"; to reassure the surgeons in Edinburgh he promised that the college would "not rival them in trying to obtain the body of even one malefactor for dissection"; and he placated the Scottish universities by "relinquishing any idea of giving degrees." The concessions were all agreed, the charter was drawn up, and royal approval given on November 30, 1681. Its development had been heavily influenced by the London College; at the end of his life Robert Sibbald was a fellow of both colleges.[68]

Efforts to Develop an *Edinburgh Pharmacopoeia*, 1681–1699

Under its charter, the College had a duty to "examine and inspect the drugs and medicines sold within the jurisdiction, suburbs and liberties of Edinburgh."[69] In order to make clear which drugs and medicines would be acceptable, a committee was appointed in March 1682 to revise and prepare a new draft of the pharmacopoeia that Sibbald, Balfour, and others had drawn up at their meetings in 1680.[70] In fact, it was the first item of business recorded in the College's minutes in 1682, and a manuscript was apparently ready for the printer in 1683.[71] It might be expected that this would quickly be completed and published. But publication was delayed until 1699, a delay attributed by Sibbald to "malice and obstruction by a faction."[72] By the 1690s, the College was divided. Cowen noted that there was some indication that "religious controversy added fuel to professional differences and jealousies."[73] The next we hear of the proposed pharmacopoeia in the minutes is that an agreed draft had not been produced by 1693.

Wider political and religious issues played an important part in explaining why publication of the *Edinburgh Pharmacopoeia* was delayed during the period from 1683 to 1699. In 1685, Charles II died and was succeeded by his brother, who became James VII of England and James II of Scotland. James was a Catholic, and within three years he had lost the support of the Scottish Parliament as he began dismantling legal restrictions placed on Catholics.[74] In 1688, his wife produced a son, raising the prospect of a succession of Catholic monarchs. A group of Protestant parliamentarians mounted a coup and James was exiled. His daughter Mary and her husband, the Protestant William of Orange, were invited to take the throne of England in the "Glorious Revolution." The Scotland Parliament was then forced to choose between an exiled Catholic or to offer the throne to William and Mary. In February 1689, they voted for a Protestant succession, but a third attempt to join England and Scotland through an act of Parliament failed.

The Glorious Revolution was followed by the Jacobite uprising of 1689 and the Glen Coe massacre in 1692.[75] Indeed, in the 1690s "Scotland experienced a level of collective misery and misfortune that was never approached again."[76] These events took a heavy toll on the college; it took the election of Sir Archibald Stevenson as president in 1693 to restore its life.[77] But when he did so his council was equally divided between Catholics and Protestants. Since 1684, the number of fellows had dropped substantially, and some of the remaining ones played little part in the life of the College. Attempts at reform introduced a decade of discord.

There were serious personal antipathies and antagonisms based on both politics and religion, and there were factional disputes over the adoption of the "new science."

Publication of the *Edinburgh Pharmacopoeia*, 1699

The longest-lasting of the College's many disputes was over the pharmacopoeia. In 1694, the responsibility for it passed to Sir Archibald Stevenson himself.[78] For the next five years under his chairmanship, a succession of committees struggled to draw up an agreed version that could go forward for publication.[79] Opposing attitudes to the "new science" appeared intractable. The medicine of Sibbald was firmly rooted in the works of Galen; he was firmly resistant to all new ideas, including iatromechanism.[80] The leading exponent of iatromechanism in Scotland was Archibald Pitcairn, a staunch Catholic and Jacobite; he was also Stevenson's son-in-law. They, along with the younger fellows of the College, regarded the Sibbald faction as reactionaries. Some of the galenicals and herbal remedies favored by Sibbald, including ones based on local tradition, were discounted by Stevenson and Pitcairn as outdated. Likewise, any remedies suggested by Pitcairn were rejected by Sibbald.

Finally, in 1699, a text was produced that met with the approval of the College, and nineteen years after the work had begun, the first edition of the *Edinburgh Pharmacopoeia* was published (figure 10.2).[81] But agreement on the contents of the pharmacopoeia had only been possible by a division within the College; Pitcairn, Stevenson, and the modernizers of their party were suspended from their fellowships. Reactionaries who supported the Sibbald faction were quickly admitted to fellowship of the College. The *Edinburgh Pharmacopoeia* decisively represented the victory of tradition and conservatism over innovation and change.

IRELAND: PHYSICIANS, COLLEGES, AND PHARMACOPOEIAS TO 1654

A college of physicians was founded in Dublin in 1654, seventeen years before the Edinburgh one. A final version of the *Dublin Pharmacopoeia* was only published in 1807, although the College produced a first specimen pharmacopoeia in 1794, 140 years after its foundation and 95 years after publication of the *Edinburgh Pharmacopoeia*. Much of the explanation for the delay lies in the political and religious contexts in which the pharmacopoeias were developed. Like Scotland, Ireland had a long political association with England. But the Irish story is one of conquest and occupation by the English over many centuries. Poynings' Law, passed in 1494 during the reign of Henry VII, declared that the Irish Parliament, founded in Dublin in 1297, could meet only with the king's permission, and that

FIGURE 10.2. First edition of the *Edinburgh Pharmacopoeia*, 1699. Courtesy of the Royal College of Physicians of Edinburgh.

it could not pass laws unless they were previously approved by the king and his English council.[82] Ireland entered into "personal union" with England in 1541, when the kingdom of Ireland was created and Henry VIII was proclaimed king of Ireland.

With the ascent of James I and passage of the Union of the Crowns Act of 1603 both Ireland and England came into "personal union" with Scotland. The founding of the College of Physicians of London in 1518 prompted a group in Dublin to propose forming one there.[83] Paul de Laune, a fellow of the London College since 1618, moved to Ireland as physician to the Lord Deputy. In 1625 Charles I succeeded James I, and the Dublin physicians took the opportunity to petition him. In 1626, de Laune and four other physicians signed a letter to the London College announcing that, following their petition, the king had agreed to found a college of physicians in Dublin "for the repression of unqualified practitioners of medicine."[84] It was to be modeled on the London one, and the Irish physicians asked the London College to send copies of their statutes.

It was to be almost forty years before a college of physicians in Dublin was finally founded, in 1654, initially without a Royal Charter. It was a period of political and religious turmoil coinciding with the English Civil War. In 1641, the Irish Rebellion had begun as an attempt to seize control of the administration in Ireland. It failed and developed into the Irish Confederate Wars from 1641 to 1653. The Commonwealth brought about more peaceful conditions and, in 1654, the initiative to found a college of physicians was taken by Dr. John Stearne, an Irish Protestant physician and a professor at Trinity College, Dublin.[85] Stearne persuaded the Dublin physicians to form a college "for the purpose of regulating the practice of medicine in Ireland." The Dublin College of Physicians was finally granted its Royal Charter in 1667 by Charles II, seven years after his restoration.

Irish Catholics remained a large majority on the island, but Irish physicians were predominately Protestants. Irish Protestants were viewed as colonists, and their Parliament was derided as a subaltern assembly.[86] James II fled to Ireland with a view to staging a return to the throne. The Battle of the Boyne in 1690, between the forces of James II and William III, resulted in victory for William, ensuring continued Protestant ascendancy in Ireland.[87] Their Royal Charter was renewed in 1692 by William and Mary, giving the College authority over physicians, apothecaries, druggists, and others engaged in the sale of medicines. In 1707, following the union of England with Scotland, the Irish Parliament petitioned Queen Anne to endorse a more comprehensive union between Ireland and England. A number of restrictions placed it subservient to the Parliament of England; the College and the country quickly realized that English law might soon apply throughout Ireland.

Initial Attempts to Prepare a *Dublin Pharmacopoeia* in 1717

There was no mention of the need to prepare a pharmacopoeia for the city of Dublin or the kingdom of Ireland in the 1692 Royal Charter, but the issue was first brought before the College in 1717. However, nothing effective was done about it for over four years. There were two main options available: they could adopt either the London or Edinburgh pharmacopoeia, or they could publish their own. The College favored the last option, and on April 13, 1719, the president and fellows again considered "the making of a dispensatory for this City and Kingdom."[88] Dr. Duncan Cumming (variously spelled Cuming or Cumyng) was asked to prepare a draft of the medicines that he thought proper for inclusion in a dispensatory.[89] But the task was a daunting one, and after two years he appears to have made little progress.

In 1721, the fourth edition of the *London Pharmacopoeia* was published, much to the relief of the College. On June 5, 1721, they agreed "till such a work could be perfected, to recommend to the Apothecaries of the City of Dublin the prescriptions of the last new *London Dispensatory* in the making up their medicines."[90] However, the resolution did not make it official in Ireland as it was in England. The apothecaries and surgeons considered that the physicians were attempting to control pharmacy and create a monopoly for themselves.[91] Adoption of the *London Pharmacopoeia* caused disagreement among the College's own members and, on October 31, 1726, a proposal was put "whether a new Dispensatory will be of use to this City and Kingdom." It was unanimously agreed.

The College had clearly given up on Cumming, for it was agreed in January 1727 "that the several Fellows at the next meeting do bring a list of the compositions in the folio edition of the *London Dispensatory* that are proper to make part of the *Dublin Dispensatory*."[92] The fellows seem to have been no more diligent than Cumming, for they did not produce their lists, and the matter was adjourned from one meeting to another for a year. However, on January 22, 1728, a committee was appointed "to prepare the Dispensatory." It was the first of six committees formed and disbanded before the *Dublin Pharmacopoeia* was published. Like the previous initiatives, this committee failed to deliver a dispensatory or even to report on progress. So, on July 31, 1729, yet another proposal was brought forward to the College to revive it, and yet again nothing was done.[93]

The next move came from the Irish Parliament. In 1735, they passed "an Act for preventing frauds and abuses in the making and vending of unsound, adulterated and bad drugs and medicines." It authorized the College to inspect the apothecaries' shops within a seven-mile radius of Dublin, and to destroy any medicines they considered unfit. The apothecaries were then part of a larger guild that included the barbers, surgeons, and wig makers.[94] They petitioned for a Charter incorporating them as a separate guild. The petition was referred to the College of Physicians, who decided that it "might be of use to incorporate the Apothecaries" provided that they were not given powers to make bylaws about the composition of medicines without the sanction of the College.[95] A Royal Charter incorporating the apothecaries of Dublin as the Guild of St. Luke was granted in 1745.[96]

The College had taken no action on a pharmacopoeia since 1729. But in 1746, the fifth edition of the *London Pharmacopoeia* was published, and on August 7, the College resolved "to use the new *London Dispensatory* in prescriptions from the first day of March next."[97] To facilitate this the College appointed a committee "to inspect the *London Dispensatory* and to report to the College the errors thereof."[98] The Dublin edition of the *London Pharmacopoeia* was published later in 1747.

The 1736 act failed to stop abuses, so in 1761 a new act was passed. This authorized the College to produce a pharmacopoeia; they were given power "to frame and publish a code or pharmacopoeia containing a catalogue of drugs or simple medicines as they shall judge necessary."⁹⁹ But the College made no attempt to do so for another twenty-three years. Britain was in a state of almost continuous war, including the Seven Years' War (1756–1763) and the American War of Independence (1775–1783). Thousands left Ireland to start new lives in America, but maintained close links with Ireland. In 1776, the Irish Parliament imposed an embargo on exporting linens and provisions to the colonies.¹⁰⁰ Surrender at Yorktown in 1781 led to concessions being made to Ireland by Britain; through the Constitution of 1782, Ireland gained effective independence. It was the high point of Protestant nationalism in Ireland, but Britain was forced to renounce any right to legislate for Ireland.

The First and Second Specimen *Dublin Pharmacopoeia*, 1784–1805

With the 1783 Peace of Paris, the College felt able to return to the issue of the pharmacopoeia. On June 14, 1784, a third committee was appointed "to take the most speedy and effectual steps toward the preparation of a dispensatory, under the title of *Dublin Pharmacopoeia*, for the general use of this kingdom." In spite of urgent requests the committee made little progress. In June of 1788, the sixth edition of the *London Pharmacopoeia* was published and in October of 1788, the College agreed to again take this as the foundation for a Dublin publication. A few changes were agreed to and a Dublin edition was published. By 1788, the Pharmacopoeia Committee had proceeded in its revision "as far as the article on vina medicata."¹⁰¹ For a time its work proceeded a little more quickly. In January 1791, the College "ordered that copies of the *Dublin Pharmacopoeia* be printed by the Committee and distributed according to their direction."

In June 1792, the Pharmacopoeia Committee reported that it had completed another revision of the text, but still there was no publication. Finally, in 1794 they reported that "they had prepared and printed 100 copies of a specimen which they had distributed among the Fellows and Licentiates of the College, and also amongst the heads of the College of Surgeons, and the Corporation of Apothecaries, and Apothecaries Hall." Until then no one other than physicians had been involved in preparing the pharmacopoeia. A new Committee, the fourth, consisting of Drs. Plunket, Harvey, and Hopkins, was then appointed "to revise the Latinity of the specimen the *Dublin Pharmacopoeia*, and to report such alterations in general as may appear to them proper on August 25 ensuing."¹⁰² The committee completed its work and reported back to the members.

Two years later, in June 1796, it was resolved that a new committee (the fifth) for framing a pharmacopoeia be appointed. The committee was to seek the cooperation of the College of Surgeons in the preparation of "such parts of the pharmacopoeia as relate to external applications." However, the surgeons declined to cooperate.¹⁰³ There was yet further delay; the fellows were engaged in an acrimonious dispute among themselves, resulting from plans for a School of Physic for clinical lectures. Outside events also played a part. The Acts of Union 1801 united the Kingdom of Great Britain and the Kingdom of Ireland to create the United Kingdom of Great Britain and Ireland, with effect from January 1, 1801.

In January 1802, a new committee (the sixth) was appointed, and Drs. Cullen and Percival were asked to cooperate. Percival suggested various changes in chemical nomenclature, and later made suggestions concerning the standardization of weights and measures. In May 1805, the committee reported to the College that the Pharmacopoeia was nearly complete. An order was made that it should be printed "for the perusal of the fellows and licentiates." Dr. Patrick Plunket was asked to write a preface for it, but it was a request "he could not comply with." Despite these setbacks, on August 28, 1805, a copy of the complete work was laid before the College for consideration. This was the second so-called "specimen" version of the *Dublin Pharmacopoeia.*

Publication of the *Dublin Pharmacopoeia*, 1807

Yet another committee (the seventh) was appointed to revise the second specimen, and on March 24, 1806, an order was made "that the College retain in their own hands the copyright of the *Dublin Pharmacopoeia,* and that 500 copies be ordered to be printed." On October 27, 1806, the book was presented to the College, along with a report dated October 18, 1806: "After regular weekly meetings, held during a year and eight months, your committee at length present to the College a *Dublin Pharmacopoeia,* with some confidence, that they have omitted no reasonable point to render it as perfect as lies in their power ... in the names and properties of ingredients for compound medicines they have principally followed the London College, whose pharmacopoeia of 1746 has been directed by our College to be the rule of compounding medicines in Ireland."[104]

This first edition of the *Dublin Pharmacopoeia* was dedicated to George III (figure 10.3).[105] A proclamation directed "all and singular apothecaries and others whose business it is to compound medicines or distil oils or waters or make other extracts within this part of His Majesty's United Kingdom called Ireland, that they and every one of them immediately after the said *Dublin Pharmacopoeia* shall be printed and published, do not compound or make any medicine ... except it shall be by the special direction or prescription of some learned physician in that behalf." It thus applied to the whole of Ireland. Its purpose was spelled out: "medicines are prepared sometimes according to the London formulae, sometimes according to the Edinburgh, and not infrequently at the mere discretion of the Apothecary; some practitioners awaiting the further directions of the College continuing to use the nomenclature of the pharmacopoeia of 1746; some adhere to the latter publication, and not a few prescribe according to

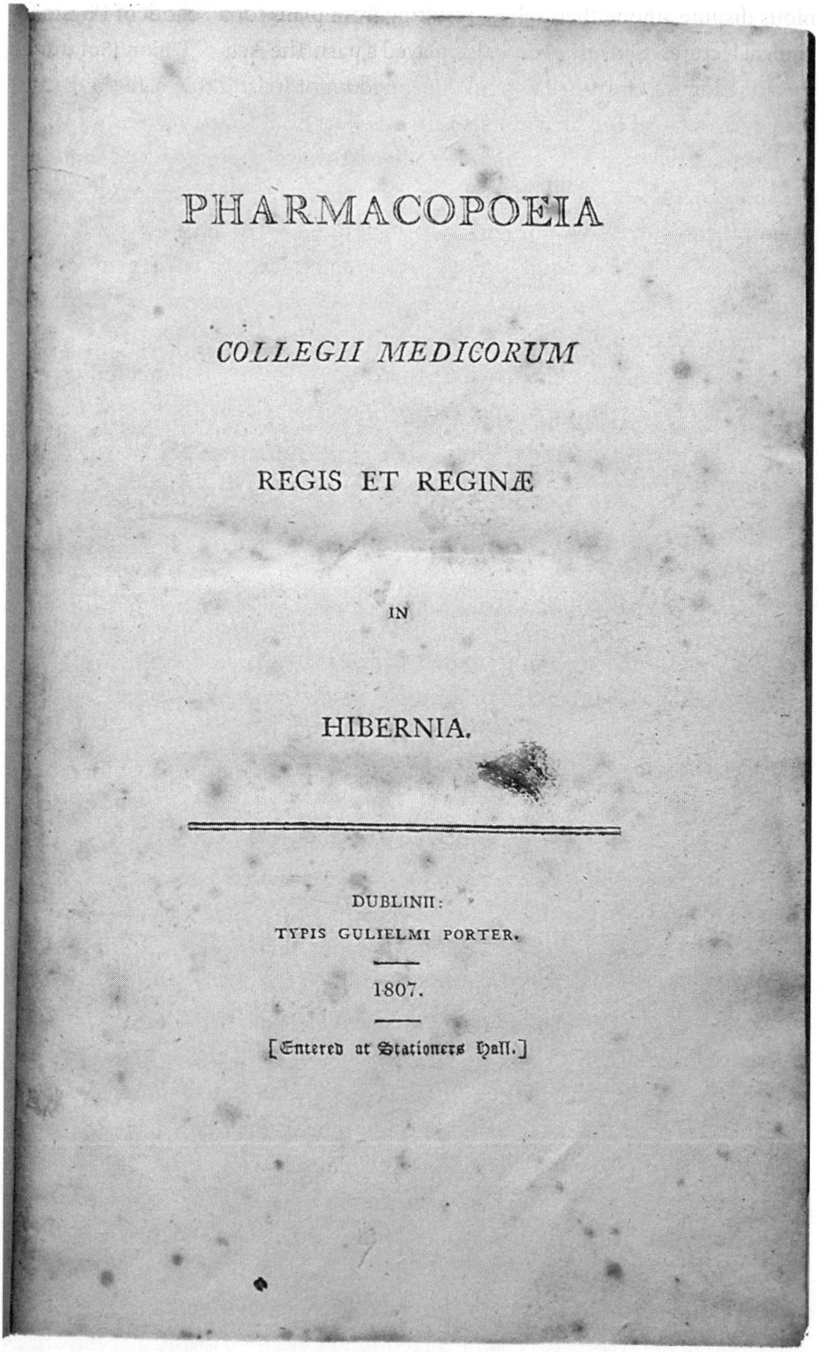
PHARMACOPOEIA

COLLEGII MEDICORUM

REGIS ET REGINÆ

IN

HIBERNIA.

DUBLINII:

TYPIS GULIELMI PORTER.

1807.

[Entered at Stationers Hall.]

FIGURE 10.3. First edition of the *Dublin Pharmacopoeia*, 1807. Courtesy of the Royal College of Physicians of Ireland.

one of the multiplied editions of the Edinburgh College." In fact, a sixth edition of the *London Pharmacopoeia* had been published in 1788 and a seventh was in preparation.

CONCLUSION

This chapter has shown how the *London, Edinburgh,* and *Dublin* pharmacopoeias were all developed against a background of political and religious turmoil. This account of the process involved illustrates the complexity of the hurdles, both internal and external, that needed to be overcome. That they were developed at all must be considered a considerable feat of perseverance and endurance. The processes had much in common; all took many years, usually decades, to develop, and there needed to be a clear purpose for their development, which was usually concern about the quality and lack of uniformity in the medicines being dispensed.

The conditions necessary for the creation of an official pharmacopoeia were formidable; the physicians had first to organize themselves into a college, which needed to be authorized through a Royal Charter; there was a need for civil and amicable relations with other medical practitioners, notably the surgeons and apothecaries, as ultimately their support was required; there needed to be people able and willing to undertake the work involved in compiling a pharmacopoeia, and someone willing to lead the process; widely differing views, beliefs, and prejudices needed to be confronted and reconciled; and the final content needed to be approved by a majority of College members. This needed to be achieved while the physicians themselves were often dependent on royal patronage.

The motives of key individuals played an important part in explaining the origins of the British pharmacopoeias. In London, Thomas Linacre was certainly influenced by the *Nuovo Receptario* he had seen in Italy, but little progress was made in his lifetime. The final push came from Sir Theodore Turquet de Mayerne, a Swiss immigrant who had been expelled from Paris. He was keen to ensure that the *London Pharmacopoeia* should include the chemical medicines he used. The *Edinburgh Pharmacopoeia* clearly owes much to Scottish nationalism and to the early life experiences of Robert Sibbald. The Scottish physicians could have avoided delays resulting from the "malice and obstruction" they experienced in the 1680s and 1690s by simply adopting the current edition of the *London Pharmacopoeia*, as the Irish physicians did for many years. That they chose not to, and that they chose not to delay publication further until after signing of the Acts of Union, tells us much about the attitudes of the Scottish physicians to the English. The Irish too played to a nationalist audience, albeit for different reasons. As an Irishman and a Protestant Robert Percival was anxious to retain a degree of Irish identity in Irish medicine even after the union of Ireland with Great Britain; it would also appeal to Irish nationalists. National identity was a key influence in the development of all three pharmacopoeias.

Medical politics too were an important contributory factor in determining the timing and nature of the pharmacopoeias. While relations with apothecaries and surgeons was a constant theme in the development of all three, internal disputes among the physicians themselves played a huge part in shaping their content and in explaining the delays in reaching agreed versions. Despite the fact that both the surgeons and the apothecaries were organized into guilds or companies the physicians rarely felt it necessary to consult either group about the content of the pharmacopoeias. The *Edinburgh Pharmacopoeia* was entirely the work of physicians.[106] For the first edition there was no input from either apothecaries or surgeons, or any recognition that such advice might be helpful. Indeed, the College continued to be highly critical of the pharmaceutical practitioner for some years. The preface to the second edition, published in 1722, stated that it was being revised "lest . . . through the unskilfulness of apothecaries the life of the patient should be endangered and the hopes of the physician frustrated."[107] Cowen notes that although similarities between the pharmacopoeias were to be expected, there were also considerable differences.[108]

Local traditions and medical philosophies brought to the discussion by physicians had a substantial effect on both the content of the pharmacopoeias and the process by which they were created. But these tensions were different for the three pharmacopoeias, occurring as they did many decades apart. While early differences emerged between followers of Galen and the chemical therapeutics of Paracelsus, later disputes centered on early attempts at rationality embodied in iatrochemistry and iatromechanics. Indeed, the tensions between the different medical philosophies, between tradition and innovation, and between local and "foreign" knowledge, which had to be addressed in developing pharmacopoeias, illustrate a key theme of this book, that of "ways of knowing."[109] They transcend communities separated not only by the Atlantic but also by other barriers to communication and exchange, as noted by Timothy Walker in chapter 5. And the contents of the pharmacopoeias themselves reflect global exchanges of medical materials and healing knowledge that was facilitated by European engagement with the Atlantic World as described in several of the contributions to this volume.

The tensions also neatly illustrate the difficulties of transitioning from one "way of knowing" to another. In his analysis, John Pickstone argues that science, medicine, and technology are linked in mutually dependent ways, and that there have been five "ways of knowing" that have shaped medical practice from the seventeenth century onward.[110] Each has been dominant at a specific point in history, although not necessarily to the exclusion of others. The first, hermeneutics or "world readings," prioritizes nature and its symbolism as a way of understanding illness. This is characterized by the ideas of Galen, which were the established truth taught in medical schools and reinforced in medical practice. This approach was displaced during the late sixteenth and seventeenth centuries by a second "way of knowing," Enlightenment rationality, where illnesses were viewed as explainable and manageable rather than supernatural.

The clash between the supporters of Galen and those of Paracelsus with his revolutionary ideas about chemical therapeutics in developing the *London Pharmacopoeia* is an early example of the tensions associated with the transition between the first and second ways of knowing. Pickstone's third "way of knowing," analysis, represents increasing rationalization in medicine, and attempts to explain the working of the human body and compounds used to treat illness. In the case of the *Edinburgh Pharmacopoeia* the supporters of Galen triumphed over the exponents of iatromechanism and other attempts at rationalization. The fourth way of knowing, which Pickstone terms "experimentalism," involves trial and error in developing optimal ways of treating disease. We see the beginnings of this in the development of the *Dublin Pharmacopoeia*. Thus, a focus on the origin and processes of pharmacopoeia development offers new insights into the creation of healing knowledge in the early modern world.

Recognizing the complexities of bringing an official pharmacopoeia into existence throws light on the huge significance of social and political context. The development of pharmacopoeias can be seen as attempts to coopt or assimilate knowledge as well as a mechanism for the control of one medical profession by another. Contingency played a crucial role in the production of official pharmacopoeias, and perhaps not too much "grand design" should be read into these texts. The *London, Edinburgh,* and *Dublin* pharmacopoeias were products of their time, and ultimately all three were replaced with publication of the first edition of the *British Pharmacopoeia* in 1864.[111]

11

THE CODEX NATIONALIZED

Naming People and Things in the Wake of a Revolution

ANTOINE LENTACKER

> *Where does the dialect spoken in your province depart most clearly from the national idiom? Is it not particularly for the names of plants, diseases, the terms of the arts and trades, of tilling instruments, of the varieties of grains, of commerce and of customary law? We would like to have this nomenclature.*
>
> ABBÉ HENRI GRÉGOIRE,
> *Enquête sur les patois*, Question 6 (1790)

THIS CHAPTER is about France's first national pharmacopoeia, the *Codex Medicamentarius, sive Pharmacopoea Gallica* of 1818. The "Codex," as it came to be known, was by no means the first pharmacopoeia published in France, but it was the first to be given the full force of law—of a law that, in the wake of the Revolution, was to apply uniformly across the nation's territory. As Alexis de Tocqueville famously argued, the centralizing and standardizing agenda of revolutionary governments had deep roots in the bureaucratization of the early modern monarchical state. Tocqueville saw the Revolution as the culmination of, rather than a clean break with, the deep tendencies of the regime it abolished. Nonetheless, the half-century between 1770 and 1820 remains a moment of crucial inflection in the long history of practices and technologies of state power in France. The story of the Codex of 1818 offers a lens onto the nature and logic of this inflection. It encapsulates in uniquely transparent ways the changes that the Revolution brought about in the governing of people and things.

THE CODEX NATIONALIZED

By tying together the words and substances of pharmacy in firm and explicit ways, pharmacopoeias aimed to create standardized pharmaceutical vernaculars. As such, the genesis of the French Codex must be situated within the Revolution's far-reaching project of reforming language. Forging a democratic polity required reliable communication channels between the people and their representatives. Seamless communication between center and periphery was needed to publicize the nation's laws, to explain revolutionary policies and gain support for them, to consult the people, conduct trials by popular jury, and more generally to build a country administered by and for the people. A reformed language, in this respect, was always more than a mere tool to govern the new Republic; it was the medium of its existence. A precise and concrete idiom grounded in the empirical world, firmly tethered to nature and to thought was an ideal of the Enlightenment. On the other hand, of the many habits and customs revolutionaries sought to reform, language was perhaps the most intimate and the most deeply ingrained. Since language, too, governs the way people relate to each other, to themselves and to the world, it is not itself governed—or not easily. For this reason, the first few years of the Revolution witnessed the heyday of a form of linguistic federalism, as local societies translated the laws voted by the National Assembly, reported on the events in Paris, and carried out in local languages the political pedagogy needed to breathe life into the Revolution. Translation offered itself as a method to overcome the obstacle that idiomatic diversity posed to the revolutionary project.

Following the radicalization of the Revolution, however, the nation's leaders changed course. Under the Terror, multilingualism and the uncertainties of translation were recast as a political problem. On 16 Prairial Year II (June 4, 1794), the Abbé Grégoire presented his "report on the necessity and the means to eradicate local dialects [*patois*] and to universalize the use of the French language" to the nation's representatives. The report was based on responses to a national survey that the Jacobin abbot had been conducting among local notables since 1790. As Michel de Certeau showed, both the inquisitor and his informants shared a vision of local dialects as languages of nature and passions, expressive rather than descriptive, unmoored by a written corpus and hence tethered to local particularisms. As mirrors of local idiosyncrasies, regional languages had to be extinguished and a standardized national language imposed in their place. That way, the relation between words and things would be fixed, misunderstandings made impossible, and the government given a firm grip on the people, things, and situations it named. Political unification was equated with semiotic unity and a rational Republican government conceived as a universal semiotic communion.[1]

In the end, the Revolution's actual accomplishments fell short of these grand ambitions. The project to open French-speaking schools in every commune was not carried out until the 1830s. No state-commissioned dictionaries or grammars saw the light of day. Yet the vision remained influential, and, if not for the language as a whole, in a number of more circumscribed domains the revolutionary and postrevolutionary state did go on to legislate over the name of things, claiming a monopoly over the power to give, or at least to fix and to police, appellations. One such domain was pharmacy—the domain located at the intersection of what Grégoire called "the names of plants, diseases, the terms of the arts and trades"—where the government claimed an exclusive right to name the drugs that would be allowed to circulate within France's borders. The Codex of 1818 became for pharmacy the binding and state-sanctioned dictionary that revolutionary governments had envisioned, but never written, for the French language. Another, however, was people, whose names were admittedly not given by the state, but legislated and regulated in new and stricter ways than they had ever been before.

In what follows, I link the measures taken to organize the circulation of people and drugs before and after the publication of the first Codex. Both people and drugs were remarkably elusive entities, hard to track and to pin down, inscrutable and often deceptive. Since in both cases the link between names and the named was so notably tenuous, unstable and always ready to unravel, bringing them together illustrates how control over the naming of the governed was deployed as a tool to govern them. To do so, I draw on a rich literature on the role of paper and paperwork in the building of modern state power.[2] The policing of both people and drugs involved the production of what I propose to call "paper doubles"; that is, of documents that identified and allowed them to circulate lawfully. As will become clear, however, the two systems of registration met with different fates. By comparison with the state's effort to account for people, its effort to account for drugs was largely a failed one. Thus, if the story of the French Codex provides a lens onto the nexus of language, print, and territory in modern state projects, it is one that highlights the limits of state power as well as the scope of its reach.

LANGUAGES AND TERRITORIES OF PHARMACY IN EARLY MODERN EUROPE

A critical reflection on the language of pharmacy is in some ways consubstantial to pharmacy itself. In Christian Europe, this reflection dates back to the early centuries of the second millennium, as apothecaries reappeared in the contact zones with the Muslim world and spread northward to most medieval cities of the continent. The division of labor between prescribers and preparers of medicines created the need for a shared vocabulary ensuring that prescriptions be read and executed as their authors intended. There were moral conditions to the purity of drugs. Early pharmacy regulations required the apothecary to be a reputable member of his community, "of clean appearance," according to Michael Stainpeis, "not given to drinking, gluttony, carousing, or whoring, but a man of decency, not too young or too vain, no gambler [. . .] diligent and meticulous, obliging, honest, god-fearing and scrupulous, righteous, just, and pious, humble in his dealings with the poor and conscientious with everyone."[3] But there were semiotic conditions as well. Given the instability of mineral and botanical classifications, the wide local variations in the tools and terms of the trade, and the special dangers of ingesting powerful poisons, common codes of pharmaceutical communication written out and agreed upon between those who prescribed and those who executed prescriptions were of equal importance in making the delegation of drug preparation transparent and reliable. A pure character was not enough; the purity of drugs also depended on a purified language.

Where dictionaries used definitions, pharmacopoeias relied on recipes. Explicit rules for the preparation of drugs were meant to firm up the relation between signifiers and signified substances, guaranteeing that drugs bearing the same name be made of the same ingredients and following the same procedures. Yet, as Paula De Vos's survey of the "textual tradition of Western pharmacy" makes clear, the multitude of texts that composed it both served and undermined that purpose (chapter 1).[4] As a tradition, it was never reproduced without some degree of revision and reinvention, so that many of these texts ended up giving different names to similar drugs or, even worse, similar names to different drugs. In certain cities, the medical faculty, apothecaries' guild, or municipal council designated one classic dispensatory or ordered the drafting of a new one to serve as sole admissible reference. Whether Florence's *Nuovo Receptario* of 1499 or Nuremberg's *Dispensatorium* of 1546 deserves the title of the first modern pharmacopoeia in the sense of a legally binding dispensatory intended to ensure the uniformity of drugs within a certain jurisdiction—long a passionately debated question in the history of pharmacy—matters less than their common anchorage in the context of the Renaissance. Both were authored by humanistic physicians intent on reforming pharmaceutical language through a return to the purity of origins rediscovered in ancient Greek sources; and both were printed under the auspices of municipal authorities. In the era of the manuscript copy, the notion of a same text guaranteeing the uniform production of drugs over an extended territory was bound to remain elusive. A technology for the reliable mass production of identical texts was from the beginning a condition of possibility for the mass production of identical drugs.[5]

The term "pharmacopoeia" itself was a creation of the Renaissance, a Latin neologism made from Greek roots whose first known occurrence was in the title of a 1548 dispensatory published by Jacques Dubois in the French city of Lyon.

Dubois's opus was not an official publication. Calls for official formularies in France, especially in the kingdom's capital, became more insistent after the middle of the sixteenth century. In 1590, the Parlement of Paris ordered the medical faculty to draft such a formulary for use in the city's pharmacies. In 1638, the first edition of the *Codex Medicamentarius seu Pharmacopoea Parisiensis* came off the press. The book's preface lamented pharmacists' indiscriminate reliance on the medieval dispensatories of Mesue or Nicholas, the pharmacopoeias of Bauderon or Valerius Cordus, or on Guybert's *Médecin charitable*, all of which offered different recipes for the same medicaments. Others, "without discernment and judgment," it added, "executed their operations according to their own whim, some having grown so bold and impudent as to add or remove certain substances to or from recipes of traditional remedies accepted everywhere and by everyone, and make for themselves and their aides their own special and personal pharmacopoeia."[6] However good the recipes, a pharmacopoeia could not be special and private. Only as a shared and public reference would it fulfill its goal of codifying a stable lexicon in place of endlessly varying local pharmaceutical dialects.

A code in the semiotic sense—a system of signs—became a code in the legal sense—a system of rules—by being bound to a particular territory. The horizon of early modern pharmacopoeias was as broad as it was vague. Elaborated by humanists through a critical appropriation of Greek sources selected by tradition and endowed with the authority of ancient texts, they were intended to possess a validity transcending time and place. Nonetheless, their concrete implementation remained narrowly local. As William Ryan showed earlier in this volume, the publication of *materia medica* entailed a process of compilation, incorporation, translation, edition, and therefore also erasure of a multiplicity of heterogeneous pharmaceutical traditions, both learned and vernacular (chapter 6). The printing of a pharmacopoeia ratified the final selection, conferring upon it the character of timeless coherence that official sanction could bestow. This process was eminently political, and in each locality the decision to adopt an existing dispensatory or to produce a new one reflected a local balance of power between scholarly and artisanal economies and epistemologies.[7] When the Parisian Codex came out in 1638, several cities across the French kingdom had already adopted their own official dispensatories. Lille, Lyon, and Blois had theirs by 1573, 1628, and 1634, respectively; while Bordeaux's and Toulouse's came out in 1643 and 1648, shortly after Paris's.[8]

The fragmented structures of the early modern economy constricted the power of official pharmacopoeias to reform and standardize pharmaceutical practice. The Parisian pharmacopoeia was originally commissioned after a royal ordinance of 1573 required that an official formulary be drawn up in every city that was home to a faculty of medicine. This represented only a handful of cities in early modern France. In others, the dispensation of drugs was overseen by one of the kingdom's three hundred apothecaries' guilds. City limits generally coincided with the limits of the market claimed by university-trained physicians and apothecaries. In rural areas, drugs were prepared at home, by local healers, or sold by peddlers. Even within cities, the services of physicians and pharmacists were affordable only to a few, so there was a general tolerance, for instance, for drugs provided charitably by religious institutions. Finally, as Justin Rivest's chapter in this volume discusses, the king granted privileges and protections to various makers of secret remedies, allowing a growing number of medicines to exist with official sanction outside the purview of official formularies (chapter 4).[9] In these circumstances, pharmacopoeias mapped only fragments of the world of drugs, most of which remained unaccounted for in authorized formularies or medical prescriptions.

The Babel of pharmaceutical codes resulting from the multiplication of municipal dispensatories prompted the design of new kinds of pharmacopoeias. Nicolas Lémery's *Pharmacopée universelle* of 1698 aimed for universal applicability precisely by relinquishing any pretention to official character: "Some will certainly find fault in the scope of this pharmacopoeia, which includes many descriptions uncommon or unused in Paris; yet since my intention was to make this work useful in all countries in which medicine is practiced, I deemed it proper to describe as much as could be done the preparations encountered in dispensatories, so that every man could find in it what suits him without having to consult other pharmacopoeias in search of what he needs. Since sentiments vary on these subjects, certain compositions gain currency in certain cities that are not found in others."[10] Being a pharmacopoeia tied to no particular place, in other words, it was to be fit for use anywhere. Meanwhile, the gradual transfer of power from local jurisdictions to the central administrative apparatuses of absolutist monarchies gave rise to the idea of a national pharmacopoeia, a single official formulary that would take the place of the myriad, more or less overlapping formularies of cities, provinces, and other self-governed communities that made up early modern kingdoms. Stuart Anderson's chapter in this volume linked the production of the *London Pharmacopoeia* of 1618—the first to be effectively promulgated (if not enforced) for a nation as a whole—to the assertion of monarchical sovereignty in the wake of the English Reformation (chapter 10). On the continent the first national pharmacopoeias were commissioned and published by the so-called enlightened despots of the eighteenth century: in Brandenburg in 1698, the same year as the publication of Lémery's universal pharmacopoeia, or in Austria in 1774.[11] The Madrid pharmacopoeia of 1739 was declared binding throughout Spain. In France, things followed a different course. In some areas such as taxation and justice, the monarchy pursued a centralizing agenda but met with growing resistance. In many others, it opposed rather than embraced the reformatory projects inspired by the political philosophy of the Enlightenment. As such, many of the kinds of reform pushed through by neighboring monarchies—including the promulgation of a national pharmacopoeia—were not undertaken in France until the time of the Revolution.

The French Codex was thus a latecomer, but therein lies its value as a case study. Enlightened despotism was an effort by Old Regime monarchs to reform their rule in order to perpetuate it. By contrast, the French Revolution was perceived both in and outside of France as the most far-reaching attempt yet to tear down the social and political structures of the Old Regime. Moreover, the Revolution opened up the doors of power to scientists. At no time in French history, before or since, did scientists become so directly involved in the government of their country. As Antoine François de Fourcroy, the future architect of the postrevolutionary "medico-pharmaceutical police," put it in a 1797 address to the newly founded Free Society of Parisian Pharmacists: "One could not encounter more fortuitous circumstances and a more perfect opportunity than the reorganization of our society under the auspices of liberty and of a republican constitution to give pharmacy a new luster." The overhaul of pharmacy was to be part and parcel of the remaking of the social order as a whole. And both were to be guided by the lights of science. In the same speech, Fourcroy linked the rebirth of pharmacy to a new way of communicating defined by "a style inherent in the things themselves, consisting solely in the purity of language and the precision of words."[12]

ENCLOSING THE WORLD OF PHARMACY IN A BOOK

In March 1791 France's National Assembly legislated the guilds that traditionally oversaw the practice of urban crafts and trades out of existence. To the nation's new representatives, guilds fit squarely within the architecture of Old Regime society, corseting the nation's economy in a system of arbitrary privileges. Like all other trades medicine and pharmacy were pronounced free. Titles, licenses, and hierarchies were struck down, and with them all the rules and regulations they imposed on their members. Upon payment of the tax required to set up shop, everyone was henceforth entitled to call themselves physician or pharmacist, and to sell or advertise medical goods and services, including drugs of one's own invention. Quickly, however, revolutionary lawmakers harbored second thoughts about the benefits of unrestrained freedom of enterprise in medical matters. Pharmacy was exempted from the antiguild legislation as early as April 1791, a mere month and a half after its passage. In the following years a series of ad hoc measures were taken to ensure continuity in the training of apothecaries until new rules could be agreed upon.[13]

By then, Fourcroy had ascended to the top of France's scientific and political worlds. The son of a pharmacist, Fourcroy was trained as a doctor and rose to prominence as a chemist. Lavoisier's death at the guillotine in 1794 made him the leading exponent of the new chemistry in France. The same year, he became a member of the revolutionary government (the Comité de Salut Public) and remained chief advisor to subsequent governments on matters of medical legislation and public health policy. It was as state councillor under Napoleon's newly established regime that he authored the laws of 19 Ventôse Year XI (March 10, 1803) on medicine and of 21 Germinal Year XI (April 11, 1803) on pharmacy. In typical Napoleonic fashion, both laws restored order and hierarchy by means of a compromise between Old Regime customs and new revolutionary principles. In a return to prerevolutionary practice it required physicians and pharmacists to be trained and licensed men of the art. But instead of being coopted by one's masters according to the special rules of the guild, aspiring professionals were to graduate from national universities under control of a central Ministry of Public Instruction. Fourcroy's legislation provided for the training, licensing, and oversight of the personnel needed to extend healthcare services to the entire population. As such, it also came with a new resolve to prosecute those who practiced without the required degree. Unlicensed practice, once widely tolerated, was from then on equated with unauthorized practice and explicitly criminalized.[14]

Under the law of Germinal, there were two ways for a drug to be legal. Any medicine in circulation had to be traceable to a formula contained in the prescription of a physician or in a soon-to-be-published national pharmacopoeia. Drugs compounded according to a prescription were called magistral; those prepared according to the pharmacopoeia were called officinal.[15] Either way, drugs were to be dispensed with a label bearing the medicine's name (or, for magistral drugs, the number under which the prescription had been copied in the pharmacy's prescription ledger) alongside the name and address of the pharmacist, so that each individual drug in circulation could be connected back to the place where its formula was archived. The drafting of the pharmacopoeia was entrusted to a commission of nine members of Paris's medical faculty and two members of its pharmaceutical faculty, Nicolas-Louis Vauquelin and Edmé-Jean-Baptiste Bouillon-Lagrange, both former students of Fourcroy. The commission's mandate was to draw up a list of standard remedies that pharmacists would be required to make and hold in stock, alongside instructions on how to prepare them as well as test their purity. The commission's work took longer than expected, but in 1818 the national formulary finally came out.

The new Codex was the true cornerstone of the new pharmaceutical regime. In form and content, little distinguished it from the Parisian Pharmacopoeia that preceded it. Its title was almost unchanged—it had become "The Book of Drugs, or, The French Pharmacopoeia" in lieu of "the Book of Drugs, or, the Parisian Pharmacopoeia"—and so too was its general appearance. Both volumes were thick, leather-bound quartos written in Latin. They opened with the traditional preface celebrating the progress of chemistry and medicine since the publication of the prior pharmacopoeia, depicting the evils of variation and inconsistency in pharmacists' preparations, allowed to spread once an outdated pharmacopoeia had fallen into disuse, and promising a new and up-to-date formulary that would restore "security for the sick, certainty for physicians, and a consistent standard to follow for pharmacists," in the words of the Parisian Codex, or ensure "that drugs be prepared according to the same method and remain identical to themselves always and everywhere," in those of the 1818 Codex. As the genre of the pharmacopoeia required, they consisted of two main sections: a list of materia medica (the mineral, vegetal, or animal substances that form the basic ingredients of drugs), followed by a compendium of a few hundred formulas for the preparation of approved medicines combining these ingredients. Each volume ended with an alphabetic index so as to allow users to retrieve the authorized recipe associated with the name of any common remedy.[16]

Yet continuity in the Codex's appearance and contents masked an entirely new relation between the book and the space it governed. The Parisian pharmacopoeia was made by the local medical faculty for the local guild of apothecaries, one of approximately three hundred self-policed corporations of apothecaries in the kingdom. Royal edicts ordered Parisian and only Parisian pharmacists conform to it, so that the Parisian Codex covered only a fragment of the nation's territory, and within that local space only a fraction of the drugs on the market. The Codex commissioned, sanctioned, and promulgated by the postrevolutionary state, on the contrary, cast its net over the entire national territory. With the requirement inscribed in the law of 1803 that no drug circulate within France's borders that could not be traced to an authorized and recorded formula, it aimed to produce uniformities on an entirely new scale. As was the case with language more broadly, the vernacular used and codified in the capital was imposed as the national vernacular, the standard against which diverse practices across a large territory had to be measured and could be recast as orthodox or divergent, or even as devious and hence punishable. The instrument for the transformation of the capital's idiom into the national idiom was a centrally governed schooling system—whether elementary school for everyday language or university for the language of pharmacy—and although the king's court in Old Regime France had already served to disseminate Parisian standards of language and taste throughout the country's nobility, the building of a national schooling system in order to spread these standards down the social scale was a typically revolutionary project.

The publication of the Codex was itself a carefully regulated and ritualized process. A royal ordinance issued on August 8, 1816, in anticipation of the Codex's release informed pharmacists that eight thousand copies of the new volume would be printed, each bearing the stamp of the medical faculty of Paris, the original signature of the dean of the school of medicine, and the cipher of the editor. "Copies not bearing these authenticating marks," the ordinance added, "will be deemed counterfeited."[17] Pharmacists were given six months from the time of publication to acquire their copy and conform to its standards. Each copy authenticated by the authorities and purchased by the pharmacist was to function as a paper mirror of the pharmacy itself, with all drugs stocked in the pharmacy also registered in the Codex, and conversely all drugs described in the Codex also stocked in the pharmacy's store. Additional copies were deposited in the Royal Library, in local public health offices, and in pharmacy schools.

Pharmaceutical education, in some sense, was about inscribing the book of pharmacy in the body of the pharmacist. If written instructions resulted in the production of uniform drugs across pharmacies, it was only because pharmacists shared a common embodied understanding of how to act on them, a tacit knowledge acquired in practical exercises, in the choreographed repetition of the gestures involved in the execution of recipes. The efficacy of the new pharmacopoeia, therefore, was inseparable from the standard, university-based pharmaceutical training introduced to replace the shop-based apprenticeships through which apothecaries used to be made and recruited into the trade. As a result, the pharmacopoeia's influence reached far beyond the realm of officinal remedies per se. By describing the basic ingredients and codifying the basic theoretical and practical vocabulary of the profession, it also determined how pharmacists read and executed physicians' magistral prescriptions. Its planned distribution and compulsory presence in the places that mattered maintained a link between central state, academic elite, and pharmacies dispersed across the nation, tying the hands of practicing pharmacists in the provinces to the academic minds that governed the profession from Paris.

Despite the fifteen years of work that went into its preparation, the publication of the first national pharmacopoeia in France did not escape controversy. Decisions to add or exclude particular remedies in the Codex carried significant practical and symbolic implications. They determined which drugs would be allowed into free circulation, but also declared with official sanction what counted as essential or inessential, what was worth knowing in the pharmaceutical arts and what was not. With physicians representing nine out of eleven members of the pharmacopoeia commission, many of these decisions were met with pharmacists' discontent. The *Journal de Pharmacie* published a scathing review of the commission's work signed by, among others, Bouillon-Lagrange, one of the two pharmacists to have sat on the commission. It noted, for instance, that the recipe of the sarsaparilla syrup, "one of the most commonly used," was nowhere to be found, whereas four different recipes were given for opium extract, one of which had fallen out of use and the three others produced different drugs among which pharmacists would be unable to choose. Overall the article fastidiously inventoried over one hundred inconsistencies of classification and inaccuracies of description, betraying pharmacists' resentment at the fact that the Codex— "their code, their law, their gospel," as the *Journal* called it—had been written mostly by physicians too far removed from the practice of their art.[18] Foreign reviewers in particular noted that, while several members of the Codex commission were distinguished scientists who had contributed to the new chemical science initiated during the French Revolution by Lavoisier, the formulary they produced failed to break with the antiquated traditions of baroque polypharmacy that characterized early modern formularies such as the Parisian Codex.[19]

The practical circumstances of the Codex's publication aroused protests as well. In June 1816, the government sold the rights to the Codex for forty thousand francs to the Parisian printing house of André-François Hacquart. The *Journal de Pharmacie* deemed the eighteen francs Hacquart asked for the volume excessive, denouncing it as an illegitimate tax to be paid to "the favorite printer of an ex-minister." The pharmacopoeia was a public document, prepared precisely to delineate a new public domain of drugs. How, then, could it have been sold away to a private entrepreneur bent on exploiting it for his personal profit? Tensions came to a head when Hacquart filed a police complaint against Julien-Joseph Virey in 1819, following the publication of the new edition of Virey's *Traité de pharmacie* in which a number of the Codex's formulas were reproduced. Upon Hacquart's request the police seized all copies of Virey's handbook pending trial. In court, however, Virey produced multiple examples of formulas that the Codex itself had borrowed from older books, including prior editions of his own treatise. Hacquart's case was thrown out in October 1819. In 1837, the rights to the second edition of the Codex, thoroughly revised and written directly in French rather than Latin, were ceded instead to J.-B. Baillière, the unofficial publishing house of the Paris School of Medicine.[20]

DOCUMENTING PEOPLE AND THINGS

In being nationalized by the new regime, the pharmacopoeia followed the same fate as a number of other paper tools to which it can be compared. As James Scott demonstrated in *Seeing Like a State*, the formation of the modern state involves a subtle dialectic between knowing and reforming. To make the disorderly social world legible and manageable, the government produces simplifying representations of it, but it also seeks to produce a simpler and more orderly social reality that conforms to the schematic representations it makes to account for the governed. In this dialectic, the giving and policing of names was the most basic of all operations. In order to firm up its hold over things and people, the government claimed control over the documents in which the governed registered and represented themselves—maps and cadasters for land, birth and death records for people, or pharmacopoeias in the case of drugs. Tightening the state's grip on the naming of the governed was the first and most fundamental step in extending the reach of the law over the land and its inhabitants.[21]

The state's efforts to register and track drugs circulating in its territory overlap in remarkable ways with measures taken to register and track people present within the country's borders. Shortly after guilds were abolished, the Constitution of September 3, 1791, suppressed passports. During the Old Regime, passports were documents issued not solely by the royal government but also by cities, guilds, or other recognized corporations in order to authorize, protect, and keep track of the movements of their subjects.[22] Deputies of the constitutional assembly denounced them as a symbol of the system of arbitrary privileges interfering with people's fundamental rights, the inability to cross territorial borders functioning as a powerful metaphor for the inability to move across social boundaries. Yet, exactly as had been the case with drugs a few months prior, the unrestrained freedom to circulate was curtailed almost as soon as it was promulgated. With growing numbers of vagrants gathering at cities' gates, passports were reinstated on February 1, 1792. On September 20 of the same year, a system of civil registration entrusted to local elected officials (the *état civil*) was instituted to record the births, marriages, and deaths of all individuals present in the national territory. The new legislation on identification and surveillance underwent multiple revisions in the context of the revolutionary wars. The law stabilized under Napoleon's rule, requiring not only of foreign nationals but of anyone traveling away from their department of birth to carry identification documents. Finally, the law of 11 Germinal Year XI (April 1, 1803) on the immutability of names—adopted ten days prior to the pharmacy law of 21 Germinal—fixed for decades to come the rules governing the attribution, transmission, and—only in rare exceptions—modification of personal names.[23]

The état civil was in effect a nationalized version of parish records. In 1667, a royal ordinance gave Catholic parish records official status, but in 1685 Louis XIV also abolished the Edict of Nantes, which had instituted a regime of tolerance toward French Protestants. Thereafter, Protestants were no longer being registered. Jews, whose numbers had greatly increased following Alsace's annexation to the kingdom in the same period, never had been. Taken over by the central state, then, national civil records were to recover and register the unaccounted for, the groups that used to slip into the interstices of the checkered jurisdictions of early modern France. Under Napoleon's rule, leaders of Jewish communities, in particular, were asked to assign surnames to their members.[24] No resident was to be left without a first and last name properly attributed and archived in official documents. In a world in which identification still relied on mutual acquaintance in relatively stable communities—and in which alien and stranger meant essentially the same thing—papers were required of everyone who traveled out of their communities of origin. That way, they could be set apart from the *sans aveu* (literally, the "unvouched for"), persons attached to no community and whom no one could or would speak for.

In the decade that separated the beginning of the Revolution from its conclusion under Napoleon's solidifying rule, thus, laws on the naming and registration of people and drugs evolved in step and exhibited a same technopolitical logic. As civil records and passports did for residents, prescriptions and the pharmacopoeias created a paper double of each drug in circulation, ensuring that no drug would be found in France that could not be linked to an archive that identified it and legalized its presence. As a French prosecutor put it in the 1850s: "A remedy's passport, so to speak, is found in the Codex."[25] Birth records, passports, prescriptions, and pharmacopoeias all had their roots in the Old Regime. They were documents that used to be produced by self-policing corporations such as churches, guilds, or communes in order to stitch together the partly overlapping and partly disjointed authorities, territories, and jurisdictions of a vast early modern kingdom. As a centralized state bureaucracy seized the powers once vested in these self-policing communities, it appropriated these paper technologies and transformed their meaning. Henceforth, the authority of official documents was to derive from the same source and extend their rule homogeneously throughout a seamless national space.

In the case of drugs, the sans aveu to be registered or removed included the many drugs made and peddled by unlicensed healers, whether lay or religious, local or itinerant, to those who had no access to urban medical markets. But most of all, they included the so-called secret remedies that the monarchy had continued to tolerate or to authorize by special privilege. As Colin Jones demonstrated, the trade in such drugs, whose formulas were to be found neither in pharmacopoeias nor on prescriptions, was flourishing as the Revolution started. Jones linked the budding trade in secret remedies on the eve of the Revolution to the rise of a new entrepreneurial spirit, the vitality of the written press in which secret remedies were advertised, and to a web of prevalent metaphors associating the regeneration of the individual body with that of the body politic. The collapsing scaffold of privileges that used to govern trade in the Old Regime facilitated their spread in the early 1790s. In some broad sense, the same disruptions were behind the multiplication of undocumented drugs and undocumented people in the chaotic early years of the Revolution.²⁶

Secret remedies loomed large on the agenda of medical reformers during the Revolution. Félix Vicq d'Azyr, secretary of the Royal Society of Medicine and Fourcroy's onetime mentor, insisted in the *Nouveau plan de constitution pour la médecine* he submitted to the National Assembly in 1790 that

> nothing is more dangerous than secrecy in the realm of remedies. The most valuable among them remain noxious as long as they are shrouded in the veils of mystery. Mystery excites enthusiasm and maintains the credulity of the people; it brings uncertainty in the knowledge of circumstances and inaccuracy in the application of a remedy that is used without being known. The Society's records contain abundant evidence of the harmful effects and, we do not shy from saying so, of the poisonings caused by remedies whose benefits were in some cases touted by famous physicians and citizens of all ranks.²⁷

Fourcroy, too, made the metaphor of transparency the dominating theme of his reform of pharmacy. In order to distinguish themselves from mere traders and craftsmen, he urged, pharmacists had to stop offering "mysterious remedies [. . .] to the credulous anxiety of the sick," and commit instead to the open process of experiment, publication, and replication that defined membership in a scientific community.²⁸ A sweeping ban on secret remedies, first recommended by the National Assembly's Public Health Committee (Comité de salubrité publique) in May 1791, eventually became a key provision of the law of Germinal. As had been the case with people, it was decided that any loophole through which unidentified and undocumented drugs came into circulation had to be closed. The new pharmaceutical police would be founded on firmly tightening the links between names, things, and bodies.

In both cases, the governmental logic was the same; naming drugs or people amounted to holding them under the grip of the law. Self-given, unregistered, and thus easily changeable names were always likely to conceal rather than reveal true identities. They functioned as effective shields against the searching gaze of the state. Names regulated and registered in official archives, by contrast, attached the governed to fixed identities and genealogies. Wresting the power to police the naming of people and things from Old Regime intermediary bodies was therefore key to the formation of the modern state. Provided they were officially fixed, names accounted for the named and made them accountable, ensnaring things and bodies in a net of signs firmly in the hands of a central power. Inscrutable, elusive, and potentially dangerous substances demonstrated this with unique clarity. Drugs had to be given their proper names to be kept in their proper places, to ensure that they remained where they belonged.

KNOWING DRUGS BY THEIR NAMES?

Of course, there remained significant differences in the paper technologies used to register people and drugs, and these differences accounted in large part for the diverging fates of the two registration systems. The identities that the state ascribed to people were individual. They were captured in proper nouns whose function was to establish the singularity of each person and distinguish her from others. The identities ascribed to drugs, on the other hand, were generic. While each drug in circulation was in some ways individualized by its label, the pharmacopoeia assigned a same name to drugs of the same kind—that is, drugs prepared with the same ingredients according to the same replicable recipe. Indeed, only secret remedies tended to be known by proper nouns, usually those of their inventor or proprietor, and were prohibited precisely because such names were regarded as dissimulating rather than disclosing a drug's true nature. This fundamental difference was reflected in the material form of the archives produced to regulate the names of people and drugs. Civil records were by nature decentralized and open-ended; they grew in step with the number of individuals they recorded, a paper double being created for each new individual born on the national territory. The pharmacopoeia, on the contrary, was a closed nomenclature containing a predefined number of common nouns, a finite set of categories that ought to contain all drugs in circulation on the national territory.

The closure of the Codex decisively altered the relations between word and world. During the fifteen years it took to draft it, members of the pharmacopoeia commission strove to produce a faithful inventory of the shared customs and usages of French pharmacy, noting in their preface that they decided against eliminating those names or formulas that were less than fully adequate but had become entrenched in the habits of physicians, pharmacists, or the public.[29] Once printed and published, though, the customs inventoried became norms as pharmacists were expected to abide by them. In this way, the Codex contributed in fact to producing the world it purported to describe. Like other official maps, grammars, or dictionaries, pharmacopoeias were at once descriptive and prescriptive; they reformed a messy social world as they recorded it, making it at once more orderly and more legible to the remote eye of an outside observer or ruler. Drugs, in other words, were expected to mirror their paper doubles rather than the other way around.

Given the closure of the Codex, the question arose of how to admit newly invented drugs into the circle of approved remedies. This was a problem of a very different nature from that which arose from the creation of civil records. Gérard Noiriel has shown how long it took to get local officials to record births, marriages, and deaths in accordance with the requirements of the decree of 1792. For decades after the Revolution, ministry officials complained of the lack of diligence with which the registers were kept. In remote rural areas, illiterate mayors failed to keep them altogether; in other places, many residents—especially young men subject to military draft—were being omitted; the writing was often indecipherable; witnesses' signatures were missing; entries did not follow one another in chronological order, or were overwritten in ways that rendered good-faith corrections impossible to distinguish from fraudulent ones. In the second half of the century, however, complaints faded away as the progress of literacy, coupled with decades of tightening administrative supervision, started to bear fruit. With time, civil records provided an increasingly accurate archive of the population.[30] Meanwhile, time worked against the pharmacopoeia. As a still frame of knowledge in flux, it was at its most accurate when it was published but slid gradually into obsolescence as new drugs made their appearance.

In 1820, for example, a mere two years after the publication of the Codex, Joseph Pelletier and Joseph-Bienaimé Caventou isolated quinine. Periodic revisions of the pharmacopoeia provided a means to incorporate new drugs, yet it had taken a decade and a half to finish the first pharmacopoeia and it took almost two to complete its first revision. Quinine remained seventeen years in legal limbo before the second edition of the Codex came out. Courts faced the question of whether drugs like quinine should cease to count as "secret remedies" once they had been described by chemists. In 1828, the Tribunal of Lyon ruled that a blanket prohibition on all drugs introduced after the last revision of the Codex would constitute an inadmissible impediment to the progress of science. This ruling consecrated physicians' right to prescribe new and as yet unregistered drugs. In doing so, it implicitly required pharmacists to keep their cabinets supplied with other medicines than those listed in the pharmacopoeia.[31]

When the law of Germinal went into effect in 1803, the trade in secret remedies was still thriving. Eliminating it by fiat, while the Codex's first edition remained years away from publication, did not seem workable. In 1805, the government allowed secret remedies approved before the Revolution to remain temporarily on the market. In 1810, it set up a commission within the Ministry of the Interior to review new proprietary medicines. If the commission deemed a drug useful, the government would purchase it from its inventor and publish its formula until it could be included in the next edition of the Codex. That procedure ensured that all legal drugs would remain within the public domain. However, members of the Ministry's review commission were cut from the same cloth as the Codex's authors. Almost all were academic physicians who shared a deep pharmaceutical conservatism, a belief that most truly valuable drugs had already been discovered, and that future discoveries would likely come from within the medical or pharmaceutical profession, not from profit-seeking industrialists. Of the countless applications submitted, only a handful received official approval even as unapproved remedies continued to clutter the shelves of pharmacies and the advertising columns of newspapers.[32]

In an effort to preserve the integrity of the public domain of drugs, academic physicians demanded that drugs be excluded from the list of patentable inventions. Lawmakers granted their wish as they passed a new law on the protection of inventions in 1844. In 1850 the government ceased to compensate inventors whose drugs it approved. Drugs deemed new and useful became legal upon publication of their formulas in the bulletin of the Academy of Medicine.[33] Thus all incentives or rewards for submitting a newly invented drug to the government had been suppressed. Without patents, trade secrets became the only way to defend a product against counterfeiting. As a result, most manufacturers simply ceased to seek the government's approval for their products. Of the few formulas that were submitted for review in the 1850s a mere six were approved and published in the academy's bulletin, even as hundreds of new proprietary compounds flooded the market.[34]

The decree of 1850 was the last official revision of the law of Germinal, a last-ditch effort to uphold the rule of the Codex, and the one that most clearly revealed its failures. The reluctance of medical authorities to expand the narrow circle of licit remedies motivated the courts' leniency toward drugs that remained outside of it. Of the many unapproved products on the market few became the object of prosecutions, and fewer still of convictions. In October of 1856 a pharmacist of the city of Metz by the name of Édant was convicted of selling seven different secret remedies. Early in 1857, the appeals court of Metz overturned his conviction, ruling that drugs that could be seen as "minor alterations" or "mere improvements" of Codex formulas were to be admitted even if "given different names than those under which they were usually known."[35] This ruling definitively undermined the Codex's attempt to regulate drugs by attaching them to known names and formulas. No longer were drugs circulating under different names from those recorded in the Codex ipso facto illegal. Only in court was a manufacturer expected to disclose the formula of the incriminated drug and let a judge decide if it was sufficiently similar to one of the pharmacopoeia's preparations. The door was open for the gradual replacement of medicines prepared by pharmacists under the rules and names of the pharmacopoeia by proprietary drugs produced in factories and sold under brand names.

The name under which France's national pharmacopoeia was known—the Codex—was a simple diminutive, an abbreviated version of the official Latin title of the document. But that diminutive was a meaningful one, for "Codex" is also the name of a particular medium—namely, that of the book. As became clear in the ever-deferred project to include all drugs within a public domain charted and ruled by a single book, the choice of a medium also entailed an implicit philosophy. The existence of a unique volume that delineated the outer bounds of the national pharmaceutical commons at once presupposed and affirmed the stability of pharmaceutical knowledge. Revised and updated versions of the Codex every other decade or so were meant to register the slight, and hence manageable, shifts that were likely to occur at the margins. Transparency was the condition of this public domain of drugs. The fact that the names of all drugs (or all ingredients of drugs in the case of "magistral" drugs formulated by physicians) were contained in an authorized handbook present in every pharmacy guaranteed that drugs' formulas would be collectively known and owned.

Yet, in the nineteenth century, that philosophy proved increasingly at odds with the realities of the nascent chemical and pharmaceutical industries. The very same scientists involved in the drafting of the Codex's first edition contributed greatly to accelerating the pace at which novel medicinal substances were discovered and isolated. Meanwhile, pharmacists developed new tools and methods to produce and package drugs on an industrial scale. Shifting spaces of reference, in other words, were inseparable from shifting temporalities. As French pharmacy became national, it also ceased to define itself in terms of a stable, unchanging pharmaceutical heritage. Progress and inventiveness became its self-proclaimed values. Its newest remedies, not its oldest, were to attest to its character.[36]

The new drugs of the nineteenth century, and especially of its later decades, tended to be proprietary drugs. As drugs produced in factories and sold under trademarked names began displacing medicines compounded in pharmacies according to pharmacopoeia recipes, brand names imposed themselves, not only in France but across Europe and North America, as the accepted way to refer to drugs. Meanwhile, pharmacopoeias continued to be updated and republished in most countries, but governed an ever-shrinking segment of the drug market. If national pharmacopoeias continue to this day to serve as reference works and compendia of standards (though for the testing rather than the preparation of pharmaceuticals), the vision they once carried of a stable and transparent public domain of drugs enclosed in one official volume vanished as the center of gravity of drug research, production, and distribution shifted from physicians and pharmacists to a powerful private industry.

12

INDIAN SECRETS, INDIAN CURES, AND THE *PHARMACOPOEIA OF THE UNITED STATES OF AMERICA*

JOSEPH M. GABRIEL

IN 1806, the American physician John Redman Coxe published the first edition of his influential *American Dispensatory*. Coxe had earned his medical degree from the University of Pennsylvania in 1794, where he had studied under Benjamin Rush, and had then studied at London, Edinburgh, and Paris. He began private practice in Philadelphia in 1796, and in 1809 became chair of chemistry at the University of Pennsylvania. Much of his book was a reproduction of Arnold Duncan's 1804 edition of the *Edinburgh New Dispensatory*, but Coxe edited Duncan's text liberally and rearranged it into what he considered an easier format, noting that "by the alteration which is here attempted... the whole is condensed into one view, and greater simplicity is thereby attained."[1] Duncan's text had, not surprisingly, included descriptions of numerous botanicals from the Americas and the Caribbean alongside those of European origin (including wild cinnamon and Jamaica pepper, both discussed by William Ryan in chapter 6 of this volume).[2] Coxe reproduced most of this material, but he also included a significant amount of information drawn from other sources, including Benjamin Smith Barton's *Collections for an Essay Toward a Materia Medica of the United States* (1801 and 1804), in order to fully document the "indigenous medicinal productions" of his home country. In doing so, he included numerous references to the medicinal uses of plants by the Indigenous peoples of North America. Fluxroot, spotted geranium, and the bark of the wild cherry tree were all used by Indians to treat venereal diseases, he noted, while alum root was used by Indians to treat ulcers, sweet fern to stop bleeding, and bone-set as an emetic. Other plant descriptions that included references to Indian use included cassena, water flag root, Carolina pink, mayapple, common dogwood, yellow root, lobelia, and grouseberry, which was used as a stimulant and was "one of the principal articles of the *materia medica* of some Indian tribes."[3]

All this meant that Coxe moved significantly beyond simply reproducing Duncan's version of the *Edinburgh New Dispensatory* and instead produced what one later observer called "the first distinctly American publication" in this genre.[4] Coxe certainly saw his work in these terms. As he noted in the preface, "the present edition of an American Dispensatory is the first attempt which has been made towards the introduction of a Standard for the United States."[5] Yet Coxe was also acutely aware that his effort to establish a "standard" for the young country had significant limitations. In part, this was due to problems related to nomenclature that he was unable to resolve; it was also due to the fact that the formulas he copied from Duncan's text were only sometimes used in the United States, where prescription and dispensing practice varied widely and, as a result, "injurious variety is introduced into the shops." More generally, Coxe recognized that his work was unable to function as a truly normative text in the way that official pharmacopoeias in Europe were supposedly able to because his book was based only on "opinion" and had no institutional authority behind it. What was needed was a truly authoritative "American pharmacopoeia" that would standardize both nomenclature and prescribing and dispensing practices across the young nation.[6]

Coxe was not alone in these types of concerns. His work was part of an ongoing effort among elite physicians to systematize knowledge about the diverse body of medicinal substances used in the early United States. At the same time, it was also a part of the broader effort to distinguish American medical science from its European counterparts and to build the institutional structures necessary to support the development of a distinctly American medical profession. One result of these efforts was the publication of the first edition of the *Pharmacopeia of the United States of America* (USP) in 1820. Produced by a small group of elite physicians, the USP was an explicitly normative work intended to standardize prescription and dispensing practices in the young nation, in part by resolving conflicts over how formulas should be prepared.[7] Both Coxe's *Dispensatory* and the first edition of the USP were important milestones in the history of medicine and pharmacy in the United States. They were followed by numerous other texts that sought to organize knowledge about medicinal goods, including subsequent editions of both volumes and new texts such as George B. Wood and Franklin Bache's tremendously influential *Dispensatory of the United States of America*, first published in 1833.[8]

Not surprisingly, indigenous plants—or what were taken to be indigenous plants—played an important role in these works, and white physicians documented and explicitly drew on Indian therapeutic knowledge as they sought to investigate the medicinal bounty of their young country. The fact that Indians sometimes kept their healing practices secret meant that such knowledge would need to be discovered and revealed in order to be made useful; at the same time, physicians such as Coxe believed that it would need to be stripped of superstitious—or even demonic—Indian beliefs in order to be rendered into useful information. As Benjamin Breen suggests in his chapter in this volume, during the early modern period the distinction between medical science and what we now consider supernatural practices and beliefs was not at all clear cut (chapter 7). In the United States, the divergence between the two grew out of—and helped to create—the institutional, normative, epistemological, and racial foundations of orthodox medical science. This included the establishment of an intellectual and scientific framework that explicitly rejected secrecy as a backward and dangerous form of irrationalism. Yet, perhaps ironically, this secrecy might also conceal something truly useful, if only it could be discovered. This possibility helped fuel not just the development of American medical science but also a populist critique of medical authority and a concomitant market in medicinal goods that practitioners of orthodox medicine considered a form of quackery. Deeply seated ideas about Indian healing forged in the colonial encounter thus helped structure the fundamental organization of pharmaceutical markets over the course of the nineteenth century and beyond.

MEDICAL SCIENCE AND INDIAN KNOWLEDGE IN THE EARLY REPUBLIC

"The Nottoway tribe of Indians have a town not far from this," wrote an anonymous doctor from a small town in Virginia to the physician and botanist Benjamin Smith Barton in 1805. "I am told, from respectable authority, that some of them have cured syphilis, with vegetable remedies. As yet, I have not been able to obtain the secret, or find out the plant, though I have been anxious on the subject. Whenever I do, I shall inform you of it."[9] Barton was a member of the scientific elite in Philadelphia who served on the faculty of the University of Pennsylvania and published extensively. His *Collections for an Essay Toward a Materia Medica of the United States* was an important and influential effort to document the "indigenous vegetables" of the young country that could be used to heal. References to Indian use appear scattered throughout the text, including a description of the treatment of syphilis with lobelia among Indians and a discussion of whether or not the disease had been introduced by white colonialists. Given this, it seems likely that Barton would have been quite interested in discovering the secret that his unknown correspondent referred to. Whether or not he ever did we do not know. Yet in this brief exchange, we see both the importance of Indian knowledge to early American medical science and the way that white physicians conceptualized Indian secrecy as a barrier to be overcome.

Barton's efforts to document indigenous medicinal plants was one small part of a much broader process through which elite physicians in the early United States forged a professional identity for themselves distinct from both their peers in Europe and their competitors at home. Americans had a remarkable and growing diversity of therapeutic options to choose from in the decades immediately following the Revolution: the gardens, wilderness, and markets of the young country provided a tremendous bounty of healing goods. Most of these were in the form of raw botanicals, but there were also plenty of tinctures, elixirs, chemicals, and other manufactured products—many of which were made by a small but rapidly expanding domestic chemical and pharmaceutical industry. At the same time, knowledge about healing, and about the use of plants and other remedies, was itself tremendously diverse and varied greatly across lines of race, class, geographic and regional difference, and numerous other factors. Elite physicians, such as Barton, were thus one group of healers among many in the early Republic. Following the Revolution, they worked to build the institutional structures necessary to support the development of distinctly American medical identity, one that was both loyal to European traditions and met the unique needs of their own geographic, social, and intellectual locations. They did so according to their own evolving ideas of what constituted both proper science and proper conduct, including the establishment of medical societies, medical schools, and domestic medical publishing. Barton was thus one among many actors who helped build the intellectual and institutional framework of what historians refer to as "orthodox" medicine in the young country.

As Barton's work indicates, a dynamic interaction between whites and Indians was an important part of this process. Other chapters in this volume have shown how knowledge about healing plants among Indians and enslaved Africans in the Americas and the Caribbean was transmitted to colonial physicians, pharmacists, and others who then incorporated it into a developing European scientific cosmology. As William Ryan suggests in his discussion of Hans Sloane (chapter 6), however, this was not always a straightforward process and European medical science in the early modern period was less epistemologically rigid than we might assume. The introduction of New World remedies into Europe thus transformed European medical science in complex ways that have not yet been fully explored by historians; at the same time, colonialists developed their own traditions that were in some ways distinct from those of Europe, in part as a result of their daily interactions with Indigenous and enslaved peoples. These traditions eventually formed the basis for the emergence of distinct medical and pharmaceutical communities in the emergent nations of the New World, including the young United States. Not surprisingly, then, references to Indian cures are scattered through medical, pharmaceutical, botanical, and popular texts that were either written in or about the North American colonies and later the young United States. Indeed, the importance of this dynamic can be seen simply based on the number of plants with the name "Indian" in their colloquial name that appeared in medical texts printed in both the colonies and the early United States, including Indian physic, Indian pink, Indian poke, Indian lettuce, Indian turnip, Indian paint, Indian cucumber, Indian sage, and Indian tobacco (or lobelia).

The history of Seneca snakeroot is illustrative (figure 12.1). In 1736 a physician from Virginia named John Tennent published an essay on the treatment of pleurisy, in which he introduced the use of a plant he called "rattle snake root" because of its apparent effectiveness in treating rattlesnake bites. Tennent was well known to other physicians in Virginia at the time—in 1734 he had published a biting critique of the medical community titled *Every Man his Own Doctor*, in which he denounced the high prices physicians charged their patients and laid out a series of treatments that common people could use to cure themselves.[10] Perhaps to escape the reaction to his essay, Tennent then studied medicine in Edinburgh from 1735 to 1736 before returning and publishing his treatment for pleurisy. Tennent claimed to have learned of the use of the plant to treat snakebites from the Seneca Indians, noting that the plant "resemble the Rattles of a *Rattle-Snake;* from which, I suppose, the *Indians,* who discovered the Quality of the Root, deduced its Efficacy."[11] He reasoned that since the symptoms of rattlesnake bite and pleurisy are similar, the plant in question could be used to treat both. Soon after, he gave the plant the name "Seneca Snake Root" and published a series of essays defending his discovery and outlining his theories.[12] He also petitioned the Virginia legislature for a reward for introducing the remedy to science rather than keeping the knowledge of it secret, which he argued would have allowed him to profit from his discovery. In 1739, the Virginia House of Burgesses granted him one hundred pounds for his efforts.[13] News spread rapidly, although most of it focused on the use of the plant for snakebite rather than pleurisy. "The Indians long made a Secret of the Herb they used in curing the Bite of that venomous Reptile a Rattle-Snake," noted Richard Saunders in the 1737 edition of *Poor Richard's Almanack,* "but since some curious Persons among the English have fully discover'd and are now well acquainted with it, I hope it will be an

FIGURE 12.1. Jacob Bigelow. "Polygala senega," *American Medical Botany*, vol. 2 (Boston, 1818), plate XXX. Courtesy of University of Wisconsin Digital Collections and the Biodiversity Heritage Library under Creative Commons License CC BY 2.0.

acceptable Service to these Parts of the World, if I make it more publick by the following Description."[14]

Tennent's effort to expand the use of Seneca snakeroot from the treatment of snakebite to another condition was mirrored by numerous other physicians over the next seven decades. The plant became an important part of orthodox medical practice, both in the colonies and in Europe, and was widely used as a diuretic, emetic, cathartic, and expectorant.[15] The plant was included in the *Edinburgh Pharmacopoeia* as early as 1744, for example, and was described in influential European texts such as Duncan's 1804 *Edinburgh Dispensatory*, who noted that it "grows wild in North America" and is used by "the Senegaro Indians" to treat rattlesnake bite.[16] At the same time, in North America the plant began to lose its reputation as an effective treatment for snakebite. In a 1778 formulary apparently intended for use by the American military, for example, Seneca pills were listed as a remedy for intermittent and persistent fevers, but not for snakebite, and by the early nineteenth century, most elite physicians in the United States appear to have doubted its effectiveness in this area.[17] Barton thus described the plant in his *Collections*, but he doubted its effectiveness in snakebite and instead focused on its use as a diuretic; he also briefly noted that "our Indians" used it for the treatment of syphilis and sore throat. As a result, although Coxe reproduced much of the *Edinburgh Dispensatory* verbatim in his 1806 *American Dispensatory*, he ignored Duncan's comment about its use for snakebite and instead reproduced Barton's comments.[18]

As the story of Seneca snakeroot illustrates, the incorporation and transformation of plants indigenous to North America into European, colonial, and early American medical science was often based on the discovery and disclosure of secret Indian knowledge. Historians of Native American medicine have persuasively argued that Indian medicine was both effective and deeply spiritual in nature, in part because the plants they used as part of their healing practices were, by today's standards, effective, and in part because disease processes—such as epidemics—and efforts to respond to them cannot be understood as somehow outside the broader context in which they take place.[19] Historians have also noted that Indians often kept their healing knowledge secret from outsiders—in this, they were not that different from the Jesuit explorers that Timothy Walker describes in his chapter in this volume (chapter 5), or from the Contugi family that Justin Rivest describes in his (chapter 4). Indian knowledge of nature's therapeutic bounty, or what John Clayton had referred to in 1687 as their "great Apothecary,"[20] had to be discovered and revealed if it were to be made useful to medical science. Secrecy was a barrier that had to be overcome, whether through persuasion, force, or some other means.[21]

Yet white physicians were only really interested in a small portion of Indian knowledge. As Martha Robinson notes, colonialists such as Clayton recognized that Indian secrecy concealed knowledge of cures that were undoubtedly useful—indeed, some, such as sassafras, had already been incorporated into colonial medical practice by the time he made his comments.[22] Yet neither Clayton nor the numerous white physicians, botanists, gardeners, and others who followed him believed that Indians understood why their cures worked. Indians held secrets that could benefit medical science, but the explanatory frameworks in which Indian healing took place was, in general, of little interest to colonialists—except, perhaps, as examples of the strange beliefs of a people trapped in an earlier stage of human development. Indian medical knowledge was thus understood as being shrouded in superstition and, in general, useful for civilized people only if adapted to current conditions through modern science. Indeed, although Indians were known to have made useful discoveries—and sometimes to have successfully treated colonial whites in cases where European cures had failed—over the course of the eighteenth and early nineteenth centuries, their knowledge was increasingly understood as having been eclipsed by the scientific work of botanists, pharmacists, and, of course, physicians. Seneca snakeroot might have initially come to the attention of medical science as a cure for snakebite, for example, but as European and colonial physicians investigated its use it was gradually transformed into a different type of cure.

Indian knowledge of healing plants was thus incorporated into the development of both European and colonial medicine in a way that extracted plants out of Indian cosmologies, reified them as discrete physical things, and transformed them into therapeutic objects that were knowable through the evolving epistemological practices of Western medical science. Equally important, in the decades following the American Revolution, such efforts took on a distinctly nationalist bent. As Philip Pauly has made clear, Europeans hostile to the Revolution argued that the environment of the New World was not conducive to either health or successful political organization; animals and plants in America, they claimed, degenerated over time, and any effort to establish an independent country would undoubtedly result in its inhabitants reverting back to the savage state of its aboriginals.[23] Citizens of the young country responded in part by documenting the utility of indigenous plants, including plants that had medical uses, in terms that blended practical, scientific, and political needs. Thomas Jefferson's *Notes on the State of Virginia* (1781), for example, included a list of indigenous medicinal plants as a part of his broader effort to document the resources of the young country in the face of European skepticism.[24] By the first decades of the nineteenth century, such efforts had resulted in systematized programs to document and classify the indigenous medical plants of the young country, including Barton's *Collections* and Jacob Bigelow's three-volume *American Medi-*

cal Botany (1817–1820). These were simultaneously scientific, political, and thoroughly practical projects—as Bigelow noted in the first volume of his work, "it is the policy of every country to convert as far as possible its own production to use, as a means of multiplying its resources, and diminishing its tribute to foreigners."[25]

This process also resulted in early efforts to establish a truly American materia medica that combined remedies long known in Europe with indigenous medicinal plants unfamiliar to physicians outside the United States. Although preceded by a handful of other texts, including the 1778 military formulary mentioned above, Coxe's 1806 *American Dispensatory* was probably the first effort to establish a systematized and truly comprehensive account of useful remedies that directly suited the needs of the young nation.[26] It was followed by a number of similar efforts, most notably the 1808 publication of the *Pharmacopoeia of the Massachusetts Medical Society* and, two years later, James Thacher's companion volume *American New Dispensatory* (1810). Knowledge of indigenous "vegetable productions," at least some of which was derived from Indian sources, was an essential part of this process. "The Indians used the root of water dock with great success in cleansing foul ulcers," noted Thacher in one of the many references to Indian practices in his work. "It is said, they endeavored to keep it a secret from the Europeans."[27]

There was also a distinct class element to all of this. European interest in natural history had long been the province of elite members of society, but in both the colonies and the young United States the study and collection of plants was pursued by a wide variety of people. As Andrew J. Lewis has argued, elite botanists "confronted a society of artisans and farmers . . . who sought economic and political freedoms at odds with the dreams of their social betters and who possessed their own ideas about plants derived from experience, gardening, folk knowledge, Indians, and local usage." As Lewis notes, elite botanists thus "encountered a democratizing republic reluctant and at times resistant to adopting a hierarchical botanical practice."[28] The same can be said of the effort to organize medical knowledge. Herbals, recipe books, newspaper columns, and other texts produced in both the colonies and the young republic sought to provide useful knowledge with which ordinary people could treat themselves. Echoing John Tennent's 1734 *Every Man His Own Doctor*, these texts sometimes struck a populist tone and, at times, were explicitly antagonistic toward elite physicians due to their high prices and esoteric theories.[29] Not surprisingly, elite physicians often considered this type of information to be irrational and even dangerous. Like the nation itself, both medicine and botany in the early republic was characterized by a tension between its democratic possibilities and elite efforts to guide it in what they understood to be a properly scientific direction. The presentation of therapeutic claims about plants that had, supposedly, been derived from Indian sources played out along these lines.

Samuel Henry's *A New and Complete American Medical Family Herbal* (1814) is a good example.[30] Henry claimed that his book was the result of more than thirty years of study among the Creek Indians, as well as extensive travels through the southern states where he made "botanic discoveries on the real medical virtues of our indigenous plants." Henry was poorly educated, but critics were initially willing to give him the benefit of the doubt since, as one reviewer noted, "it has long been received opinion, that the Indian Natives are well acquainted with the medical virtues of many indigenous plants of which the descendants of Europeans are perfectly ignorant," and that as a result anyone claiming to have extensive experience with them had the potential to contribute something useful. Despite such possibilities, Henry's work was roundly criticized for misclassifying plants and using inaccurate scientific names, for including exotic species in what was supposed to be a catalogue of indigenous plants, and for numerous other problems. The net result of all this was that the work was simply untrustworthy and, as a result, dangerous. "Much mischief may be done by masters and mistresses of families attempting to identify American with European plants, which are preposterously figured by the herbalist," noted one critic. "The result [may] be fatal... and cause the masters and mistresses of families, as well as those who have *rejoiced* at this publication, to repent of their folly when too late."[31]

The organization of knowledge about materia medica in the early Republic was thus a project in both epistemology and nationalism. It was an effort to draw limits around what counted as legitimate scientific inquiry, both by separating medical science from supposedly irrational Indian beliefs and by its dismissal of supposedly scientific claims made by the uneducated. Participation in the effort to organize the world of healing goods was thus to be limited through the constraints of proper scientific practice, and this practice, in turn, would result in reliable knowledge that would protect ordinary people from the dangers of fraud, superstition, and error. Henry's work, in other words, demonstrated the need for authoritative texts grounded in proper scientific practice that would distinguish between trustworthy knowledge and irrational claims. Of course, his work was dangerous not only because it threatened the health of the public. It was also dangerous because his sloppy methods produced unreasonable claims that threatened the reputation of the young nation and, in doing so, seemed to threaten the nation itself. One reviewer thus explained why he had written a highly critical review of the book by noting that he did so "not only for the sake of exposing a work which is evidently calculated to do mischief, but also with a hope that it may have the effect of disarming those European critics, who would be happy to avail themselves of such an opportunity to disparage our literary and scientific character, should the work fall into their hands."[32]

THE PHARMACOPOEIA OF THE UNITED STATES OF AMERICA

In 1811 the physician Valentine Seaman assembled a small compilation of written formulas for use in the New York Hospital. Working with Samuel L. Mitchill, over the next several years Seaman expanded the list and then published it under the title *The Pharmacopoeia of the New-York Hospital* (1816). Seaman's volume drew heavily on the London, Edinburgh, and Dublin pharmacopoeias, but it also included a variety of articles indigenous to the United States that were not included in those texts. It was written with the intention of "correcting abuses, and of effecting reform in the pharmaceutical department" by replacing the "numerous and varied Pharmacopoeias and Dispensatories in common use" at the hospital. Seaman also recognized that his volume had broader significance; as he noted in the preface, "apothecaries, who reside in parts of the United States where no regular Pharmacopeia has been established, must have seriously felt the inconvenience of not having any uniform standard for compounding their medicines; to those particularly [it] cannot but be intrinsically useful."[33] Seaman's pharmacopoeia was, at heart, an effort to improve the practice of medicine and pharmacy in the United States by establishing a "standard" that could be used by apothecaries across the country.

Seaman's pharmacopoeia does not appear to have been distributed widely, but it did spark Mitchill's interest in the issue. Mitchill was well suited to advance the cause of a national pharmacopoeia; one of the founders of the country's first medical journal in 1797, he served as its principal editor for much of his career and maintained a wide network of professional contacts among the elite orthodox medical community.[34] Following the publication of the *Pharmacopoeia of the New-York Hospital*, Mitchill began corresponding about the issue with a young physician named Lyman Spalding, whom he took under his wing and helped settle in New York. In 1817, with Mitchill's support, Spalding proposed a plan to the New York Medical Society in which medical societies in different sections of the country would hold regional meetings and elect delegates to a national convention in order to formulate a truly national pharmacopoeia. The idea proved appealing, and the convention was held in early 1820. Only a small number of representatives actually participated, and those who did were mostly from New England and New York. Still, imagining themselves representative of the nation as a whole, this small group of physicians created a text that they designated a national pharmacopoeia. It was, they believed, a testament to the abilities of the young nation to contribute to the advancement of medical science on its own terms. As Spalding noted at the end of the convention, "the great national work is spoken of by the President [James Monroe] as an undertaking which will assist in giving us a National character."[35]

The first edition of the USP appeared soon after. It was an explicitly normative work, designed to combat "the evil of irregularity and uncertainty in the preparation of medicines" by resolving disputes about the proper compounding of remedies and replacing existing texts on the market.[36] It was also self-consciously modeled on the republican ideals of its elite authors. "A National Pharmacopoeia, which should be established and adopted by the consent of all the medical corporate bodies throughout the United States," noted the introduction, is "evidently the only mode by which a uniform system could be introduced at once into all parts of the American territory."[37] At the same time, if the writing of a national pharmacopoeia grew out of the ideals of the young republic—or, at least, out of the ideals of some of its most elite members—then its production also helped create the discursive conditions of the new nation, demonstrating to the world that a free and democratic people could produce a scientific work equal to, or better than, that of Europe. As one reviewer of the first edition put it:

> This work forms an era in the history of the profession. It is the first one ever compiled by the authority of the profession throughout a nation. Collections of this sort have been made in other countries, but none, so far, under the impressive sanction which distinguishes this. Many of the authorities of the Past compiled similar works, later still, the Colleges of Great Britain have followed their examples. France by command of her Monarch has furnished her "Codex," but it has remained for American Physicians to frame a work which emanates from the profession itself, and is founded on the principles of Representation. It embodies a Codex Medicum of the free and independent United States.[38]

The USP was thus organized according to the ideological constructs of the young republic, and, in turn, reinforced the hierarchical nature of its distinctive brand of medical science. After all, Samuel Henry's text had no place in the creation of this document. In the coming years, new editions flowed from the pens of elite physicians on a regular basis, both confirming and reproducing the connection between science, class, and national identity.

Not surprisingly, the first edition of the USP drew heavily on the efforts of botanists, physicians, and others to document the indigenous healing plants of the young country. A small number of plants that were used medicinally by Indians were included in its pages, including Seneca snakeroot, skunk cabbage, wild cherry, water dock, dogwood, mayapple, Carolina pink, Virginia snakeroot, and thoroughwort.[39] However, the first edition of the USP also remained deeply indebted to its European antecedents. The text was heavily influenced by the Edinburgh pharmacopeia, and many of the indigenous remedies were only included on a secondary list of articles deemed "of secondary or doubtful efficacy."[40] Perhaps more importantly, many indigenous plants known to have medical uses were simply left out. This did not go unnoticed. Critics of the first edition regretfully denounced it as filled with errors, overly reliant on European remedies, and as having failed to sufficiently take into account the indigenous "vegetable productions" of the land from which it had come. As one reviewer bitterly put it, "is an American pharmacopoeia deserving the name, which betrays a servility to the influence of foreign example in its arrangement of matter, and which shows as a *secondary list* of articles of the materia medica, the *products of our own territory*?"[41] For many critics, a truly American medical science would fully include the natural bounty of the young country. Left unsaid, but obvious to everyone concerned, was the fact that this inclusion would be based, in part, on the efforts of American botanists, physicians, and others to disentangle useful Indian knowledge from supposedly irrational belief.

As Stuart Anderson reminds us (chapter 10), local political considerations can play an important role in how pharmacopoeias linked to national identity are developed. The same can be said of individual personalities and disputes between small numbers of people. Indeed, depending on the scale of analysis, explanations for the historical trajectory of national pharmacopoeias and other similar texts can range from individual behaviors to local political and professional disputes to national ambitions. The USP was no exception. Due to tensions between the New York and Philadelphia medical communities, when it came time to issue a second edition of the USP in 1830, competing conventions were held in New York and Washington. The medical community in New York issued a second edition later that year, while the Philadelphia medical community issued its own second edition in early 1831. Among the principal authors of the Philadelphia edition were George B. Wood and Franklin Bache, both affiliated with the Philadelphia College of Pharmacy. Shortly after the publication of their volume, however, Jonathan Redman Coxe criticized both revisions in the preface to the 1831 edition of his *American New Dispensatory*. Coxe argued that neither of the "so called Conventions," as he put it, had followed the directions for establishing a national convention outlined in the original 1820 edition. Neither, therefore, had a rightful claim to being duly constituted or national in scope, with the unfortunate result that two competing works had been issued that disagreed in important ways about even common remedies. "From this general outline," Coxe noted, "it will easily be seen, that, unless the principles of *free* representation be fully maintained, we shall here sanction an irregularity of infinite injury to the medical profession."[42] Declaring both texts illegitimate, Coxe recommended that physicians rely on the 1820 edition of the USP until the difficulties could be resolved.

From Wood's perspective this made little sense. In a detailed rebuttal published the following year he outlined numerous problems with Coxe's critique. Wood paid virtually no attention to Coxe's argument about the process through which the two revisions had been produced, instead focusing on scientific issues related to nomenclature, taxonomy, the preparation of formulas, and related topics.[43] The dispute appears to have had its roots in an earlier controversy between the two men over Coxe's suggestion that the University of Pennsylvania medical school issue a degree in pharmacy; Wood appears to have rejected the idea and instead helped to establish the Philadelphia College of Pharmacy in 1821. A decade later, the two men were still competitors, and in 1833 Wood and Bache issued the first edition of their *Dispensatory of the United States of America*. More than a thousand pages in length, this was a massive text that reproduced and synthesized a huge amount of material. It was also a direct competitor to Coxe's volume and quickly became the standard reference work of its type.[44] The 1831 version of the USP that had been drafted by Wood and Bache also went on to serve as the basis for subsequent editions, issued once every ten years, while both the New York version and Coxe's critique were forgotten.

Over the next several decades, the USP and the *Dispensatory of the United States of America* played increasingly important roles in the practice of American medicine and pharmacy. Perhaps not surprisingly, the transformation of Indian practice into Enlightenment science can be seen in their pages. Except for the occasional use of the term "Indian" in the common name of a simple, the Indigenous influence on medical science is virtually invisible in the pages of the USP. Wood and Bache's text is more complicated. Numerous, albeit brief, references to Indian practice are included in the multiple editions of the *Dispensatory* issued before the outbreak of the Civil War in 1861. Such practices, however, belonged to the past. They were not a part of medical science itself. In a long description of Seneca snakeroot, for example, the 1833 edition includes only a single sentence describing the origins of white knowledge about its medicinal use: "It was introduced into practice about a century ago by Dr. Tennant of Virginia, who recommended it as a cure for the bite of the rattlesnake and in various pectoral complaints." Wood and Bache do not recommend the plant for the treatment of snakebite, of course, and instead describe it as a useful diuretic, stimulating expectorant, emetic, and cathartic. They also discuss the fact that the root appears to contain an "active principle" called "senegin," and they describe a number of experiments on the supposed principle conducted by other researchers. The direction of things to come is clear. The rise of laboratory science and the eclipse of raw botanicals in favor of manufactured goods—these and other scientific, commercial, and epistemic transformations would help create a growing distance between Indian practice and orthodox medical science.[45]

Still, the persistent traces of Indian medicine in Wood and Bache's text are important. They suggest that efforts by American physicians to differentiate themselves from their European counterparts relied, in part, on the recognition of the hybrid origins of their practice, if only as a fleeting gesture to an earlier time. They also point to the violent processes on which American medical science was partially based. In a period of time characterized by national expansion and brutal violence toward Indigenous peoples, the separation of these traces from the broader cosmologies in which the Native use of healing plants took place illustrates how the formation of a distinctly American medical identity was also based on the eradication of Indian ways of being and knowing. Like the nation as a whole, orthodox physicians in the early United States formulated and reformulated their identity through the process of self-invention; the simultaneous incorporation and eradication of the Indian was an important part of this process. It was, to use Richard Slotkin's memorable phrase, a form of regeneration through violence.[46]

Subtle changes in the discussion of Carolina pink are illustrative. The medical use of the plant had long been traced to Indian origin, and in the 1836 edition of the *Dispensatory* Wood and Bache note that "it is collected by the Creek and Cherokee Indians, who dispose of it to the white traders. By these it is packed in casks, or more commonly in large bales, weighing from three hundred to three hundred and fifty pounds."[47] However, the 1843 edition notes the following:

> The drug was formerly collected in Georgia and the neighboring States by the Creek and Cherokee Indians, who disposed of it to white traders. The whole plant was gathered and dried, and came to us in bales or casks. After the emigration of the Indians, the supply of spigelia from this source very much diminished, and has now nearly if not quite failed. The consequence was for a time a great scarcity and increase in the price of the drug; but a new source of supply was opened from the Western and South-western States, and it is now again plentiful.[48]

The switch from the present to the past tense is telling. Wood and Bache's use of the delicate term "emigration" in the 1843 edition obscures the violent history through which the Creek and Cherokee peoples were forced out of the region. In 1832, after decades of brutal confrontation with the states of Georgia and Alabama, the Creek had begun to move to what is now Oklahoma. Four years later, the federal army—assisted by state militias—forcibly removed the remainder. In 1838 the Cherokee were also forcibly removed from Georgia and surrounding areas and driven west along what came to be known as the "Trail of Tears." Wood and Bache's brief gesture toward this brutal process was, of course, just a tiny part of their discussion of Carolina pink. Their interest in the fate of the Cherokee and

Creek people was passing at best, embedded in a scientific discussion of a therapeutically useful plant. Yet the gesture was still meaningful. This and other similar echoes reflected the changing fate of Indian peoples in the expanding nation. They also reminded the readers of the text of the distinctiveness of a truly American medical science. They reminded American physicians of who they were.

THE ATTACK ON QUACKERY

"There is a kind of superstition among our people, as to Indian skill in physic," noted Alexander Coventry, the president of the New York Medical Society, in 1825. "This is a fallacy.... You may as well tell of an Indian mathematician whose knowledge of numbers extends to his ten fingers, as of an Indian physician." Like many of his colleagues, Coventry was concerned about what he saw as rampant quackery in the medical market. In addition to Indian physicians, for example, he also denounced "root doctors," "poison-curing Negros," and the use of patent medicines, which he characterized as "irreconcilable to common sense." The problem here, in part, was that Indians had no understanding of modern science and their cures were primitive at best. Popular reverence for Indian medicine was thus akin to a belief in "witchcraft and demonology," and those who practiced it should be suppressed. Coventry thus also criticized the 1813 New York licensing law that regulated the practice of medicine in the state; among other stipulations, the law exempted "those using roots, bark, or herbs ... from the regulations of said act," an exemption that Coventry derided as "a disgrace to our statute book, and a stigma on a civilized people" since it allowed "Indian doctors" and other quacks to practice without interference.[49]

Coventry was not alone in such views. The general framework he articulated was widely shared among the physicians who worked to establish the institutional structures of American medicine in the early decades of the nineteenth century, including state and local medical societies, medical schools, domestic medical publishing, and early licensing laws. Although these efforts were only modestly successful in the early decades of the nineteenth century, especially when compared to later achievements, they laid the groundwork for the transformation of orthodox medicine into the powerful profession that it is today. As Coventry's lecture suggests, the rejection of what orthodox physicians considered irrational forms of quackery was central to this process. It was one of the defining features of orthodox medical identity.

The rejection of secrecy was an important part of all this. As I have argued elsewhere, early American orthodox medical science was grounded on the assumption that medical knowledge should be freely shared among physicians for both benevolent and scientific reasons. The monopolization of knowledge through secrecy—and other means, including patents—seemed deeply contrary to the advancement of a rational medical science. The Connecticut Medical Society, for example, was first organized in 1792 under a state law granting the society licensing authority over the practice of medicine. In 1793, the society passed a resolution expelling any member who might "assume or hold the knowledge of any nostrum, or palm any medicine or composition on the people as a secret." Over the next several decades a wide variety of local and state medical societies incorporated similar prohibitions into their codes of ethics; conformity to the dictates of these codes was, in turn, linked to the legal ability to practice medicine through the institution of state and municipal licensing laws. Orthodox physicians, in other words, distinguished their own supposedly rational practice from what they considered dangerous forms of quackery partially through the institutionalization of a prohibition on secrecy.[50]

As Coventry's lecture suggests, this process overlapped with a growing concern among elite physicians about so-called Indian doctors and Indian cures. By the 1820s, it was widely assumed among orthodox physicians that although Indians might, in fact, know something valuable about indigenous remedies, this knowledge needed to be both revealed and disentangled from superstitious beliefs if it were to be made useful. Indeed, the reliance on secrecy was itself understood as one of the many superstitious and backward practices among Indians that stood in contrast to the rationalism of medical science.[51] Yet elite physicians also confronted a medical market in which Indian medicine appeared to be increasingly popular—or, at least, purported Indian medicine. Cures and remedies that supposedly had Indian origins, for example, were printed in newspapers and other popular texts, including one "Indian prescription" for curing cysts that was made from baking a frog in butter.[52] More generally, as Joshua David Bellin has argued, popular culture in the early nineteenth-century United States often included portrayals of Indian healing and descriptions of Indian cures. Bellin, for example, discusses the performances of a man named Henry Tufts who, in his 1807 autobiography, claimed to have learned "the Indian practice of physic" and to have established a great reputation for himself as a purveyor of "extraordinary cures." We do not know what benefit Tuft's patients derived from their interactions with him, but his apparent popularity suggests that whatever he was selling did indeed mean something important to his customers. "Playing Indian," as Philip Deloria has termed the process through which white Americans incorporated representations of Indians into their own social and cultural practices, included "taking the Indian cure." Representations and performances of Indian healing such as those offered by Tufts were one of the many resources that white Americans used to build their identity.[53]

The rhetorical uses of secrecy and disclosure were an important component of the popular representations of Indian medicine. On the one hand, during the early decades of the nineteenth century, a growing number of itinerant healers followed in Tuft's footsteps and presented themselves as so-called Indian doctors who had access to secret medical knowledge. The claim to esoteric knowledge was a key part of how these people justified their supposed expertise and established legitimacy for themselves as they plied their wares. At the same time, however, the disclosure of what were purportedly secret Indian practices and remedies—often for supposedly benevolent reasons—was also an important rhetorical strategy in these types of representations. Take, for example, Peter Smith's *The Indian Doctor's Dispensatory* (1813). Smith was a self-trained "Indian doctor" who claimed to have traveled extensively through the country learning secret Indian remedies. Although his book was expensive, in the preface Smith argued that it would actually save the purchaser money over time because the remedies he disclosed in its pages would eliminate the need to pay high fees of a physician. He also noted that "the author is well aware that the public mind has long been impressed with these ideas: vis. '*The natives of our own country are in possession of cures, simples & c. that surpass what is used by our best practitioners.*'" Such language echoed longstanding populist critiques of elite medical knowledge and high physician prices; like numerous other so-called Indian doctors that traveled the land, Smith sought to make his fortune by combining claims to secret Indigenous knowledge with a populist critique of medical elites.[54]

The growing popularity of Indian medicine—or what purported to be Indian medicine—was also linked to the rapidly developing trade in patent medicines. As James Harvey Young noted, remedies made from secret ingredients and imported from England had long been sold in the American colonies.[55] Following the revolution, American manufacturers began to enter the market, and by the early decades of the nineteenth century a robust and rapidly growing trade in domestically produced patent medicines had developed. Secrecy was an important part of the business strategy of these companies. Keeping their ingredients secret allowed patent medicine manufacturers to protect their formulas from their competitors; secrecy also allowed them to change ingredients as they deemed necessary and to advertise their goods as they saw fit.[56] At the same time, many of these manufacturers began to rhetorically link their products to supposedly secret Indian knowledge through their advertising and naming practices. In 1800, for example, a manufacturer named Richard Lee introduced his "Patent Indian Vegetable Specific" as a cure for venereal disease.[57] Lee's remedy went on to become one of the most successful patent medicines of the first decades of the nineteenth century, widely advertised in newspapers across the country. Other manufacturers followed suit, selling products such as "Indian Tooth-Ache Drops" and "Romlindorf Digestive Lozenges—An Indian Specific," which was supposedly made from a recipe "presented to the proprietor as a special mark of favor by a chief of the Ogageoattos, highly esteemed for his knowledge of the virtues of the vegetable kingdom."[58]

Orthodox physicians had little patience for such things. Although they recognized that many of the remedies used in orthodox medicine had originally been introduced from Indian sources, by the 1820s orthodox physicians generally regarded so-called Indian cures and Indian Doctors with attitudes ranging from skepticism to outright hostility. In part, this was because they were increasingly skeptical that Indians had much to offer anymore—after all, could a backward people really contribute something truly new to the advanced state of medical knowledge? "It is hardly credible," noted one physician in 1823, "that a good farmer, or an Indian, or an illiterate quack doctor will be likely to discover useful compounds which have escaped the researches of learned and experienced practitioners."[59] More generally, so-called Indian doctors seemed to be little more than one among a variety of quacks who duped the gullible public, while patent medicines—whether advertised as supposedly based on secret Indian knowledge or not—were increasingly denounced as a dangerous threat that needed to be suppressed. Given the advanced state of medical science, and the unlikelihood that Indian remedies could help the diseases of civilization, practices and goods that explicitly referenced supposed Indian sources of knowledge were increasingly understood as iterations of therapeutic irrationality. Physicians thus warned the public to be wary of the "uneducated, presuming imposter, who, under the name of an Indian, or a Cancer Doctor, or Doctor by instinct, tampers with the health, the limbs, and the life of the patient."[60]

Beginning in the 1830s this antagonism became decidedly more pronounced. In part, this was undoubtedly because the brutal process of Indian removal meant that Indigenous practices were increasingly remote from the experiences of elite physicians interested in materia medica. Equally important, orthodox physicians continued to build the institutional structures that underlay an emergent profession and to distinguish themselves from other forms of healing practice through the consolidation and institutionalization of a binary between "medical science" and "quackery." As Owen Whooley argues, debates about medical professionalism in the nineteenth-century United States were deeply intertwined with epistemological concerns about what constituted legitimate medical knowledge.[61] They were also closely connected to both the process of institution building and efforts among orthodox practitioners to suppress competing visions of medical practice. In 1849, for example, the American Medical Association was established, which formalized both ethical and epistemic norms among elite orthodox physicians into an institutional structure that supported an emergent professional identity. The formalization of a sharp distinction between orthodox medicine and supposedly irrational practices such as Thomsonian medicine, homeopathy, mesmerism and "Indian physic" was an important part of this process. Popular texts such as Jas. W. Mahoney's *The Cherokee Physician, or Indian Guide to Health* (1849), which purported to reveal the secrets of Cherokee medicine, seemed to orthodox physicians to be a part of the dangerous medical hucksterism of the day.[62] At the same time, even when apparently sincere, Indian healing practices themselves were described as superstitious and backward in the medical literature, when they were mentioned at all. Indian medicine—or what was claimed to be Indian medicine—was associated with ignorance and superstition on the one hand and outright quackery on the other. Both were considered irrational practices that had no place in medicine and while one could perhaps be tolerated as an antiquated curiosity, the other needed to be actively suppressed.

Early efforts to restrict the sale of patent medicines should be understood in these terms. During the 1840s and 1850s the patent medicine industry grew rapidly, in part by frequently including Indian motifs in its advertising. Elite orthodox physicians looked on the trade with horror, and in response began to work with their colleagues in organized pharmacy to advocate for laws to suppress their manufacture and sale. These efforts were rarely successful during the antebellum period, but there were occasional victories: In 1849, for example, the state of Pennsylvania passed a law requiring manufacturers and vendors of patent medicines to be taxed and licensed, while exempting "regular apothecaries for the sale of simple medicines, the prescriptions of physicians, and the compounds of the pharmacopoeia and the several dispensatories of the United States."[63] The conceptual distinction between science and quackery underlying the law is clear—patent medicines were considered outside the domain of medical science, and while reformers in the medical and pharmacy communities in Pennsylvania did not have the political strength at the time to fully ban their use, the law was a clear effort to suppress their manufacture and sale. Legitimate compounds, on the other hand, were those defined by the USP, the *Dispensatory of the United States of America*, and other reputable texts. The suppression of quackery would thus be pursued, in part, through the establishment of legally enforceable taxonomies of knowledge. The sale of "Dr. Cullen's Indian Vegetable Remedy," with its secret ingredients and supposedly fraudulent origin, would be suppressed as a dangerous form of irrationalism; the sale of a decoction made from Seneca snakeroot, official in the USP for more almost three decades, would be allowed as one small part of the broader practice of scientific medicine.

Indeed, the sale of medicines that conformed to the shifting epistemic and institutional norms of the medical community was not just tolerated—it was actively encouraged by an emergent wing of the manufacturing industry that self-consciously adopted these norms and, simultaneously, began to seek out new remedies to introduce. Indian medicine held an important possibility for these companies. Despite their association with irrationality, Indian practices might also include useful kernels of truth that could be investigated and perhaps exploited as a means of advancing both medical science and private profit. Doing so, of course, required that the "gross superstitious and magical beliefs" of the Indian, as one observer put it in the 1850s, be stripped away so that their knowledge of "the valuable articles of the materia medica" might be revealed.[64] Indian secrecy, once again, acted as a barrier to the progress of civilization. Manufacturing pharmacists undoubtedly considered it the backward practice of an irrational people, something to be overcome in the effort to develop pharmaceutical goods and markets.

These complex dynamics continued in the decades following the Civil War. On the one hand, patent medicine manufacturers continued to incorporate Indian motifs into their advertising and, at times, explicitly presented themselves as selling remedies based on secret Indian knowledge (figure 12.2). At the same time, efforts by orthodox physicians and pharmacists to eliminate the "evil of irregularity" in prescribing and dispensing practice continued through subsequent editions of the USP, as did efforts to compile systematic and comprehensive descriptions of the therapeutic armature, including subsequent editions of the *Dispensatory of the United States of America*. The effort to suppress quackery continued as well, including efforts to restrict the sale of patent medicines and otherwise rationalize the drug market. These efforts eventually resulted in the passage of the 1906 Pure Food and Drug Act, which, among other things, was a direct assault on patent medicines and other supposed forms of irrationalism in the drug market. The law linked the definition of drug purity to standards established in USP; it also required that pharmaceuticals made from certain dangerous substances disclose this fact by listing their ingredients, and required that nonofficinal drugs (i.e., products made according to formulas that were not included in either the USP or the *Dispensatory*) be sold under "distinctive names" with which they could be identified. Finally, the law empowered the federal government to require manufacturers to demonstrate that the claims they made for their products conformed to the epistemic norms of medical science by prohibiting deceptive statements on product labels. Although not adequate to the challenges of the day, the 1906 law was a key moment in the emergence of federal power over the therapeutic market in the United States. It was a direct assault on what physicians considered one particularly dangerous form of quackery, and one in which the USP played a central role. The fact that it was also an attack on manufacturers who claimed to be in possession of secret Indian knowledge was no coincidence.

The USP continued to play a central role in the effort to rationalize the drug market over the course of the twentieth century, linking the epistemic norms of

FIGURE 12.2. Advertisement for Old Sachem Bitters. Wm Goodrich, New York, ca. 1859. Courtesy of the Library of Congress.

the medical community to the development of state power. There was little room for "hocus pocus" in this vision of the world. Still, the idea that Indians and other Indigenous peoples possess secret knowledge that, if disclosed and disentwined from irrational beliefs, might lead to the successful development of new products continued to be a tremendously important assumption among drug manufacturers. Pharmaceutical manufacturers thus engaged in systematic efforts to discover new medicinal plants in distant lands; interrogating Native peoples about their customs was an important part of this process, as was the effort to overcome what was often seen as a reticence to reveal their practices. Following World War II, for example, American manufacturer Merck & Company recognized that their efforts to "obtain a complete knowledge of native medicinal plants and native remedies" in Central America had not been adequate given the efforts of their competitors. In response, they planned to send a team of researchers to Guatemala and other areas in order to establish a research program on local medicinal plants. In preparation for doing so, the director of the enterprise corresponded with a variety of people knowledgeable about "Guatemalan Indian customs."[65] The results of the survey were not promising. "The difficulty is going to be to get the basic information from the Indians," he noted in a letter to a company director. "In fact they said the only practical approach is for possibly one of the Merck executives to marry and live with an Indian woman for five years. Perhaps this is somewhat exaggerated but no doubt the field personnel have to be closely associated with the Indians and have their complete confidence."[66]

Indigenous peoples have long been understood as a source of knowledge about healing plants, and their efforts to conceal this knowledge from outsiders has long been understood as a barrier to the advancement of medical science. In the early decades of the nineteenth-century United States, this secrecy was associated with a wide variety of other practices that were thought to be irrational. By the mid-twentieth century, beliefs that had once been dismissed as superstitious were in the process of being redefined as a form of tradition that should, perhaps, be preserved. Yet secrecy still operated as an obstacle to the advancement of medical science when it was used to conceal the therapeutic possibilities of so-called traditional cures. Indian customs were thus both a resource and a barrier, something to be investigated and overcome in the search for new drugs—drugs that would, in turn, be incorporated into the developing regime of pharmaceutical knowledge of which the USP was an important part. The "advance" of medical science during the nineteenth and twentieth centuries was thus based, in part, on the logic of incorporation, transformation, distortion, and erasure of Indian knowledge and practice. Indian medical knowledge had a place in this new world, but only to the extent that it could be stripped from the cosmologies in which it had once been embedded and reified as "tradition" devoid of actual metaphysical significance. Like plants themselves, Indian knowledge had to be harvested and cleaned, and thus destroyed, to be made useful in the construction of pharmaceutical markets.

AFTERWORD

THE POWER OF UNKNOWING

Early Modern Pharmacopoeias and the Imagination of the Atlantic

Pablo F. Gómez

EARLY MODERN pharmacopoeias are peculiar objects whose taxonomies and formulas were born out of myriad paradoxes.[1] These lists of medicines ostensibly aimed to classify and organize, yet they relied for their very functioning on what they could not explicitly establish. They claimed to carry authoritative knowledge while betraying profound anxieties about epistemological hierarchies. Pharmacopoeias' authors' stern dictums demanded traditional prescriptive modes for the crafting of substances with bodily effect, but they also recognized and promoted innovation and improvisation. The multiple individuals and institutions in charge of crafting lists of medicinal substances in the early modern Atlantic shaped a relationality to nomenclatures designed for the enactment of ideas about belonging, value, organization, and intellectual hierarchies. They did so, however, through the active creation of multiple "others."[2]

It is precisely because of their contradictory and entangled nature, as the chapters in this volume show, that pharmacopoeias are fertile and productive spaces of historical inquiry and unexpected revealers of complex social, political, and economic dynamics that were linked, crucially, to a mercurial materiality. Early modern lists of medicines carry in them traces of the multiplicity of aspirations, social boundaries, fears, and learned and unlearned histories of the Atlantic. They are potent symbols of the classificatory energy of Western modernity.[3] As engines of illusory hierarchies, pharmacopoeias channeled shared imperial, national, scientific, and professional medical aspirations. In doing so, they worked through the purported organization of the mores of manifold societies whose medicines—including those of Europeans—worked on the basis of occult properties that pharmacopoeias could not discern, and whose practitioners thrived in secrecy and obfuscation.[4]

Pharmacopoeias, thus, point out to the multiplicity of lifeworlds that coproduced the Atlantic and its fictional categorical divisions. But they also signal to the crucial material exchange that was at the center of the making of the multi-originated and imprecise idea of the West. Early modern Atlantic lists of medicines as parts of active assemblages of people and things signal to the concrete processes of value creation that emerged from the very spaces where the vaunted divide between nature and culture, "us" and "others" appeared in Europe during the early modern era.[5] Practitioners and organizers of these early texts of fetishization proclaimed the superiority of their medical taxonomies and explained the power of their new ways of organizing the substances and objects of the Atlantic for specific material concerns. Under this light, the process of creation of the fetish is not uniquely a development emerging out of intellectual rationalizing and naturalizing impetuses, or racialized thought. Instead, creating fetishes evinces the means through which practitioners, professional societies, regulators, and empires created commercial worth in fiery processes of exchange and commercial competition related to *materia medica*.[6]

Pharmacopoeias did not simply provide descriptions but also in a very active manner created the things they were supposed to detail.[7] Lentacker's chapter, for instance, reminds us of the transmuting power of pharmacopoeias.[8] These codices, to put it bluntly, created the worlds they aimed to describe and the value of the categories they conveyed. Pharmacopoeias, clearly, then, tend to be by their nature originators and simultaneous validators of groupings and vocabularies. Anderson's chapter makes explicit how these lists primarily functioned as tools of professional control over specific legal and intellectual territories, and that rather than being the result of unique moments of inspiration, their crafting followed long and contentious historical trajectories.[9]

But such attempts at control were limited. Florentine ricettario[s], as Beck shows, reveal how early modern lay practitioners had different manners of interpreting and acting upon the learned authority deployed in pharmacy texts and regulations.[10] This is also evident in Gabriel's chapter, where he exploits the tension between pharmacopoeias as instruments for reification and standardization of medical practice, and for the filtering of Native American herbal knowledge in the United States. Such a process, as widely discussed in the literature, involved a procedure of "sanitization" that subtracted from the material culture being listed the very elements that made it work. In this way, the pharmacopoeias of the nineteenth century aimed to bring into printed traditions secretive knowledge without, in reality, moving the social and religious power that made medicines powerful into the public realm of the new nation-states.[11]

It would be unproductive to rehearse here the well-known pitfalls of examining early modern pharmacopoeias under modern biomedical or nationalistic frames of analysis—a type of approach that the chapters in this volume assiduously avoid.[12] Early modern practitioners, patients, apothecaries, and pharmacists, after all, were continuously reinventing connections between the flesh of human bodies, the natural world, the celestial sphere, and the social order that had little to do with the explanatory epistemologies that emerged in the nineteenth century about the workings of bodies and medicines. As the chapters in this volume show, instead, the impetus for the development of early modern pharmacopoeias came from an appetite for other types of innovations.

A close reading of the materiality of the processes involved in the making of early modern Atlantic medicinal substances uncovers rich dynamics of historical change. Such transformations, however, did not only come about through the desacralization of canonical medical texts, often seen as bounded by attachment to rigid institutional and corporative controls. Pharmacopoeias, as De Vos shows, can make evident strategies for empirical innovation within Galenic or Hippocratic frameworks that had little to do with notions of modernity, or with the prevailing narrative of experimental revolution associated with the very uneven process that came to be known as the Scientific Revolution.[13] The *new* also came from revolutionary processes of linguistic, epistemological, and ontological translation. Increasingly, from the sixteenth century onward, ventriloquists of nature living in Atlantic locales found themselves immersed in a veritable Tower of Babel. This era saw the emergence of vibrant and fluid global markets where people speaking languages as diverse as Castilian, Portuguese, French, Dutch, English, and dozens of Amerindian, Ewe/Gbe/Fon, and Bantu languages, among many others, came together to discuss the nature and properties of medicinal substances.

Atlantic lists of materia medica also provided specific, if illusory, boundaries of the very materiality they described—limits that were a product of the ascription of provincial, European mostly, ontological and epistemological frameworks posing as universal to explain how medicines functioned. Material culture related to healing thus allowed for the articulation of distinct notions of belonging and separation. The examination of the writings about medicine and materia medica of early modern Europeans like Hans Sloane, as Ryan's chapter argues, can reveal the "intimacy of the colonial relation." These writings evidence "the epistemological, representational, and social instabilities of the colonial space" and open windows for historians to analyze not only colonial ambitions and commercial drives but also the fears and intense competition driving the modeling of early modern Atlantic medical worlds.[14] Medicines also functioned as a tool for the elaboration of languages and geographies of embodied similarities and the claiming of new imagined landscapes—like those of New France, as Parsons shows. The case of the *capillaire* is particularly suggestive of how otherwise "unremarkable plant[s]" were closely associated with the conceptualization of early modern European empires' nature.[15]

But for all that they cemented, early modern Atlantic materia medica and their imagination also created worlds through disruption—through "disturbances" and "unstable geographies." Such instability is apparent in the uneven processes of adoption and attempts at classification that drove metropolitan Spanish attempts at codification of New World's materia medica, as Crawford demonstrates in his piece.[16] Lists of medicinal substances in the early modern era, after all, classified not only physical but also numinous therapeutic agents—Mohegan, Montaukett, Congo, and Bran, for instance—and were robust indexes of social relationships. These were vibrant, unstable compilations that could perform reflective, classifying, and prescribing labor. Lists of materia medica allow historians to map filial, commercial, religious, and political networks, and the multiplicity of social constraints that are brought upon by them.

As a result, issues of proprietary recipes, privacy, and fierce competition for profit also figure prominently in these histories. Pharmacopoeias make transparent attempts of elite groups at producing strategies for professional control, and the answers that those outside of circles of power developed for the negotiation of specific modes of securing financial and social standing associated with the trade of specific medicinal substances. This is apparent in the anxieties behind the publication of the "secret" recipe of *orviétan*, as Rivest eloquently explores in his chapter.[17] Attempts at secrecy also figure prominently, as Walker shows, in the history of the medicinal trade developed by Jesuits, and in the similar dynamics modeling the actions of Portuguese crown agents angling for economic advantages out of proprietary materia medica coming from Asia and the Americas.[18]

Pharmacopoeias were, thus, not surprisingly, closely aligned with the articulation of imperial designs and plans, and reflect idiosyncratic preoccupations about the often-cited "imperial body." Medicinal substances' value as strategic imperial objects relied little on the "scientifically" identifiable effectiveness they carried, and depended, rather, on the epistemologies created around them—be they related to creation of colonial others, reification of ideas about the value of ancient knowledge, or to the production of imaginaries about valuable networks of commercial exchange.[19] These "suturing objects," such as Walker, Crawford, and Parsons show, braid together networks of belonging and recognition.[20]

Lists of plants and medicines also contain clues to otherwise invisible encounters and the intensely personal nature of early modern materia medica's crafting. Wisecup, for instance, makes a strong case for considering lists of herbals—at least some of them—as formal media of communication that serve for the examination of Native American histories. The lists she discusses link histories, languages, and lifeworlds.[21] They also make apparent another paradox of these lists: as ultimate arbiters of what supposedly circulated, they reaffirmed the impossibility for the dissemination of the power that ultimately made the substances they itemized work.

The power of early modern pharmacopoeias, indeed, relied on what went unsaid in them. Much like Michael Taussig's "public secrets"—an intricate linkage of half-truths, fantasies, and omissions articulated in social codes sustaining political, social, religious, and cultural regimes—pharmacopoeias made invisible the power of the unsaid that supported their creators' attempts at structuring medicinal-substance usage both in Europe and in other Atlantic locales.[22] This is because the pharmacopoeias themselves were based on the very make-believe processes that Europeans, paradoxically, increasingly used to describe and classify substances outside these lists as coming from insalubrious places, inhabited by "primitive" peoples that worked not on "marvelous" or "miraculous" terms but on "magic" ones. Such processes, as Breen proposes in his contribution to this volume, might be behind the scarce incorporation of the medicinal knowledge of people of African descent in Western European early modern Atlantic pharmacopoeias—even if in practice, many of these medicines were in abundant use. However, as Breen argues, the specific material, commercial and social circumstances surrounding the different commercial and colonial projects that Europeans developed in different parts of the Atlantic (the slave trade in particular), were also crucial, perhaps even more so than other factors, in defining what substances would become part of the therapeutic armamentariums circulating in European medical circles. In other words, access through trade to materia medica, possibilities for profit, anxieties about emerging ideas about race, and the presence or absence of well-connected—academically, socially, and commercially—historical actors were definitive shapers of these lists.[23]

Early modern historical actors involved in the process of creating authoritative lists of substances with bodily effect, either in print, or transmitted through oral traditions, depended, thus, on the purposeful use of a series of "unknown knowns."[24] These historical actors, in other words, actively chose to unknow, through a series of practices related to the production of value, certain kinds of healing practices and elements to which they had access, and prominently, the origin, social customs, spiritual, and cultural practices that gave the early modern medical substances they incorporated power—and not only to heal but also to kill—in the very local contexts in which they were deployed. Furthermore, the process of the creation of authority also required the active process of ignoring the similarities between explanatory frameworks for the working of substances with bodily effect in "others'" medicinal or ritual practices. In doing so, the methods that allowed practitioners, and later imperial and nation-state apparatuses, to create enlightened taxonomies and primitive lists of magical substances involved both practices of active learning and strategies of deliberate forgetting.

Finally, it is crucial to recognize that African, Amerindian, and Atlantic creole practitioners were as invested as their European counterparts in processes of classification, borrowing, and adoption of medicinal substances of all origins. They were, in other words, similarly interested in crafting economic, social, and political power from the meteria medica circulating in the Atlantic.[25] They searched for materials with powers over the body and explored the natural and social landscape of the Atlantic to imagine therapeutic and diseasing arsenals. The curiosity and artifice of "the others" of the Atlantic—which were also immersed in the reimagination of the early modern world—puts them in the company of a large international community preoccupied with the "the secrets of nature."[26] Such appetite, as was the case with Europeans, was primordially related to the concrete results that shaping materia medica's usage would have on their economic and social lives. The fact that their historical record of non-European health practitioners cannot be easily excavated does not mean that they were not interested in projects of exchange, cataloging, and obscuring of medicinal substances' origins as much as Europeans were.

Pharmacopoeias carried with them the resemblance of representation—a depiction that was by its nature and inception only partial. The purpose of the pharmacopeia was to create an idea of order and structure. A framework of codified knowledge that both practitioners and regulators knew was aspirational. A mirage, the pharmacopeia was, nevertheless, powerful. It was the physical representation of a regime sustained in what could not be said—aspirational attempts at bringing to light what functioned on the basis of obscurity.

NOTES

INTRODUCTION: Thinking with Pharmacopoeias

1. Ole Worm, *Museum Wormianum: seu historia rerum rariorum, tam naturalium, quam artificialium, tam domesticarum, quam exoticarum, quae hasniae danorum in aedibus authoris fervantur* (Lugduni Batavorum: Apud Iohannem Elsevirium, 1655).

2. Paula Findlen, *Possessing Nature: Museums, Collecting, and Scientific Culture in Early Modern Italy* (Berkeley: University of California Press, 1994); Oliver Impey and Arthur MacGregor, eds., *The Origins of Museums: The Cabinet of Curiosities in Sixteenth- and Seventeenth-Century Europe* (Oxford: Clarendon, 1985).

3. Royal College of Physicians of London, *Pharmacopoeia Londinensis: of 1618*, reproduced in facsimile with a historical introduction by George Urdang (Madison: State Historical Society of Wisconsin, 1944).

4. Jole Shackelford, "Documenting the Factual and the Artifactual: Ole Worm and Public Knowledge," *Endeavour* 23 (1999): 65–71; see also: H. D. Schepelern, "The Museum Wormianum Reconstructed: A Note on the Illustration of 1655," *Journal of the History of Collections* 2 (1990): 81–85.

5. Worm, "Praefatio ad Lectorem," in *Museum Wormianum*, nn. A4v.

6. Worm, "De Vegetabilibus Rarioribus," in *Museum Wormianum*, 137–228.

7. Francisco Ximénez, *Quatro libros. De la naturaleza y virtudes de las plantas y animales que estan recividos en el uso de Medicina en la Nueva España* (Mexico: Casa de la Viuda de Diego Lopez Davalos, 1615); see also: Miguel Figueroa Saavedra and Guadalupe Melgarejo Rodríguez, "La *Materia Medicinal de la Nueva España* de Fray Francisco Ximénez. Reapropiación y resignificación del conocimiento médico novohispano," *Dynamis* 38 (2018): 219–41; Simon Varey, Rafael Chabrán, and Dora B. Weiner, eds., *Searching for the Secrets of Nature: The Life and Works of Dr. Francisco Hernández* (Stanford: Stanford University Press, 2002).

8. Michael Flannery, "Introduction," in *The English Physician*, ed. Michael Flannery (Huntsville: University of Alabama Press, 2007), 4.

9. Nicholas Culpeper, *The English Physician, or An astrological Discourse on the vulgar herbs of this Nation being a compleat method of physick, whereby a man may preserve his body in health, or cure himself being sick for three pence charge, with such things only as grow in England, they being most fit for English bodies* (London: Printed by William Bentley, 1652).

10. For a detailed examination of the publishing history of Culpeper's texts in London in the latter half of the seventeenth century, see Mary Rhinelander McCarl, "Publishing the Works of Nicholas Culpeper, Astrological Herbalist and Translator of Latin Medical Works in Seventeenth-Century London," *Canadian Bulletin of Medical History* 13 (1996): 225–76.

11. Nicholas Culpeper, *The English Physician. Containing admirable and approved remedies for several of the most usual diseases* (Boston: Reprinted for Nicholas Boone, 1708). Historians sometimes claim that this book was the first medical text printed in British North America. For example, see Flannery, "Introduction," 12; David L. Cowen, "The Boston Editions of Nicholas Culpeper," *Journal of the History of Medicine and Allied Sciences* 11 (1956): 156–65.

12. Culpeper, *English Physician* (Boston: 1708), 58.

13. A cursory search on JSTOR reveals that this usage of the term "pharmacopoeia" is most common in anthropological and ethnobotanical scholarship; see: Robert Voeks, "Disturbance Pharmacopoeias: Medicine and Myth from the Humid Tropics," *Annals of the Association of American Geographers* 94 (2004): 868–88; Kevin D. Janni and Joseph W. Bastien, "Exotic Botanicals in the Kallawaya Pharmacopoeia," *Economic Botany* 58 (2004): S274–S279; Bradley C. Bennett and Ghilean T. Prance, "Introduced Plants in the Indigenous Pharmacopoeia of Northern South America," *Economic Botany* 54 (2000): 90–120.

14. Pablo Gómez, *The Experiential Caribbean: Creating Knowledge and Healing in the Early Modern Atlantic* (Chapel Hill: University of North Carolina Press, 2017).

15. Matthew James Crawford, *The Andean Wonder Drug: Cinchona Bark and Imperial Science in the Spanish Atlantic World, 1630–1800* (Pittsburgh: University of Pittsburgh Press, 2016).

16. The term "pharmacopoeic forms" comes from the title of Kelly Wisecup's essay in this volume (chapter 9).

17. A. C. Courtwright, *The British Pharmacopoeia, 1864–2014: Medicines, International Standards, and the State* (Farnham: Ashgate, 2015); David L. Cowen, *Pharmacopoeias and Related Literature in Britain and America, 1618–1847* (Burlington: Ashgate, 2000); Jacques Saller, *La pharmacopée française, dans l'évolution scientifique, technique et professionnelle* (Metz: Maisonnueve, 1969).

18. Lee Anderson and Gregory J. Higby, *The Spirit of Voluntarism: A Legacy of Commitment and Contribution: The United States Pharmacopoeia, 1820–1995* (Rockville, MD: United States Pharmacopeial Convention, 1995); Joseph M. Gabriel, *Medical Monopoly: Intellectual Property Rights and the Origins of Modern Pharmaceutical Industry* (Chicago: University of Chicago Press, 2014).

19. N. Jardine, J. A. Secord, and E. C. Spary, eds., *Cultures of Natural History* (Cambridge: Cambridge University Press, 1996); Michel Foucault, *The Order of Things: An Archaeology of the Human Sciences* (New York: Vintage, 1994).

20. Harold J. Cook and Timothy D. Walker, "Circulation of Medicine in the Early Modern Atlantic World," *Social History of Medicine* 26 (2013): 337–51.

21. Daniela Bleichmar, "Books, Bodies, and Fields: Sixteenth-Century Transatlantic Encounters with New World Materia Medica," in *Colonial Botany: Science, Commerce, and Politics in the Early Modern World*, ed. Londa L. Schiebinger and Claudia Swan (Philadelphia: University of Pennsylvania Press, 2005), 83–99; see also Kelly Wisecup, *Medical Encounters: Knowledge and Identity in Early American Literature* (Amherst: University of Massachusetts Press, 2013).

22. Londa Schiebinger, *Secret Cures of Slaves: People, Plants, and Medicine in the Eighteenth-Century Atlantic World* (Stanford: Stanford University Press, 2017); James H. Sweet, *Domingos Álvares, African Healing, and the Intellectual History of the Atlantic World* (Chapel Hill: University of North Carolina Press, 2011).

23. Dániel Margócsy, *Commercial Visions: Science, Trade, and Visual Culture in the*

Dutch Golden Age (Chicago: University of Chicago Press, 2015); Daniela Bleichmar, *Visible Empire: Botanical Expeditions and Visual Culture in the Hispanic Enlightenment* (Chicago: University of Chicago Press, 2012); Neil Safier, *Measuring the New World: Enlightenment Science and South America* (Chicago: University of Chicago Press, 2008); Susan Scott Parrish, *American Curiosity: Cultures of Natural History in the Colonial British Atlantic World* (Chapel Hill: University of North Carolina Press, 2006).

24. Abena Dove Osseo-Asare, *Bitter Roots: The Search for Healing Plants in Africa* (Chicago: University of Chicago Press, 2014); Kristin Peterson, *Speculative Markets: Drug Circuits and Derivative Life in Nigeria* (Durham: Duke University Press, 2014); Gabriela Soto Laveaga, *Jungle Laboratories: Mexican Peasants, National Projects and the Making of the Pill* (Durham: Duke University Press, 2009); Jeremy A. Greene, *Prescribing by the Numbers: Drugs and the Definition of Disease* (Baltimore: Johns Hopkins University Press, 2008); Marcy Norton, *Sacred Gifts, Divine Pleasures: A History of Tobacco and Chocolate in the Atlantic World* (Ithaca: Cornell University Press, 2004); Clifford Foust, *Rhubarb: The Wonder Drug* (Princeton: Princeton University Press, 1992)

25. Alison Games, "Atlantic History: Definitions, Challenges, and Opportunities," *American Historical Review* 111 (2006): 741–57; David Armitage, "Three Concepts of Atlantic History," in *The British Atlantic World, 1500–1800*, 11–27, ed. David Armitage and Michael J. Braddick (New York: Palgrave Macmillan, 2002).

CHAPTER 1: Pharmacopoeias and the Textual Tradition in Galenic Pharmacy

1. Archivo General de la Nación, México (hereafter AGN/M), Civil, V. 72, Exp. 9, 1774. "Autos de Concurso de Acreedores a Bienes de Don Jacinto Herrera Dueno de Botica en esta Ciudad." All descriptions of the pharmacy and its contents come from two different inventories included in this document beginning September 14, 1775, and June 7, 1776, on fs. 340v–362v and 397r–411v, respectively.

2. Other Spanish pharmacopoeias include that of Barcelona (1511) and Zaragoza (1546). See *Concordie Apothecariorum Barchinone 1511, la primera farmacopea española* (Tarragona: Colegio Oficial de Farmacéuticos de Tarragona, 1941); Rafael Folch Andreu, 1948, "La Concordia o Farmacopea de Zaragoza de 1553"; "Las Farmacopeas Nacionales españolas," *Actas del XV Congreso Internacional de Historia de la Medicina* 1 (1956): 247–67.

3. *Recopilación de leyes de Indias*. See also María Soledad Campos Díez, *El Real Tribunal del protomedicato castellano, siglos XIV–XIX* (Cuenca: University de Castilla La Mancha, 1999), 76. Philip II had made a call for such a standard in 1592 but it did not come to fruition until 1739—see Matthew Crawford's essay (chapter 3). However, Charles Davis and María Luz López Terrada, "Protomedicato y farmacia en Castilla a finales del siglo XVI: Edición crítica del *Catálogo de las cosas que los boticarios han de tener en sus boticas*, de Andrés de Zamudio de Alfaro, Protomédico General (1592–1599)," *Asclepio* 42 (2010): 579–626, have uncovered a catalogue that did result from these orders and formed the basis for a 1791 Petitorio.

4. *Pharmacopea Matritense*, "Decreto del Tribunal del Real Prot-Medicato" (Madrid: Typographia Regia D. Michaelis Rodriguez, 1739).

5. They also had to buy each new updated edition when it appeared—in 1762, 1794, 1797, 1803, and 1817. "Decreto del Tribunal del Real Prot-Medicato," *Pharmacopoeia Matritense*, "Despacho del Real Tribunal del Proto-Medicato" (Madrid: Typis Antonii Perez de Soto, 1762); and *Pharmacopea Hispana* (Madrid: Typographia Ibarriana, 1794).

6. For Galenic medicine, see Oswei Temkin, *Galenism: Rise and Decline of a Medical Philosophy* (Ithaca: Cornell University Press, 1973); essays in *'On Second Thought' and Other Essays in the History of Medicine and Science* (Baltimore: Johns Hopkins University Press, 2002); and *The Double Face of Janus and Other Essays in The History of Medicine* (Baltimore: Johns Hopkins University Press, 2006); Luis Garcia Ballester et al., *Galen and Galenism: Theory and Medical Practice from Antiquity to the European Renaissance* (Aldershot: Ashgate, 2002); and *Galeno en la sociedad y en la ciencia de su tiempo* (Madrid: Ediciones Guadarrama, 1972).

7. I discuss this in my book manuscript, tentatively titled "The Foundations and Development of Galenic Pharmacy in the Western Tradition." This is not to say that the New World context had no impact on Galenic pharmacy in Mexico. American substances such as hipecac, guaiacan, and cinchona were listed in pharmacy inventories and outside the cities a hybrid kind of "rustic pharmacy" brought together Indigenous and Galenic traditions more fully. In the cities, however, these influences remained minimal: only 12% of substances in pharmacies inventoried were of American origin.

8. Pamela Smith, *The Body of the Artisan* (Chicago: University of Chicago Press, 2004), 142. See also Smith, "Science on the Move: Recent Trends in the History of Early Modern Science," *Renaissance Quarterly* 62 (2009): 345–75, 353, where she discusses the "pseudosciences" of alchemy, astrology, and magic, all of which "possessed long disciplinary histories with both practical and textual components reaching back to antiquity."

9. Esteban de Villa, *Examen de boticarios* (Burgos: por Pedro de Huydobro, 1632), chapter III, 13v and 14r (page is misnumbered as 12).

10. Martínez de Leache, *Discurso pharmaceutico, sobre los Canones de Mesue*, Proemio (no pagination), "los Empiricos, que applican medicinas conforme vieron sin mas fundamento que averlo sonado, o averlo visto aplicar a alguno."

11. Martinez de Leache, *Discurso Pharmaceutico*, 74, 78.

12. Martinez de Leache, *Discurso Pharmaceutico*, 74.

13. I compiled this corpus of works first noting the books that were listed in archival inventories of eighteenth-century Mexican pharmacies. I then noted the authors referenced in those sources to gain a broader sense of the pharmaceutical corpus. I corroborated these findings with bibliographic searches in modern library catalogues as well.

14. This habit of referencing, however, carried only through the seventeenth century for the pharmaceutical genre. Only one eighteenth-century work, Francisco de Brihuega's *Examen Pharmaceutico Galeno-Chimico Teorico-Practico* (1776), regularly referred to other authors for authority.

15. Chiara Crisciani, "History, Novelty, and Progress in Scholastic Medicine," *Osiris*, 2nd ser., 6 (1990): 118–39. See also Leigh Chipman, *The World of Pharmacy and Pharmacists in Mamlūk Cairo* (Leiden: Brill, 2010). Juan Antonio Castels, *Theorica y pratica de boticarios en que se trata de la arte y forma como se han de componer las confectiones ansi interiores como Exteriores* (Barcelona: en casa de Sebastian de Cormellas, 1592) was particularly helpful as it contained a list of 91 "names of authors to whom this book refers."

16. The author in this case most probably used a pseudonym of the ninth-century physician and medical author Yuhanna ibn Masawaih and is often referred to as pseudo-Mesue or Mesue Junior. For more information on the complexities of this author's identity, see Paula De Vos, "The 'Prince of Medicine,' Yūhannā ibn Māsawayh and the Foundations of the Western Pharmaceutical Tradition," *Isis* 104, no. 4 (2013): 667–712.

17. I have shown elsewhere the lasting influence of *De materia medica* on compilations of European materia medica through the nineteenth century, where the simples named by Dioscorides continued to show a correlation of as much as 60%. See Paula De Vos, "European Materia Medica in Historical Texts: Longevity of a Tradition and Implications for Future Use," *Journal of Ethnopharmacology* 132, no. 1 (2010): 28–47. Similarly, I have discussed Mesue's place in the foundations of the western pharmaceutical tradition in "'Prince of Medicine.'"

18. Oswei Temkin, "Byzantine Medicine: Tradition and Empiricism," *Dumbarton Oaks Papers* 16 (1962): at 97, Philip van der Eijk, "Principles and Practices of Compilation and Abbreviation in the Medical 'Encyclopaedias' of Late Antiquity," in *Condensing Texts—Condensed Texts*, ed. M. Horster and C. Reitz (Stuttgart: de Gruyter, 2010) 519–54.

19. Only one medieval pharmaceutical author, Gilberto Anglico, came from a non-Mediterranean region, and since he was trained in Italy I have included him in southern Europe.

20. Of course, these findings derive from a survey of Spanish works that may very well have privileged a southern tradition, so further research is needed to corroborate or challenge these findings. A comparative survey of references in northern European pharmacy texts would reveal the extent to which this was a shared tradition. At the same time, it is well established that the translation of Arabic medical and pharmacological texts greatly influenced developments in medieval Latin medicine and pharmacy, especially in the medieval medical schools of Montpellier, Paris, Padua, and Bologne. The translations largely occurred in and disseminated from Salerno and Toledo, so in essence this tradition entered Europe from these southern centers.

21. Saladino da Ascoli, *Compendio de los boticarios*, trans. Alonso Rodríguez de Tudela (Salamanca, 1515; facsímile edition, Extramuros Edición, 2008), 3r.

22. Ascoli, *Compendio*, 3r.

23. Pedro Benedicto Mateo, *Loculentissimi viri . . . Liber in examen apothecariorum . . .* (Barcelona: Joan Rosembach, 1521), CIIv (f. 102v).

24. Cristobal Suarez de Figueroa, trans., *Plaza universal de todas ciencias y artes* (Madrid: por Luis Sanchez, 1615).

25. Villa, *Examen*, "Al Lector," 5v, and chapter 1, fs. 1r–12v.

26. Villa, *Examen*, 12r.

27. Joël Coste, "La médicine pratique et ses genres littéraires en France à l'époque moderne," Bibliotheque interuniversitaire de Médicine, Paris; Manfred Ullmann, *Islamic Medicine* (Edinburgh: Edinburgh University Press, 1978), 103; and Carmélia Opsomer, "La Pharmacopée Medieval: Images et Manuscrits," *Journal de Pharmacie de Belgique* 57 (2002): 3–10.

28. Danielle Jacquart, *La medicine médiévale dans le cadre parisien: XIV–XV siècle* (Paris: Fayard, 1998), 97; Crisciani, "History, Novelty, and Progress in Scholastic Medicine"; Irma Taavitsainen, and Paivi Pahta, "Corpus of Early English Medical Writing, 1350–1750," *ICAME Journal: Computers in English Linguistics* 21 (1997): 71–78; and Paivi Pahta and Irma Taavitsainen, "Vernacularization of Scientific and Medical Writing in Sociohistorical Context," in *Medical and Scientific Writing in Late Medieval English*, ed. Irma Taavitsainen and Paivi Pahta (Cambridge: Cambridge University Press, 2004), 1–2. See also Linda Ehrsam Voigts, "Scientific and Medical Books" in *Book Production and Publishing in Britain 1375–1475*, Jeremy Griffiths, Derek Pearsall, eds. (Cambridge: Cam-

bridge University Press, 2007): 345–402. See also Monica Azzolini, "In Praise of Art: Text and Context of Leonardo's *Paragone* and its Critique of the Arts and Sciences," *Renaissance Studies* 19, no. 4 (2005): 487 for a brief discussion of the importance of rhetoric in medical and scientific works. See also Jole Agrimi and Chiara Crisciani, *Edocere medicos: medicina scolastica nei secoli XIII–XV* (Naples: Guerini e associati, 1988), 237–73.

29. Taavitsainen and Pahta, "Corpus of Early English Medical Writing, 1375–1750," 71–72.

30. For histories of herbals, see Charles Singer, "The Herbal in Antiquity and Its Transmission to Later Ages," *Journal of Hellenic Studies* 47, no. 1 (1927): 1–52; Vivian Nutton and John Scarborough, "The *Preface* of Dioscorides' Materia Medica: Introduction, Translation and Commentary," *Transactions and Studies of the College of Physicians of Philadelphia* IV (1982): 187–227; Agnes Arber, *Herbals: Their Origin and Evolution*, 3rd ed. (Cambridge: Cambridge University Press, 1987, orig. pub. 1912); Graeme Tobyn, Alison Denham, and Margaret Whitelegg, *The Western Herbal Tradition: 2000 Years of Medicinal Plant Knowledge* (London: Elsevier Health Sciences, 2010); and Anne Van Arsdall and Timothy Graham, eds., *Herbs and Healers from the Ancient Mediterranean through the Medieval West: Essays in Honor of John M. Riddle* (Ashgate Publishing, Ltd, 2012).

31. See *The Seven Books of Paulus Aegineta*, trans. Francis Adams (London: Printed for the Sydenham Society, 1846), John Riddle, "The Introduction and Use of Eastern Drugs," and an article on a treatise written by Ibn Māsawaih /Mesue Senior, Martin Levey, "Ibn Māsawaih and His Treatise on Simple Aromatic Substances," *Journal of the History of Medicine and Allied Sciences* 16 (1961): 394–410. Levey's research shows that Romans knew of many of these plants, but they are generally not named by Dioscorides and are attributed to Arab-Islamic introductions. See also Dimitri Gutas, *Greek Thought, Arabic Culture: The Graeco-Arabic Translation Movement in Baghdad and Early 'Abbasaid Society (2nd-4th/5th-10th c.)* (London: Routledge, 1998), 25.

32. Nutton, "Matthiolo and the Art of the Commentary."

33. See Otto Raubenheimer, "History of Substitutes and Substitution," paper prepared for the Section on Commercial Intersts, Symposium on Substitution, American Pharmaceutical Association, Atlantic City meeting, 1916, 50–55. Raubenheimer, 52, gives three main reasons for using drug substitutes: "1) As a matter of convenience for physician, pharmacist, and patient. 2) As a matter of necessity. 3) Because frequently certain drugs were unobtainable, as commerce to the Orient, which was the source of most drugs, became interrupted."

34. Raubenheimer, "History of Substitutes and Substitution."

35. Vivian Nutton, "Ancient Mediterranean Pharmacology and Cultural Transfer," *European Review* 16, no. 2 (2008): 213. For a modern example of the continued problem of precise plant identification with regard to the plant "rue," see Pollio et al., "Continuity and Change in the Mediterranean Medical Tradition: *Ruta* spp. (rutaceae) in Hippocratic Medicine and Present Practices," *Journal of Ethnopharmacology* 116, no. 3 (2008): 469–82.

36. Nutton, "Ancient Mediterranean Pharmacology," 213.

37. Interestingly, the need for drug synonyms still exists, though the context is very different and due not to language differences but to various names given to the same chemically synthesized drug by different drug companies. See G. W. A. Milne, ed., *Drugs: Synonyms & Properties* (Burlington, VT: Ashgate, 1999).

38. Bernardino de Laredo, *Modus faciendi cum ordine medicando* (Sevilla: por Jacobo Cromberger, 1527), facsimile edition, ed. Milagro Laín and Doris Ruiz Otín (Madrid:

Ediciones Doce Calles, 2001), 225. For information on the etymology of the term, see Chipman, *World of Pharmacy and Pharmacists in Mamlūk Cairo*, 13, and Sābūr ibn Sahl, *Dispensatorium parvum (al-Aqrabadhin al-saghir)*, ed. and trans. By Oliver Kahl (Leiden: Brill, 1994), 3n8.

39. On Scribonius, see Vivian Nutton, "Scribonius Largus, The Unknown Pharmacologist," *Pharmaceutical Historian* 25 (1995): 5–8.

40. On these treatises, see Sabine Vogt, "Drugs and Pharmacology," in *The Cambridge Companion to Galen*, ed. R. J. Hankinson (Cambridge: Cambridge University Press, 2008); Cajus Fabricius, *Galens Exzerpte aus älteren Pharmacologen* (Berlin: Walter de Gruyter, 1972), and Laurence Totelin, "And to end on a poetic note: Galen's Authorial Strategies in the Pharmacological Books," *Studies in History and Philosophy of Science* 43 (2012): 307–15. These texts were written between 180 and 193, probably simultaneously, and were rewritings of earlier works that had been destroyed in a fire. Although they were of substantial length, they included a relatively narrow set of types of compounds, with only sporadic discussion of technique or rationale for compounding. Despite the promising focus in *Kinds*, discussion of the different types of preparations is largely unsystematic. The chapters do not each consistently treat a different kind of compound but rather focus on individual recipes randomly with a prevalence of chapters on plasters; and some chapters are still organized around the diseases to be treated.

41. See Paula De Vos, "'Prince of Medicine,'" for these arguments. For Arabic formularies, see Chipman, *World of Pharmacy*, 13: and Sabur ibn Sahl, *Dispensatorium parvum (al-Aqrabadhin al-saghir)*, 3n8. See these and the following works regarding the structure and characteristics of the Arabic formulary discussed here: Oliver Kahl, *The Dispensatory of Ibn at-Tilmīḏ* (Leiden: Brill, 2007); Sābūr ibn Sahl, and Oliver Kahl, *Sābūr ibn Sahl's Dispensatory in the Recension of the 'Aḍudī Hospital* (Leiden: Brill, 2009); and *Sābūr ibn Sahl, The Small Dispensatory* (Leiden, Brill, 2003). See also Martin Levey, *Early Arabic Pharmacology* (Leiden: Brill, 1973) (though see Kahl's serious reservations as to its reliability in *Dispensatory of Ibn at-Tilmīḏ*, 3); Abū Yūsuf Yaʾkūb ibn Ishāk ibn Subbāh, al-Kindī, and Martin Levey, *The Medical Formulary or Aqrabadhin of Al-Kindī* (Madison: University of Wisconsin Press, 1966); and Martin Levey and Noury al-Khaledy, *The Medical Formulary of Al-Samarqandī and the Relation of Early Arabic Simples to Those Found in the Indigenous Medicine of the Near East and India* (Philadelphia: University of Pennsylvania Press, 1967).

42. Laurence M. V. Totelin, "Mithradates' Antidote: A Pharmacological Ghost," *Early Science and Medicine* 9, no. 1 (2004): 1–19; and Gilbert Watson, *Theriac and Mithridatium: A Study in Therapeutics* (London: Wellcome, 1966); Michael McVaugh, "Theriac at Montpellier 1285–1325" *Sudhoffs Archiv* 56 (1972): 113–44; and Gary Leiser and Michael Dols, "Evliya Chelebi's Description of Medicine in Seventeenth-Century Egypt," *Sudhoffs Archiv* 71 (1987): 197–216, and 73 (1988): 49–68.

43. Chipman, *World of Pharmacy*, 107. Chipman explores the relationship between cookbooks and formularies and finds similar descriptions of how to prepare syrups.

44. The twelve method-based categories included decoctions, infusions, electuaries, confections, lozenges, lambatives, ointments, oils, plasters, pills, powders, and syrups.

45. For professionalization of pharmacy in the late Middle Ages, see Walton O. Schalick, "Add One Part Pharmacy to One Part Surgery and One Part Medicine: Jean De St. Amand and the Development of Medical Pharmacology in Thirteenth Century Paris" (PhD diss., Johns Hopkins University, 1997).

46. This may be similar to what Crisciani has discussed as the "operative" knowledge for medieval alchemy, in which alchemy was excluded from university curricula due to its operative nature, or its "understanding through effect." However, she argues, along with M. Pereira, that this knowledge was not without principles, theory, or philosophy and was, in fact, what amounted to "developing a philosophy rooted in the contact with matter and its activity." See Chiara Crisciani, "Alchemy and Medieval Universities: Some Proposals for Research," *Universitas* 10 (1997). James Bennett uses the term "operative knowledge" in discussing mathematics and scientific instruments but it is very useful here—a "knowing by doing." See J. A. Bennett, "Practical Geometry and Operative Knowledge," *Configurations* 6, no. 2 (1998): 195–222, especially 220–22.

47. *Servidor de albuchasis*, f. ii.

48. *Servidor de albuchasis*, Tratado primero, fs. iii verso–xiiii verso.

49. *Servidor de albuchasis*, Tratado segundo, fs. xiiii verso–f. xliiii recto.

50. *Servidor de albuchasis*, Tratado tercero, fs. xliiii verso–xlviii verso. For more on this, see De Vos, "Rosewater and Philosophers' Oil: Thermo-Chemical Processing in Medieval and Early Modern Spanish Pharmacy," *Centaurus*, accepted for publication, forthcoming.

51. Sami Hamarneh and Glenn Sonnedeker, *A Pharmaceutical View of Abulcasis al-Zahrawi in Moorish Spain* (Leiden: Brill, 1963). See also E. Cordonnier, "Sur le Liber Servitoris d' Aboulcasis," *Janus* 9 (1904): 425–87.

52. See McVaugh, "Practical Pharmacy and Medical Theory at the End of the Thirteenth Century," 3.

53. Luis de Oviedo, *Método de la colección*, f. 32r.

54. Oviedo, *Método*, f. 32v.

55. See Sami Hamarneh, "The Rise of Professional Pharmacy in Islam," 59–66, for an overview of that development.

56. Antonio Aguilera, *Exposicion sobre las preparaciones de Mesue* (Alcalá: en Casa de Juan de Villanueva, 1569), Prefacion, 7v.

57. Alonso de Jubera, *Dechado y reformacion de todas las medicinas compuestas usuales, con declaracion de todas las dudas enellas cōtenidas, assi de los simples que enellos entrā y succedaneos que por los dudosos se hayā de poner, como en el mondo de las hazer* (Valladolid: por Diego Fernandez de Cordova, 1578), "Prologo."

58. See Chipman, *World of Pharmacists*.

59. Saladino da Ascoli, *Compendium aromatarium*, 1r.

60. Saladino da Ascoli, *Compendium aromatarium*, 1r.

61. It should be noted, however, that these two were very important: they consist of the *Pharmacopea Matritense* and the *Pharmacopoeia Hispana*, which became the standard pharmacopoeias throughout the Spanish Empire and went through many editions. In addition, texts within Castilian publications were liberally sprinkled with long quotations in Latin, indicating a reading audience at least somewhat cognizant of the language.

62. Saladino, *Compendium aromatarium*, "Prolego" (second page of manuscript): "Porque los mas de los boticarios de estos reynos carecen de la lingua Latina y no se podrian aprovechar deste tan provechoso libro me parecio cosa muy util y aun necessaria trasladarlo en castellano para que consigan el fruto que dio occasion al dotor Saladino de tomar el trabajo de componerlo."

63. Aguilera, *Exposicion sobre las preparaciones de Mesue*, Prefacion, fs. 8r–v.

64. Jubera, *Dechado y reformación de todas las medicinas compuestas usuales*, prologo.

65. Oviedo, *Método*, "Luys de Oviedo al lector."
66. Castels, *Theorica y Pratica de boticarios*, Epistola al Benigno Lector.
67. Castels, *Theorica y Pratica de boticarios*, Epistola al Benigno Lector.
68. Pedro de Vinaburu, *Cartilla Pharmaceutica Chimico-Galenica* (Pamplona, 1778), Censura de Don Manuel Rodrigo y Andueza.
69. Stephen Clucas, "Galileo, Bruno and the Rhetoric of Dialogue in Seventeenth-Century Natural Philosophy," *History of Science* 46 (2008): 405–29.
70. Mateo, *Liber in Examen Apothecariorum*, fol. iir.
71. Aguilera, *Exposicion sobre las preparaciones de Mesue*, Prefacion, f. 6r. "Fabricar y componer la presente obra y dotrina en forma de dialogo por sus capitulos dividida para que assi por via de demanda y respuesta sea mentilada y rectamente entendida la dicha declaracion y dotrina."
72. Aguilera, *Exposicion sobre las preparaciones de Mesue*, Prefacion, f. 10r. (pages are misnumbered—they go from 6 to 10).
73. Aguilera, *Exposicion sobre las preparaciones de Mesue*, Prefacion, fs. 6r–v.
74. Aguilera, *Exposicion sobre las preparaciones de Mesue*, Prefacion, f. 6v.
75. George Urdang, "The Development of Pharmacopoeias: A Review with Special Reference to the Pharmacopoea Internationalis," *Bulletin of the World Health Organization* 4 (1951): 578.
76. Urdang, "Development of Pharmacopoeias," 578. See also Teresa Huguet-Termes, "Islamic Pharmacology and Pharmacy in the Latin West: An Approach to Early Pharmacopoeias," *European Review* 16, no. 2 (2008): 232. Some of the medieval Arabic formularies also included materia medica as well as recipes for compounds, information on weights and measures, and best practices. Further research is necessary to locate these works within the development of the pharmacopoeia.
77. Saladino, *Compendium aromatarium*, 1r.
78. For the discussions surrounding the establishment of a standard pharmacopoeia in the Spanish Empire, see *Estatutos del Real Colegio de Professores Boticarios de Madrid, aprobados y confirmados por su Magestad, que Dios guarde* (Marid: Imprenta Real, 1737).
79. Allen G. Debus has written extensively about medical chemistry and the chemico-Galenic compromise, especially the debates that ensued regarding these medicines in early modern England and France: see *The Chemical Philosophy: Paracelsian Science and Medicine in the Sixteenth and Seventeenth Centuries* (Dover Publications; Revised edition, 2002; orig. pub. Science History Publications, 1977), *The English Paracelsians* (Chicago: University of Chicago Press, 1968), *The French Paracelsians: The Chemical Challenge to Medical and Scientific Tradition in Early Modern France* (Cambridge: Cambridge University Press, 1977); *Alchemy and Chemistry in the Seventeenth Century* (Los Angeles: William Andrews Clark Memorial Library, University of California, 1966).

CHAPTER 2: Authority, Authorship, and Copying

1. Chollegio degli Eximii Doctorii della Arte et Medicina della Inclita Cipta di Firenze, *Il Nuovo Ricettario Fiorentino (1498): Testo e Lingua*, ed. and trans. Olimpia Fittipaldi (Pluteus, 2011), 18, accessed February 1, 2018, http://www.pluteus.it/wp-content/uploads/2014/01/nuovo%20ricettario.pdf.. For more information, see James Shaw and Evelyn Welch, *Making and Marketing Medicine in Renaissance Florence* (Amsterdam: Brill, 2011); Sali Morgenstern, "The Ricettario Fiorentino. 1567. The Origin of the Art of the Apothecary the Florentine Pharmacopoeias," *Academy Bookman* 30 (1977): 3–12; and

Ernesto Riva, "Confronto tra le prime due edizioni del Ricettario Fiorentino e l'Antidotario mantovano del 1558," *Atti e Memorie* 8 (1991): 99–104.

2. Mesue's pharmaceutical writings were foundational to premodern European pharmacy. He was referred to by "early modern European scholars . . . as 'the Divine Mesue' and 'the Prince of Medicine'"; see: Paula De Vos, "The 'Prince of Medicine': Yuhanna ibn Masawayh and the Foundations of the Western Pharmaceutical Tradition," *Isis* 104 (2013): 667–712. "Niccholao" refers to the *Antidotarium Nicolai*, a twelfth-century recipe book that became a fundamental drug text in European universities. See Maria Jose Carrillo-Linares, "Antidotarium Nicolai," *Oxford Dictionary of the Middle Ages* (Oxford: Oxford University Press, 2010), http://www.oxfordreference.com/view/10.1093/acref/9780198662624.001.0001/acref-9780198662624-e-0382?rskey=s69olU&result=1. Avicenna was a Persian physician whose *Canon on Medicine* was "the basic textbook for the study of medicine in the west." See Charles Burnett and Jonathan Yaeger, "Avicenna (Ibn Sina)," *Oxford Dictionary of the Middle Ages* (Oxford: Oxford University Press, 2010), http://www.oxfordreference.com/view/10.1093/acref/9780198662624.001.0001/acref-9780198662624-e-0651?rskey=a28nrh&result=1. Rhazes was a well-known Persian physician whose *Al Mansore* became an important drug text in Europe. See Ahmed H. al-Rahim, Alain Touwaide, "Razi, Abu Bakr Muhammad ibn Zakariyya al-," *Oxford Dictionary of the Middle Ages* (Oxford: Oxford University Press, 2010), http://www.oxfordreference.com/view/10.1093/acref/9780198662624.001.0001/acref-9780198662624-e-4912?rskey=r5lyoV&result=1.

3. See Stuart Anderson's essay (chapter 10) and Antoine Lentacker's essay (chapter 11) for discussions of later pharmacopoeias.

4. "Twelve Reformers" wrote the 1567 edition and the 1597 edition was again written by the College of Physicians.

5. It is useful to recall that doctors "employed a more limited range of standard products than was listed in the recipe books but modified these extensively in practice in accordance with the condition of the patient." Shaw and Welch, *Making and Marketing Medicine in Renaissance Florence*, 257. According to the College, apothecaries were not invited to make those adjustments themselves. Cristina Bellorini, *The World of Plants in Renaissance Tuscany: Medicine and Botany* (New York: Routledge, 2016), 157.

6. See, for example, Andrew Wear, "Medicine in Early Modern Europe, 1500–1700," in *The Western Medical Tradition 800 B.C. to A.D. 1800*, ed. Lawrence I. Conrad et al. (Cambridge: Cambridge University Press, 1995), 215–62.

7. Nancy Siraisi, *Medieval & Early Renaissance Medicine: An Introduction to Knowledge and Practice* (Chicago: University of Chicago Press, 2004), xi. Cristina Bellorini has made a useful comparison of official recommendations in the *Ricettario Fiorentino* with the evidence of Paracelsian alchemical medicine and activities in the grand dukes' foundry in Renaissance Tuscany. Bellorini, *World of Plants in Renaissance Tuscany*, 155–88. For another perspective on the utility of the *Ricettario Fiorentino*, see Teresa Huguet-Termes, "Standardising Drug Therapy in Renaissance Europe? The Florence (1499) and Nuremberg Pharmacopoeia (1546)," *Medicina & Storia—Saggi* VIII 15 (2008), 77–101.

8. Shaw and Welch, *Making and Marketing Medicine in Renaissance Florence*, 236.

9. Edgecumbe Staley, *The Guilds of Florence* (New York: B. Blom, 1906), 238.

10. Chollegio degli Eximii Doctorii della Arte et Medicina, *Nuovo Ricettario Fiorentino (1498) Testo e Lingua*, 18.

11. Shaw and Welch assert that the College did not have legal authority over apoth-

ecaries until the 1560s. Shaw and Welch, *Making and Marketing Medicine in Renaissance Florence*, 291–92.

12. The *Ricettario* was published in Latin for the first time in 1518. Morgenstern, "Ricettario Fiorentino. 1567," 8.

13. Chollegio degli Eximii Doctorii della Arte et Medicina, *Nuovo Ricettario Fiorentino (1498) Testo e Lingua*, 29.

14. Chollegio degli Eximii Doctorii della Arte et Medicina, *Nuovo Ricettario Fiorentino (1498) Testo e Lingua*, 29. See chapter 1, De Vos, "The Textual Tradition of Pharmacy in Early Modern Spain," 7–8. Rhazes is referred to in these volumes both by his name and by his book, *Almansore*. The *Antidotarium Nicolai* was a well-known pharmaceutical volume from the Salernitan School.

15. The 1498 edition finishes on page 87v. The 1567 edition is 246 pages plus a table of contents and statutes. The 1597 edition is 296 pages plus a table of contents and statutes. For more on the history of collecting and botany, see David Freedberg, *The Eye of the Lynx: Galileo, His Friends, and the Beginnings of Modern Natural History* (Chicago: University of Chicago Press, 2002); Paula Findlen, *Possessing Nature: Museums, Collecting, and Scientific Culture in Early Modern Italy* (Los Angeles: University of California Press, 1994).

16. See Daniela Bleichmar, "Books, Bodies, and Fields: Sixteenth-Century Transatlantic Encounters with New World Materia Medica," in *Colonial Botany: Science, Commerce, and Politics in the Early Modern World*, ed. Londa L. Schiebinger and Claudia Swan (Philadelphia: University of Pennsylvania Press, 2005), 83–99; and Dániel Margócsy, *Commercial Visions: Science, Trade and Visual Culture in the Dutch Golden Age* (Chicago: University of Chicago Press, 2014).

17. Collegio de Medici di Firenze, *Ricettario Fiorentino* (Firenze: Heredi di Bernardo Giunti, 1567), 43.

18. Collegio de Medici di Firenz, *Ricettario Fiorentino*, 1567, 59.

19. Even in the Spanish *Pharmacopoeia Matritensis* (1739), "only 5.2% of all the 'official simples'" were from the Americas, as noted by Matthew Crawford in chapter 3.

20. Morgenstern, "Ricettario Fiorentino," 9. Morgenstern argues that in the later editions of the *Ricettario*, "most of the added new formulas and compounds were then in use and were taken from contemporary physicians" (11–12). Ernesto Riva also argued, in terms of the 1550 edition of the *Ricettario Fiorentino* and the first edition of the *Antidotario* of Mantova, that by the mid-sixteenth century, classical medicine of the Greeks and Romans had been "hung up" in favor of experience of official physicians practicing in the city. Riva, "Confronto tra le prime due edizioni del Ricettario Fiorentino e l'Antidotario mantovano del 1558," 102.

21. When the College of Physicians was reconstituted in 1560, it "assumed most of the control over medicine which had previously been invested in the guild as a whole. It oversaw the licensing of all practice, and advised the state on matters of public health." Katharine Park, *Doctors and Medicine in Early Renaissance Florence* (Princeton: Princeton University Press, 1985), 237–38.

22. Historical catalogues of the Biblioteca Riccardiana can be found at the Fondazione Memofonte's: "Cataloghi ed Inventari della Biblioteca," Fondazione Memofonte, last modified June 11, 2010, http://www.memofonte.it/ricerche/cataloghi-ed-inventari-della-biblioteca.html.

23. See, for example, Sandra Cavallo and Tessa Storey, *Healthy Living in Late Renaissance Italy* (Oxford: Oxford University Press, 2014). For information on intertextuality

and recipe books, see Martti Makin's dissertation, "Between Herbals et alia: Intertextuality in Medieval English Herbals" (University of Helsinki, 2006).

24. Because the authors did not include information about where they were from, it is impossible to say with absolute certainty that the manuscripts were written in Florence. Dates and placement can be approximately determined based on language and paleography. The ways in which the library's collection was accumulated over time also clouds their provenance: the Biblioteca Riccardiana's collection was amassed by various members of the Riccardi family, from before the sixteenth century through nineteenth century. The vernacular Tuscan in which these manuscripts were written is regular and consistent, and there are no explicit comments in the text that would suggest they were written anywhere else.

25. Although this project focuses exclusively on recipes written in Italian, most early modern Tuscan manuscript recipe books (including these five) also include recipes written in Latin. For a complete reference on medical and pharmacological manuscripts in the Biblioteca Riccardiana, see Mahmoud Salem Elsheikh, *Medicina e farmacologia nei manoscritti della Biblioteca Riccardiana di Firenze* (Rome: Vecchiarelli Editore, 1990).

26. Collegio dei Medici, *Ricettario Fiorentino di nuovo illustrato* (Florence: G. Marescotti, 1597), 195.

27. *Segni delle Febbri / Ricettario* (MS 3057) Manuscript, ca. sixteenth century, Biblioteca Riccardiana, Florence, Italy: 45r.

28. Chollegio degli Eximii Doctorii della Arte et Medicina, *Nuovo Ricettario Fiorentino (1498) Testo e Lingua*, 140. For more on this section, see Bellorini, *World of Plants in Renaissance Tuscany*, 157.

29. Elaine Leong and Sara Pennell, "Recipe Collections and the Currency of Medical Knowledge in the Early Modern 'Medical Marketplace,'" in *Medicine and the Market in England and its Colonies, c. 1450–1850*, ed. Mark S. R. Jenner and Patrick Wallis (New York: Palgrave MacMillan, 2007), 138.

30. Leong and Pennell, "Recipe Collections and the Currency of Medical Knowledge in the Early Modern 'Medical Marketplace,'" 134, 149.

31. The author indicated that he or she copied 21 recipes from the writings of "M.o Ben.to mio zio." *Ricettario* (MS 2376), manuscript, ca. sixteenthth century (Biblioteca Riccardiana, Florence, Italy): 98r.

32. The number of recipe author references does not represent unique authors, but rather individual references to recipe authors. Not every recipe has an author, but it should also be noted that some recipes have more than one author. For example, "Mitridato d'Andromaco secondo Galeno" lists both Andromaco and Galen as authors. Il Collegio de Medici, *Ricettario Fiorentino*, 1597, 172.

33. Twenty-five of the first edition authors do not appear in the third edition, and thirty-two of the third edition authors were not included in the first edition.

34. David Gentilcore, *Medical Charlatanism in Early Modern Italy* (Oxford: Oxford University Press, 2006); David Gentilcore, "Apothecaries, 'Charlatans,' and the Medical Marketplace in Italy, 1400–1750," *Pharmacy in History* 45 (2003): 91–94; Pamela Smith, *The Body of the Artisan: Art and Experience in the Scientific Revolution* (Chicago: University of Chicago Press, 2004).

35. Roger French has discussed the increasing preference for Hellenistic sources in the Italian medical world in the late fifteenth century. Roger French, *Dissection and Vivisection in the European Renaissance* (Aldershot: Ashgate, 1999).

36. For more on Mesue, see De Vos, "'Prince of Medicine'"; Teresa Huguet-Termes, "Islamic Pharmacology and Pharmacy in the Latin West: An Approach to Early Pharmacopoeias," *European Review* 16 (2008): 229–39. Huguet-Termes's work confirms that the majority of compound medicines in the 1498 edition of the *Ricettario* were from Arabic authors (232). For more on Islamic pharmacology in general, see Danielle Jacquart, "Islamic Pharmacology in the Middle Ages: Theories and Substances," *European Review* 16 (2008): 219–27; and Vivian Nutton, "Ancient Mediterranean Pharmacology and Cultural Transfer," *European Review* 16 (2008): 211–17.

37. For more on Fuchs, see Sachiko Kusukawa, *Picturing the Book of Nature: Image, Text, and Argument in Sixteenth-Century Human Anatomy and Medical Botany* (Chicago: University of Chicago Press, 2012). For more on naming and botanical exchange, see Margócsy, *Commercial Visions*.

38. The numbers here represent the recipe authors for the recipes that are legible and in Italian in each of the manuscripts. Unfortunately, portions of many of the texts are illegible because of preservation issues, microfilm that is difficult to read, or handwriting that is simply too challenging to decipher.

39. For example, "Tutte le sopradette Ricette ho havuto da A.M.S. in S.M.N. copiate in sur uno suo libro," *Ricettario* (MS 2376), manuscript, ca. sixteenth century, Biblioteca Riccardiana, Florence, Italy, 28r.

40. "Per la cuttura del fuoco del Sr.e imbasciatore di Lucca provato da lui," *Ricettario* (MS 2376), manuscript, ca. sixteenth century, Biblioteca Riccardiana, Florence, Italy, 54r. "Cerotto mag.le per le ferite cosa ottima . . . havuto da un' padre scappuccino usato da M.o Tommaso lignani Cerusico a Citta di Castello." *Ricettario* (MS 2376), manuscript, ca. sixteenth century, Biblioteca Riccardiana, Florence, Italy, 60v–61r.

41. Chollegio degli Eximii Doctorii della Arte et Medicina, *Nuovo Ricettario Fiorentino (1498) Testo e Lingua*, 71.

42. Chollegio degli Eximii Doctorii della Arte et Medicina, *Nuovo Ricettario Fiorentino (1498) Testo e Lingua*, 130–31.

43. Chollegio degli Eximii Doctorii della Arte et Medicina, *Nuovo Ricettario Fiorentino (1498) Testo e Lingua*, 70.

44. Chollegio degli Eximii Doctorii della Arte et Medicina, *Nuovo Ricettario Fiorentino (1498) Testo e Lingua*, 70. The multiple spellings of both author names and recipe titles are common throughout printed and manuscript volumes from this era.

45. Collegio de Medici di Firenze, *Ricettario Fiorentino*, 1567, 225.

46. Collegio de Medici di Firenze, *Ricettario Fiorentino*, 1567, 228–38.

47. Collegio de Medici di Firenze, *Ricettario Fiorentino*, 1567, 127.

48. Collegio de Medici di Firenze, *Ricettario Fiorentino*, 1567, 170–71.

49. *Ricettario* (MS 2376), manuscript, ca. sixteenth century, Biblioteca Riccardiana, Florence, Italy. With a more comprehensive analysis, one might be able to correlate recipes with symptom- or illness-specific titles with author or formal name titled-recipes in the printed volume. Further analysis could find patterns in ingredients to correlate recipes that do not necessarily have explicitly common authors.

50. Bellorini, *World of Plants in Renaissance Tuscany*, 155.

51. Gianna Pomata, "Sharing Cases: The *Observationes* in Early Modern Medicine," *Early Science and Medicine* 15 (2010): 197. Gianna Pomata, "Epistemic Genres or Styles of Thinking? Tools for the Cultural Histories of Knowledge," lecture given at the University of Minnesota, May 10, 2013.

52. Bellorini, *World of Plants in Renaissance Tuscany*, 185. Christian Bec's research on writers and on merchant libraries demonstrates that writing and reading was not restricted to a particular social or economic level. See Christian Bec, "Lo statuto socio-professionale degli scrittori (Trecento e Cinquecento)," in *Letteratura italiana, Volume secondo. Produzione e consumo*, ed. Alberto Asor Rosa (Torino: Giulio Einaudi, 1983), 229–67, and Christian Bec, "I mercanti scrittori," in *Letteratura italiana, Volume secondo. Produzione e consumo*, ed. Alberto Asor Rosa. (Torino: Giulio Einaudi, 1983), 269–97.

CHAPTER 3: An Imperial Pharmacopoeia?

Part of the research for this essay was supported by a Curtis G. Lloyd Fellowship from the Lloyd Library and Museum in Cincinnati, OH, and a Doan Fellowship from the Science History Institute in Philadelphia, PA.

1. Tribunal del Real Protomedicato, *Pharmacopoeia Matritensis Regii ac Supremi Hispaniarum* (Matriti: E Typographia Regia Michaelis Rodriguez, 1739).
2. "Decreto del Tribunal del Real Proto-Medicato," in *Pharmacopoeia Matritensis*, nn. A1v–A2r.
3. "Decreto," nn. A2v.
4. "Decreto," nn. A2v.
5. On the Bourbon Reforms, see Allan J. Kuethe and Kenneth J. Andrien, *The Spanish Atlantic World in the Eighteenth Century: War and the Bourbon Reforms 1713–1796* (Cambridge: Cambridge University Press, 2014); Jorge Cañizares-Esguerra, "'Enlightened Reform' in the Spanish Empire: An Overview," in *Enlightened Reform in Southern Europe and its Colonies*, ed. Gabriel Paquette (Burlington: Ashgate, 2009), 33–37; Gabriel Paquette, *Enlightenment, Governance and Reform in Spain and Its Empire, 1759–1808* (New York: Palgrave Macmillan, 2008); Stanley J. Stein and Barbara H. Stein, *Apogee of Empire: Spain and New Spain in the Age of Charles III, 1759–1789* (Baltimore: Johns Hopkins University Press, 2005).
6. George Urdang, "Pharmacopoeias as Witnesses to World History," *Journal of the History of Medicine and Allied Sciences* 1 (1946): 46–70.
7. Urdang, "Pharmacopoeias," 46–47.
8. Urdang, "Pharmacopoeias," 46–47.
9. Sujit Sivasundaram, ed., "Focus: Global Histories of Science," *Isis* 101 (2010): 95–158; James Delbourgo and Nicholas Dew, eds., *Science and Empire in the Atlantic World* (New York: Routledge, 2007); Londa Schiebinger, ed., "Focus: Colonial Science," *Isis* 96 (2005): 52–63.
10. Federico Marcon has recently offered a similar observation in his excellent study of natural history and natural historical texts in Tokugawa Japan; see Federico Marcon, *The Nature of Knowledge and the Knowledge of Nature in Early Modern Japan* (Chicago: University of Chicago Press, 2017).
11. This essay is certainly not the first to ask about the relationship between pharmacy and empire; see Stuart Anderson, "Pharmacy and Empire: The 'British Pharmacopoeia' as an Instrument of Imperialism, 1864–1932," *Pharmacy in History* 52 (2010): 112–21.
12. Kelly Wisecup, *Medical Encounters: Knowledge and Identity in Early American Literatures* (Amherst: University of Massachusetts Press, 2013); Delbourgo and Dew, *Science and Empire in the Atlantic World*.

13. For various discussions on pharmacopoeias as a genre with reference to the Anglophone context, see David L. Cowen, *Pharmacopoeias and Related Literature in Britain and America, 1618–1847* (Burlington: Ashgate, 2000). Pharmacopoeias are noticeably lacking in the recent scholarship on genres of medical writing in early modern Europe; see Pomata, "Sharing Cases," 193–236; Ian Maclean, *Logic, Signs and Nature in the Renaissance: The Case of Learned Medicine* (Cambridge: Cambridge University Press, 2002), 55–62.

14. The *Pharmacopoeia Augustana* was one of the most popular and emulated pharmacopoeias in early modern Europe. It was first published in 1564 with regular updates and improvements appearing into the eighteenth century. As an example of a pharmacopoeia that provides extensive descriptions of materia medica, see Collegium Medicum, *Pharmacopoeia Augustana* (Augustae Vindicorum [Augsburg]: Ex Off. A. Apergeri : Prostant apud J. Krugerum, 1622).

15. See, for example, Royal College of Physicians of London, *Pharmacopoeia Londinensis* (London: Printed for John Marriott, 1639); Université de Paris, Faculté de médecine, *Codex Medicamentarius seu Pharmacopoea Parisiensis* (Lutetiae Parisiorum: Sumptibus Olivarii de Varennes, 1645).

16. Paula De Vos, "From Herbs to Alchemy: The Introduction of Chemical Medicine to Mexican Pharmacy in the Seventeenth and Eighteenth Centuries," *Journal of Spanish Cultural Studies* 8 (2007): 135–68.

17. De Vos, "From Herbs to Alchemy," 135.

18. Moyse Charas, *Pharmacopoée royale galenique et chymique* (Paris: Charas, 1676); Félix Palacios, *Palestra Pharmaceutica Chymico-Galenica* (Madrid: Juan García Infanzón, 1706).

19. Johann Zwelfer, *Pharmacopoeia regia* (Noribergae: Typis & Sumptibus Balthasaris Joachimi Endteri, 1693).

20. George Bate, *Pharmacopoeia Bateana* (Londini: Impensis Sam Smith, 1691).

21. Benjamin Breen, "Drugs and Early Modernity" *History Compass* 15 (2017), accessed September 1, 2018, https://doi.org/10.1111/hic3.12376.

22. "Decreto del Tribunal del Real Proto-Medicato," *Pharmacopoeia Matritensis*, nn. A2r; Michele L. Clouse, *Medicine, Government and Public Health in Philip II's Spain: Shared Interests, Competing Authorities* (New York: Routledge, 2011), 139–40; John Tate Lanning, *The Royal Protomedicato: The Regulation of the Medical Professions in the Spanish Empire*, ed. John TePaske (Durham: Duke University Press, 1985).

23. Antonio Barrera-Osorio, *Experiencing Nature: The Spanish American Empire and the Early Scientific Revolution* (Austin: University of Texas Press, 2006); Raquel Alvarez Peláez, *La conquista de la naturaleza americana* (Madrid: CSIC, 1993).

24. Antonio González Bueno, "An Account [of] the History of the Spanish Pharmacopoeias," History of Pharmacopoeias Working Group, International Society for the History of Pharmacy, accessed May 30, 2017, http://www.histpharm.org/ISHPWG%20Spain.pdf.

25. Between 1706 and 1792, there were nine editions of Palacios's *Palestra Pharmaceutica Chymica-Galenica*. Eight of these editions were published in Madrid, in 1706, 1716, 1725, 1730, 1737, 1753, 1763, 1778, and 1792. Another edition was published Barcelona in 1716; De Vos, "From Herbs to Alchemy."

26. Félix Palacios, *Palestra Pharmaceutica Chymico-Galenica* (Madrid: Juan de Sierra, 1730), part I, chapters IX–XI, 116–28.

27. González Bueno, "Spanish Pharmacopoeias."
28. *Pharmacopoea Hispana* (Madrid: Ex Typographia Ibarriana, 1794).
29. *Pharmacopoeia Matritensis*, "Index Capitum," nn.
30. *Pharmacopoeia Matritensis*, "Index Capitum," nn.
31. *Pharmacopoeia Matritensis*, Part 1, Chapter II ("De Simplicibus Officinalibus") and chapter III ("De Simplicibus Exoticis").
32. Here, I have opted to keep the categories from the original text rather than trying to classify marine substance as plant, animal, and mineral.
33. These geographical regions are my own and not from the pharmacopoeia itself.
34. Lanning, *Royal Protomedicato*, 231–36.
35. A recent study by Linda Newson suggests that some pharmacists in early colonial Lima, at least, had little difficulty stocking materia medica from Europe and show a relatively strong commitment to a largely European and conservative approach to medicine similar to the approach embodied in the *Pharmacopoeia Matritensis* in the early eighteenth century; see Linda Newson, *Making Medicines in Early Colonial Lima, Peru: Apothecaries, Science and Society* (Leiden: Brill, 2017).
36. Paula De Vos, "The Business of Pharmacy," in "The Art of Pharmacy in Seventeenth- and Eighteenth-Century Mexico," (PhD diss., University of California, Berkeley, 2001); Newson, *Making Medicines*.
37. Francisco Ximénez, "Al Lector," in *Quatro Libros. De la Naturaleza y virtudes de las plantas, y animales que estan recevidos en el uso de Medicina de la Nueva España* (Mexico: En la casa de la Viude de Diego Lopez Davalos, 1615).
38. Jorge Cañizares-Esguerra, *How to Write the History of the New World: Histories, Epistemologies and Identities in the Eighteenth-Century Spanish Atlantic* (Stanford: Stanford University Press, 2001); Antonello Gerbi, *The Dispute of the New World: From Christopher Columbus to Gonzalo Fernandez de Oviedo*, trans. Jeremy Moyle (Pittsburgh: University of Pittsburgh Press, 1985).
39. Whereas the first edition (1706) of the *Palestra Pharmaceutica* ran to 480 pages in one volume, Palacios expanded the work to 708 pages in one volume starting with the second edition.
40. Palacios, *Palestra Pharmaceutica*, part 3, chapter II, "De los simples que se piden absolutamente en las Recetas," 172–73.
41. Palacios, *Palestra Pharmaceutica* (1730), part II, chapter II, 172–73.
42. Nicolas Lémery, *Cours de Chymie* (Paris: Lémery, 1675).
43. Matthew James Crawford, *The Andean Wonder Drug: Cinchona Bark and Imperial Science in the Spanish Atlantic, 1630–1820* (Pittsburgh: University of Pittsburgh Press, 2016).
44. Anecdotal evidence from my ongoing and still incomplete survey of official pharmacopoeias in early modern Europe suggests that pharmacopoeias often borrowed heavily from other pharmacopoeias. Indeed, the structure and format of the *Pharmacopoeia Matritensis* closely resembles that of the *Pharmacopoeia Augustana*, an official pharmacopoeia from Augsburg, Germany, that was first published in the sixteenth century and served as a model for other official pharmacopoeias in early modern Europe. See *A Facsimile Edition of the First Edition of the Pharmacopoeia Augustana with Introductory Essays by Theodor Husemann* (Madison: State Historical Society of Wisconsin, 1927):
45. Crawford, *Andean Wonder Drug*, 53–57.

46. Patrick Wallis, "Exotic Drugs and English Medicine: England's Drug Trade, c. 1550–c. 1800," *Social History of Medicine* 25 (2011): 25–46.

47. Andreas-Holger Maehle, *Drugs on Trial* (Amsterdam: Rodopi, 1999), 225–33.

48. Robert Voeks, "Disturbance Pharmacopoeias: Medicine and Myth from the Humid Tropics," *Annals of the Association of American Geographers* 94 (2004): 868–88.

49. I would like to acknowledge Dr. Paola Bertucci for making this point in her question and comment on a presentation on this material that I gave at the annual meeting of the History of Science Society in 2014.

CHAPTER 4: Beyond the Pharmacopoeia?

1. On the terminology of operator vs. charlatan, see Emma C. Spary, *Translations of Potency: Taking Drugs in The Sun King's Reign* (forthcoming). On the Italian origins of the pan-European "charlatan" phenomenon, see David Gentilcore, *Medical Charlatanism in Early Modern Italy* (Oxford: Oxford University Press, 2006).

2. On the phenomenon of secret remedies in ancien régime France, see chapter one of Justin Rivest, "Secret Remedies and the Rise of Pharmaceutical Monopolies in France during the First Global Age" (PhD diss., Johns Hopkins University, 2016); Christian Warolin, "Le remède secret en France jusqu'à son abolition en 1926," *Revue d'histoire de la pharmacie* 90, no. 334 (2002): 229–38; Matthew Ramsey, "Traditional Medicine and Medical Enlightenment: The Regulation of Secret Remedies in the Ancien Régime," in *La médicalisation de la société française 1770–1830*, ed. Jean-Pierre Goubert (Waterloo, Ontario: Historical Reflections Press, 1982), 215–32; Maurice Bouvet, "Histoire sommaire du remède secret," *Revue d'histoire de la pharmacie* 45, no. 153–54 (1957): 57–63, 109–18.

3. These terminological distinctions expand on Ramsey, "Traditional Medicine and Medical Enlightenment," 215–16.

4. Joseph M. Gabriel, *Medical Monopoly: Intellectual Property Rights and the Origins of the Modern Pharmaceutical Industry* (Chicago: University of Chicago Press, 2014), esp. ix–x.

5. On these and other distinctions, see Mario Biagioli, "Patent Republic: Representing Inventions, Constructing Rights and Authors," *Social Research* 73, no. 4 (2006): 1150–172.

6. Harold J. Cook, "Markets and Cultures: Medical Specifics and the Reconfiguration of the Body in Early Modern Europe," *Transactions of the Royal Historical Society* 21 (2011): 123–45.

7. On magistral vs. officinal preparations, see for example Gabriel-François Venel, "Magistral, remède (Thérapeut.)," *Encyclopédie ou Dictionnaire raisonné des sciences* (1751–1772), 9:855; Jean-François Lavoisien, *Dictionnaire portatif de médecine* (Paris, 1771), 15, 429.

8. *Arrêt du conseil d'Etat du Roy, qui défend à toutes personnes de distribuer des remèdes spécifiques et autres sans en avoir obtenu de nouvelles permissions. Du 3 Juillet 1728* (Paris, 1728), 1–2.

9. *Arrêt du conseil d'Etat du Roy* [...] *Du 3 Juillet 1728*, 2.

10. Jean Verdier, *La jurisprudence de la médecine en France*. (Alençon, 1762), 1:150.

11. On the role of patient trials in granting drug monopoly privileges, see Justin Rivest, "Testing Drugs and Attesting Cures: Pharmaceutical Monopolies and Military Contracts in Eighteenth-Century France," *Bulletin of the History of Medicine* 91, no. 2 (2017): 362–90.

12. Verdier, *La jurisprudence de la médecine*, 1:152.

13. Verdier, *La jurisprudence de la médecine*, 1:147.
14. Verdier, *La jurisprudence de la médecine*, 1:161–62.
15. Ramsey, "Traditional Medicine and Medical Enlightenment," 219; and the sources above in note 2.
16. Bibliothèque interuniversitaire de la santé (BIUS) Médecine, Ms 2006, fol. 310r–v.
17. BIUS Médecine, Ms 2006, fol. 310v.
18. BIUS Médecine, Ms 2006, fol. 310r.
19. Fontenelle, "Éloge de Fagon," *Histoire de l'Académie royale des sciences* (1718), 99.
20. The most detailed account of the Contugi family and orviétan in France remains Claude-Stéphen Le Paulmier, *L'Orviétan. Histoire d'une famille de charlatans du Pont-Neuf aux XVIIe et XVIIIe siècles* (Paris: Librairie illustrée, 1893). See also Gustave Planchon, "Notes sur l'histoire de l'Orviétan," *Journal de pharmacie et de chimie* 26, nos. 3–7 (1892): 97–103, 145–52, 193–98, 241–50, 289–98.

The best survey of the pan-European orviétan phenomenon is David Gentilcore, *Healers and Healing in Early Modern Italy* (Manchester: Manchester University Press, 1998), 96–124, esp. 99–100, on the Contugi family. Gentilcore's more recent *Medical Charlatanism in Early Modern Italy* (Oxford: Oxford University Press, 2006) contains many brief references to orviétan, but does not supersede the synthetic account provided in chapter three of *Healers and Healing*.

21. Johann Schröder, *Pharmacopoeia medico-chymica*, 4th ed. (Ulm, 1655), 184.
22. Jean-Thomas Riollet, *Remarques curieuses sur la thériaque, avec un excellent traité sur l'orvietan* (Bordeaux, 1665); Pierre-Martin de La Martinière, *Traitté des compositions du mitridat, du thériaque, de l'orviétan, et des confections d'alkermes et d'hyacinthe et autres compositions antidotoires* (Paris, 1665).
23. On Riollet, see Gentilcore, *Healers and Healing*, 100–101, 112–13, 116; Le Paulmier, *L'Orviétan*, 30–32.
24. La Martinière, *Traitté des compositions du mitridat, du thériaque, de l'orviétan*, 48; on "professors of secrets," see William Eamon, *Science and the Secrets of Nature: Books of Secrets in Medieval and Early Modern Culture* (Princeton: Princeton University Press, 1994), esp. 134–67.
25. On La Martinière's colorful career, see Françoise Loux, *Pierre-Martin de La Martinière : un médecin au XVIIe siècle* (Paris: Imago, 1988).
26. On these antidotes, see Gilbert Watson, *Theriac and Mithridatium: A Study in Therapeutics* (London: Wellcome, 1966); Frederick W. Gibbs, "Medical Understandings of Poison circa 1250–1600" (PhD diss., University of Wisconsin–Madison, 2009).
27. Le Paulmier, *L'Orviétan*, 40.
28. See esp. Rivest, chapter three of "Secret Remedies"; Le Paulmier, *L'Orviétan*, 39–43, 130–42 for *pièces justificatives*.
29. La Martinière, *Traitté des compositions du mitridat, du thériaque, de l'orviétan*, 29–30.
30. Archives nationales (AN) V^6 346, May 25, 1657, n. 1; Le Paulmier, *L'Orviétan*, 21 and pièces justificatives, 137–39; Gentilcore, *Healers and Healing*, 98–99; Le Paulmier, *L'Orviétan*, 21.
31. Le Paulmier, *L'Orviétan*, 43–44.
32. Le Paulmier, *L'Orviétan*, 52.
33. These details are from the notarized act that ultimately resolved the dispute: see BIUS Pharmacie, Reg. 31, "Transaction faite entre le Sieur et veuve Contugi et les apothiquaires privilégiés sur l'instance à l'occasion de la saisie faite à le Sieur Boulogne" (August

14, 1685), fol. 272–75. For a summary see Planchon, "Notes sur l'histoire de l'Orviétan," 245–50.

34. *Requeste servant de Factum pour Antoine Boulogne Ayde-Apoticaire du Corps du Roy, Défendeur. Contre Roberte Richard, veuve Contugi, dit l'Orvietan, et son fils, Demandeurs et Défendeurs* (Paris, 1684), 5.

35. *Requeste servant de Factum*, 9.

36. *Requeste servant de Factum*, 9.

37. Moyse Charas, *Pharmacopée royale galénique et chymique* (Paris, 1676), 323–24.

38. Patrizia Catellani and Renzo Console, *L'Orvietano* (Pisa: ETS, 2004), 59.

39. Catellani and Console, *L'Orvietano*, 61.

40. Catellani and Console, *L'Orvietano*, 61–79. Of these, Boulogne only cites Schröder, Charas, and the pharmacopoeias of Lyon and Rome.

41. Catellani and Console, *L'Orvietano*, 59–60.

42. "Préparation célèbre de la Thériaque nouvellement faite à Paris par M. de Rouvière," *Journal des Sçavans* 13 (1685): 228–31.

43. "Préparation célèbre," 228–31.

44. Olivier Lafont, *Échevins et apothicaires sous Louis XIV : la vie de Matthieu-François Geoffroy, bourgeois de Paris* (Paris: Pharmathèmes, 2008), 47–52.

45. Catellani and Console, *L'Orvietano*, 119–34.

46. Most orviétan recipes I have seen are careful to specify the use of "old" (*vetere, vieille*) in the sense of "aged" or well-fermented theriac (not "old" in the sense of "expired"). See for instance Nicolas Lémery, *Pharmacopée universelle* (Paris, 1698), 601–2.

47. Catellani and Console follow Planchon in dividing orviétan into two general categories: the various forms of "Italian orviétan," which contain theriac; and the supposed orviétanum praestantius of Nicolas Lémery and the eighteenth-century Paris *Codex*: see *L'Orvietano*, 58–60.

48. La Martinière, *Traitté des compositions du mitridat, du thériaque, de l'orviétan*, 42.

49. Gentilcore, *Medical Charlatanism*, 112.

50. BIUS Pharmacie, Reg. 31, fol. 272–75.

51. BIUS Pharmacie, Reg. 31, 274v°.

52. BIUS Pharmacie, Reg. 31, 274v°.

53. John Styles, "Product Innovation in Early Modern London," *Past & Present* 168, no. 1 (2000): 124–69; for a survey of French drug branding from this period, see Maurice Bouvet, "Sur l'historique du conditionnement de la spécialité pharmaceutique," *Revue des spécialités* (1928): 101–43, 213–23, 297–315.

54. Roughly, "Unique as the sun, salutary as the salt," a clear reference to Louis XIV, the "Sun King" and source of the Contugi privilege, and perhaps also to the volatile salt to which the drug's effects as an antidote were attributed. Cf. the very similar Italian broadsheet of Franceso Nava, the Roman orviétan: see Gentilcore, *Healers and Healing*, 102, fig. 6.

55. Treasure-hunters and other modern-day antiquarians maintain online forums where they display their finds and help one another to identify them. Several include orviétan containers, notably la-detection.com and echange-passion.com.

56. Styles, "Product Innovation," 153–56.

57. For the letters see AN O¹ 30, fol. 397 (December 27, 1686); and the passport AN V^s 1246, fol. 209 (February 6, 1686). Le Paulmier, *L'Orviétan*, 147–50.

58. Maurice Bouvet, "Les apothicaires royaux," *Bulletin de la société de l'histoire de la pharmacie* 5, no. 58 (1928): 62.

59. On these later episodes, see Rivest, "Secret Remedies," 190–97.

CHAPTER 5: Crown Authorities, Colonial Physicians, and the Exigencies of Empire

This work was supported by funding from the US National Endowment for the Humanities, the University of California, Davis Department of Nutrition, the University of Massachusetts Dartmouth Center for Portuguese Studies and Culture, the Fundação Calouste Gulbenkian, and the Fundação Luso-Americana para o Desenvolvimento. In addition, for logistical support in Brazil, thanks are due to the Fundação Oswaldo Cruz, the Biblioteca Nacional do Rio de Janeiro, the Biblioteca da Escola de Farmácia da Universidade Federal de Ouro Preto, and the Arquivo Público de Salvador da Bahia.

1. See, for examples, Mansel Longworth Dames, trans., *The Book of Duarte Barbosa* (New Delhi: Asia Educational Services, 2002) (reprinted from the 1918–1921 London edition), vol. 1, 88–90, 90–95; and Armando Cortesão, *Suma Oriental of Tomé Pires: An Account of the East, from the Red Sea to China, written in Malacca and India in 1512–1515* (New Delhi: Asia Educational Services, 2005), vol. 1, xviii–xlv, and vol. 2, 512–18.

2. Anthony R. Disney, *A History of Portugal and the Portuguese Empire.* (Cambridge University Press, 2009), vol. 1, 281–82, 294–305; see also C. Michaud, "Un anti-jesuite au service de Pombal: l'abbé Platel," in *Pombal Revisitado*, ed. M. H. Carvalho dos Santos, (Lisbon: Editorial Estampa, 1984), 387–402; and B. Duhr, *Pombal: Sein Charakter und Sein Politik: Ein Beitrag zur Geschichte des Absolutism* (Freiburg in Briesgau: Herder Verlagshandlung, 1891), 106–41.

3. Dauril Alden, "The Gang of Four and the Campaign against the Jesuits in Eighteenth-Century Brazil," in *The Jesuits II: Cultures, Sciences, and the Arts, 1540–1773*, ed. John W. O'Malley et al. (Toronto: University of Toronto Press, 2006), 707–24.

4. Timothy Walker, "Acquisition and Circulation of Medical Knowledge within the Portuguese Colonial Empire during the Early Modern Period," in *Science in the Spanish and Portuguese Empires*, ed. Daniela Bleichmar et al. (Stanford: Stanford University Press, 2009), 266–70.

5. Timothy Walker, "The Medicines Trade in the Portuguese Atlantic World: Dissemination of Plant Remedies and Healing Knowledge from Brazil, c. 1580–1830," in "Mobilising Medicine: Trade & Healing in the Early Modern Atlantic World," special issue, *Social History of Medicine* 26, no. 3 (2013): 411–17.

6. Charles J. Borges, *The Economics of the Goa Jesuits, 1542–1759* (New Delhi: Concept Publishing, 1994), 41, 86; and *Colecção de Varias Receitas e Segredos Particulares das Principais Boticas da Nossa Companhia de Portugal, da India, de Macao e do Brazil* (Anonymous, 1766), Archivum Romanum Societatis Iesu (ARSI), Rome, Italy, Opp. NN. 17, introduction, 1–2.

7. Joaquim Verríssimo Serrão, *História de Portugal.* (Lisbon: Editora Verbo, 1996), vol. 6 (1750–1807), 268; and Rocha Brito and Feliciano Guimarães, "A Faculdade de Medicina de Coimbra," *Actas Cibas* 14, 555–56; cited in José Sebastião Silva Dias, "Portugal e a Cultura Europeia: Séculos XVI a XVIII," *Biblos* XXVIII (Coimbra: Universidade de Coimbra, 1952), 368.

8. For transatlantic slave trade mortality rates, see David Eltis, *The Rise of African Slav-*

ery in the Americas (Cambridge: Cambridge University Press, 1999), 68, 159, 185–86. For the annual number of patients treated at the Hospital Militar in Goa in the late eighteenth century, see Historical Archive of Goa, India, Livros dos Monções do Reino (HAG MR) 173, f. 168 (3476 patients in 1791); HAG MR 176B, f. 436 (3858 patients in 1793); HAG MR 176B, f. 448 (3076 patients in 1794); and HAG MR 177A, f. 218 (1932 patients in 1797).

9. See the discussion in Walker (2009), 248–62.

10. For India, see Ignácio Caetano Afonso, *Discripçoens e Virtudes das Raizes Medicinaes*, manuscript booklet (1794), HAG MR 175, ff. 219–30; references to a similar royal directive, dated April 2, 1798, are in HAG MR 178B (1798–1799), ff. 644–45. Also, for India, see Biblioteca da Academia das Ciências de Lisboa (BACL), Mss. 21 (Série Azul): *Medicina Oriental: Soccorro Indico, Aos Clamores dos Pobres Enfermos do Oriente; Para total profligação de seus males adquiridaa da varios Professores de Medicina* (anonymous; n.d. [late 18th century]) 1–632. For Brazil, see de Bento Bandeira Mello, manuscript (1788); Arquivo Nacional da Torre do Tombo (ANTT), Ministério do Reino, cx. 555, mç. 444. For Angola, see José Pinto de Azeredo, *Ensaios sobre algumas enfermidades de Angola* (Lisbon: Regia Officina Typografica, 1799).

11. For North America, see the written agreement between American merchant Peter Tilly and Portuguese minister to the United States Cipriano Ribeiro Freire: March 6, 1798, Biblioteca Nacional do Portugal (BNP), mss. 60, nr. 6, doc. 59.

12. Domingos Vandelli (1735–1816) published influential works on botany in Portugal and its colonial territories, including the *Diccionario dos termos technicos de Historia Natural, extrahidos das obras de Linneo, e a memoria sobre a utilidade dos Jardins Botanicos* (Coimbra: Real Officina da Universidade, 1788), and *Florae lusitanicae et brasiliensis specimen [...] et epistolae ab erudits viris* (Coimbra: Typografia Academico-Regia, 1788).

13. For Portuguese state-sponsored attempts to exploit medical natural resources in colonial Mozambique, Angola, and Brazil, see William J. Simon, *Scientific Expeditions in the Portuguese Overseas Territories (1783–1808)* (Lisbon: Instituto de Investigação Científica Tropical, 1983), 59–78, 79–104, and 23–58, respectively.

14. Anonymous, *Medicina Oriental: Soccorro Indico, Aos Clamores dos Pobres Enfermos do Oriente; Para total profligação de seus males adquiridaa da varios Professores de Medicina*; Biblioteca da Academia das Ciências de Lisboa (BACL), Códice 22, mss. 21 (Série Azul); n.d. [Final quarter of the eighteenth century?]), 1–640.

15. Correspondence of Francisco Luís de Meneses with the *Academia Real das Ciências de Lisboa*, dated 1786; BNP, Códice 6377. I am indebted to Fabiano Bracht of the Universidade do Porto for this reference.

16. BACL, Códice 22, mss. 21, 1 (frontispiece).

17. Fabiano Bracht, "Condicionantes sociais e políticas nos processos de produção de conhecimento: o caso da Índia portuguesa do século XVIII," in *História & Ciência: Ciência e Poder Na Primeira Idade Global*, ed. Amélia Polónia, Fabiano Bracht, Gisele C. Conceição, and Monique Palma (Porto: Biblioteca da Faculdade de Letras da Universidade do Porto, 2016), 161.

18. HAG MR 178B (1798–1799), ff. 644–64.

19. HAG MR 178B (1798–1799), f. 644.

20. HAG MR 175, ff. 219–30.

21. Letter from Dom Rodrigo de Souza Coutinho to the Portuguese Secretary of State in Lisbon dated March 14, 1798; HAG MR 177A, f. 212.

22. Letter dated April 28, 1799; HAG MR 178A, f. 272.

23. See Licurgo de Castro Santos Filho, *História da Medicina no Brazil, do Século XVI ao Século XIX*. (São Paulo: Editora Brasiliense Ltda., 1947), vol. 1, 50–51; and Conde dos Arcos, memorandum of October 8, 1757; in Eduardo de Castro e Almeida, *Inventário dos Documentos Relativos ao Brasil Existantes no Archivo de Marinha e Ultramar de Lisboa*, I: "Bahia, 1613–1762" (Rio de Janeiro: Officinas Graphicas da Biblioteca Nacional, 1913), doc. 2917, 255–56.

24. Biblioteca Nacional do Rio de Janeiro (BNRJ), unpublished manuscripts: I-47, 19, 20: 'Anotações Sobre Medicina Popular,' ff. 1–32; and I-47, 23, 5: 'Botânica Médica Vulgar Brasileira: Drogas Orgânicas & Medicina Popular,' ff. 1–17.

25. BNRJ, mss I-12,01,019; Francisco António de Sampaio, *História dos Reinos Vegetal, Animal e Mineral, Pertencente à Medicina* (manuscript compiled at Cachoeira, Bahia, Brazil, 1782 [volume I] and 1789 [volume II]). Both volumes of Francisco António de Sampaio's *História dos Reinos Vegetal, Animal e Mineral* were published together as a special issue of the journal *Anais da Biblioteca Nacional*, vol. 89 (Rio de Janeiro, 1969).

26. BNRJ, mss I-12,01,019; volumes I and II.

27. Gisele C. Conceição, "Estudos De Filosofia Natural No Brasil Ao Longo Do Século XVIII," in *História & Ciência: Ciência e Poder Na Primeira Idade Global*, ed. Amélia Polónia et al. (Porto: Biblioteca da Faculdade de Letras da Universidade do Porto, 2016), 126.

28. ANTT, *Chancelarias* de Dom José I, livro 70, f. 282v. I am indebted to Gisele C. Conceição of the Universidade do Porto for this reference.

29. Sampaio (BNRJ, 1782), vol. I, f. 1.

30. Conceição, 126–29.

31. In order, the twelve sections discuss various substances with the following medical uses: "Resolutives; Detergents; Coagulants for internal use; Astringents; Purgatives and Emetics; Decongestants; Anti-Venoms and Febrifuges; Diaphoretics; Anti-Venereals; Anti-Colics; Anti-Spasmodics; and Refrigerants and Coolants for external use."

32. Sampaio (BNRJ, 1782), ff. 117–24.

33. Thomas E. Skidmore, *Brazil: Five Centuries of Change* (Oxford: Oxford University Press, 1999), 32–40.

34. José Sebastião Silva Dias, "Portugal e a Cultura Europeia: Séculos XVI a XVIII," *Biblos* XXVIII (Coimbra: Universidade de Coimbra, 1952), 292–97.

35. Bento Bandeira de Mello, manuscript (1788); ANTT, Ministério do Reino, *cx*. 555, mç. 444. I am indebted to Bruno Barreiros of Universidade Nova de Lisboa for alerting me to this document.

36. Saul Jarcho, *Quinine's Predecessor: Francesco Torti and the Early History of Cinchona* (Baltimore: Johns Hopkins University Press, 1993), 102–4, 297–98; and Andreas-Holger Maehle, *Drugs on Trial: Experimental Pharmacology and Therapeutic Innovation in the Eighteenth Century* (Amsterdam: Rodopi, 1999), 223–33.

37. Vicente Jorge Dias Cabral, "Ensaio Botanico de algumas plantas de parte interior do Piauí (. . .)," Arquivo Histórico Ultramarino (AHU), AHU-ACL-CU-016, cx. 25, D.1311, 1801; António José de Sousa Pinto, *Materia Medica* (Ouro Preto, Brazil: Typografia de Silva, 1837), 21, 31; José E. Mendes Ferrão, *A Aventura das Plantas e os Descubrimentos Portugueses* (Lisbon: Chaves Ferreira Publicações, 2005), 157–60.

38. AHU, São Tomé and Príncipe Collection; *cx*. 55, doc. 75; and A. M. Amaro, *Introdução da Medicina ocidental em Macau e as receitas de segredo da botica do Colégio de São Paulo* (Macau: Instituto Cultural de Macau, 1992), 7–11.

39. José Pedro Sousa Dias, "Documentos sobre duas boticas da Companhia de Jesus em Lisboa: Colégio de Santo Antão e Casa Professa de S. Roque," in *Economia e Sociologia*, 88/89 (2009), 295–312. For Brazilian medicinal plants circulating in the *Estado da Índia*, see for example HAG 7926, f. 56r/v (report of medicines sent from *Hospital Real* of Goa to the Fortress of Diu, 1785); and HAG 1346: "Relação dos Medicamentos que fazem precizo para o Hospital Publico Militar dos Ilhas de Soldar e Timor" (Dili, 5 May 1838), f. 183.

40. ANTT, Ministério do Reino, cx. 555, mç. 444., f. 2.

41. Manoel Joaquim Henriques de Paiva, *Farmacopéa Lisbonense, ou Collecçao dos simplices, preparaçoes e composiçoes mais efficazes e de major uso* (Lisbon: Filippe da Silva e Azevedo, 1785), frontispiece. In English, the title is: *Lisboner's Pharmacopea, or Collection of Simples, Preparations, and Compositions Most Efficacious and of Greatest Use.*

42. Henriques de Paiva, *Farmacopéa Lisbonense...*, frontispiece.

43. In 1802 the original printer issued an expanded, corrected edition: Manoel Joaquim Henriques de Paiva, *Farmacopéa Lisbonense, ou Collecçao dos simplices, preparaçoes e composiçoes mais efficazes e de major uso* (Lisbon: Filippe da Silva e Azevedo, 1802).

44. Such expeditions were usually accomplished by gold- and slave-seeking quasi-military bands of frontiersmen called *bandeirantes*. See Joseph Smith, *A History of Brazil, 1500–2000* (London: Longman/Pearson Education Press, 2002), 10–12, 16–17.

45. Nisia Trindade Lima, director, *A Ciência dos Viajantes: natureza, populaces, e saúde em 500 anos de interpretações do Brasil* (Rio de Janeiro: Fundação Oswaldo Cruz, 2000), 40–41.

46. Lima, *Ciência dos Viajantes*, 40–41; and BNRJ, Manuscripts Division, Coleção Alexandre Rodrigues Ferreira.

47. See Alexandre Rodrigues Ferreira, "Diario da Viagem Philosophica pela Capitania de São José do Rio Negro...," in *Revista Trimestral do Instituto Histórico e Geográfico Brasileiro*, vols. 48, 49, 50, 51 (Rio de Janeiro, Brazil, 1885–1888).

48. José E. Mendes Ferrão et al., *Plantas do Brasil: Flora económica do Brasil no Século XVIII; Plantas do Maranhão-Piauí* (Lisbon: Chaves Ferreira Publicações, 2002), 9–11.

49. File of twenty-four watercolor illustrations, held in the Arquivo Histórico Ultramarino (AHU-ACL-CU-016, cx. 25, D.1311). A contemporary portfolio of thirty-one botanical illustrations from Brazil, called "Plantas do Piauhi," is held at the Library of the Museum of the Royal Botanical Garden, Lisbon (BMJB, Nr. E 166/24).

50. In Portuguese, "Ensaio Botanico de algumas plantas de parte interior do Piauí" (AHU-ACL-CU-016, cx. 25, D.1311); the work also includes information about Brazilian native woods exploitable for shipbuilding.

51. Treaty of Commerce and Friendship between Portugal and the United States of America, London, 25 April 1786 (ANTT, Ministério dos Negócios Estrangeiros, cx. 550, nr. 13). The treaty's terms were respected by both national signatories, though it was never ratified by the US Senate.

52. Jorge Manuel Martins Ribeiro, "Comércio e Diplomacia nas Relações Luso-Americanas (1776–1822)," (doctoral thesis, Universidade do Porto, 1997), vol. 1, 153–265.

53. *Gazeta de Lisboa*, January 6, 1789; January 18, 1791; March 26, 1799; and Ribeiro, vol. 1, 266–84.

54. *Ofício* of D. Rodrigo de Sousa Coutinho to Cipriano Ribeiro Freire relating to Virginia tobacco seeds, Lisbon: Palácio de Queluz, July 23, 1798 (BNP, Mss. 60, nr. 6, doc. 85).

55. See André João Antonil, *Cultura e Opulência do Brasil por suas Drogas e Minas...*

(Lisbon: Officina Real Deslandesiana, 1711) (modern edition, Rio de Janeiro: Editora Itatiaia, 1997), 149, 156–57.

56. Agreement between Peter Tilly and Cipriano Ribeiro Freire, March 6, 1798 (BNP, mss. 60, nr. 6, doc. 59).

57. José Calvet de Magalhães, *História das Relações Diplomáticas entre Portugal e os Estados Unidos da América (1776–1911)* (Lisbon: Publicações Europa-América, 1991), 26–27 and 30.

58. See the discussion in David Abernethy, *The Dynamics of Global Dominance: European Overseas Empires, 1415–1980* (New Haven, CT: Yale University Press, 2002), 209–14.

59. Lorelai Kury, "Plantas sem fronteiras: jardins, livros e viagens, séculos XVIII–XIX," in *Usos e circulaçao de plantas no Brasil; Séculos XVI–XIX*, ed. Lorelai Kury (Rio de Janeiro: Andrea Jakobsson Estúdio Editorial, 2013), 228–88.

60. Letter from George Washington to Scott, Pringle, Cheap & Company (Funchal, Madeira), dated February 23, 1768; Papers of George Washington (University of Virginia), Colonial Series (July 7, 1748–June 15, 1775), vol. 8, 68–69.

61. Pedro Calmon, *História do Brasil*, 3rd ed., 7 vols. (São Paulo: José Olympio, 1959), vol. 4, 1352–1354; cited in Serrão, 385; Serrão, 390–92.

62. HAG MR 46A (1681–1682), ff. 96r-97v.

63. AHU, São Tomé and Príncipe Collection; *cx*. 55, doc. 75.

64. Arquivo Histórico do São Tomé e Príncipe (AHSTP), *Alfandega* [Customs] records of São Tomé town port, February 1, 1899; and *Sociedade e Emigração para São Thomé e Principe* (Relatorio da Direcção; Paracer do Conselho Fiscal; Lista dos Acionistas, 2ª Anno) (Lisbon, 1914), 93.

65. AHU; São Tomé and Príncipe collection; *caixa* 54; doc. 15.

66. William Gervase Clarence-Smith, *Cocoa and Chocolate, 1765–1914* (London: Routledge, 2002), 208, 238–39; and Ministério da Agricultura, Industria e Commércio; Serviço de Informações, "Producção, Commércio e Consumo de Cacáo" (Rio de Janeiro: Imprensa Nacional, 1924), 3, 6–7, 14.

67. See *Curiosidade; Un Libro de Medicina escrito por los Jesuitas en las Misiones del Paraguay en el año 1580*, BNRJ, Manuscripts Division; Nr. I-15, 02, 026, Capítulo I, p. 1.

68. Louis E. Grivetti, "Medicinal Chocolate in New Spain, Western Europe, and North America," in *Chocolate: History, Culture and Heritage*, ed. Louis Grivetti and Howard Shapiro (Malden: Wiley, 2009), 67–88.

69. Jarcho, 102–4; 297–98; and Maehle, 223–33.

70. See the discussion in Walker (2013), 406, 411–17.

71. For comparative insight into this process and dynamic in Spanish American colonies, see António Barrera-Osorio, "Knowledge and Empiricism in the Sixteenth-Century Spanish Atlantic World," and Daniela Bleichmar, "A Visible and Useful Empire: Visual Culture and Colonial Natural History in the Eighteenth-Century Spanish World," both in *Science in the Spanish and Portuguese Empires*, ed. Daniela Bleichmar, Paula De Vos, Kristine Huffine, and Kevin Sheehan (Stanford: Stanford University Press, 2009), 219–32 and 290–310, respectively.

CHAPTER 6: Imperfect Knowledge

1. The phrase "res publica medica" first appears in a prefatory letter to Johann Schenck's widely reprinted collection of medical case studies, *Paratereseis; or Medical Observations, Rare, New, Wonderful, Monstrous* (1584–1597). The seven-volume compen-

dium collects observational medical cases from ancient sources, those from Schenck's practice as a town physician in Freiburg, and those of his contemporaries. In the preface, Swiss humanist Theodor Zwinger lauds Schenck's volumes for embodying the past, present, and future circulation of knowledge among a community of practicing physicians. On the emergence of this commonwealth of medical letters, see Gianna Pomata, "Observation Rising: Birth of an Epistemic Genre, 1500–1650," in *Histories of Scientific Observation*, ed. Lorraine Daston and Elizabeth Lunbeck (Chicago: University of Chicago Press, 2011), 45–80. For more on Sloane's publication history, see Michael Hunter, "Introduction," *Magic and Mental Disorder: Sir Hans Sloane's Memoir of John Beaumont* (London: The Robert Boyle Project, 2011); and Michael Hunter, Alison Walker, and Arthur MacGregor eds., *From Books to Bezoars: Sir Hans Sloane and His Collections* (London: British Museum, 2013).

2. Hans Sloane, "A Description of the Pimienta or Jamaica-Pepper Tree," *Philosophical Transactions* [*PT*] 16 (1686): 463.

3. Sloane, "Description," 463. On Sloane's emblematic method of specimen description, see Christoper Iannini, *Fatal Revolutions: Natural History, West Indian Slavery, and the Routes of American Literature* (Chapel Hill: University of North Carolina Press, 2012); and James Delbourgo, "Sir Hans Sloane's Milk Chocolate and the Whole History of the Cacao," *Social Text* 29, no. 1 (2011): 71–101.

4. Hans Sloane, *A Voyage to the Islands of Madera, Barbados, Nieves, S. Christophers, and Jamaica, with the Natural History of the Herbs and Trees Four-Footed Beasts, Fishes, Birds, Insects, Reptiles, etc. of the Las of those Islands . . .*, 2 vols. (London, 1707–25). Hereafter, Sloane's text will be cited parenthetically as *A Voyage to . . . Jamaica*. Sloane's popular and successful natural history was originally published in two lavish folio volumes and included 274 copperplate engravings. For more on the print history of *A Voyage . . . to Jamaica*, see Iannini, *Fatal Revolutions*, 72–73; and James Delbourgo, *Collecting the World: Hans Sloane and the Origins of the British Museum* (Cambridge, MA: Harvard University Press, 2017), 168–72.

5. A number of authors in this volume share my interest in identifying the diverse strategies developed by medical authors, readers, and practitioners for bridging the divergent epistemic systems that governed healing knowledge in the early modern world. In particular, Timothy Walker's analysis of bioprospecting in the Portuguese empire highlights the complex ways in which Europeans, responding to the exigencies of life in the colonial space, integrated non-European healing knowledge into pharmacopoeia, be they official or informal (chapter 5). Kelly Wisecup demonstrates through her careful attention to the rhetorical strategies in Samson Occom's herbal how an apparently taxonomical list can, for an epistemologically attuned readership, "delineate objects' connections to interpersonal and other-than-human contexts as well as objects' roles in exchanges of medical knowledge" (chapter 9).

6. Schiebinger builds on Philip Curtin's concept of the "plantation complex": a social, political, and economic order centered on the plantation that underpins the Atlantic World in the seventeenth and eighteenth centuries. James Delbourgo also describes Sloane's editorship as making him "one of the pivotal information brokers" in the Atlantic World. Londa Schiebinger, *Secret Cures of Slaves* (Palo Alto: Stanford University Press, 2017), 3, 13; Philip Curtin, *The Rise and Fall of the Plantation Complex* (Cambridge: Cambridge University Press, 1990); and James Delbourgo, *Collecting the World: Hans Sloane and the Origins of the British Museum* (Cambridge, MA: Harvard University Press, 2017), 159.

7. Gianna Pomata, "The Medical Case Narrative: Distant Reading of an Epistemic Genre," *Literature and Medicine* 32, no. 1 (2014): 2; and Pomata, "Sharing Cases: The *Observationes* in Early Modern Medicine," *Early Science and Medicine* 15, no. 3 (2010): 193–236.

8. Recent scholars have focused on Sloane's innovative *Voyage to . . . Jamaica*, a multivolume work notable for the combination of empirical precision with pietistic interpretation. See Kay Dian Kriz, "Curiosities, Commodities, and Transplanted Bodies in Hans Sloane's *Voyage to . . . Jamaica*," in *An Economy of Colour: Visual Culture and the North Atlantic World*, ed. Geoff Quilley and Kay Dian Kriz (Manchester: Manchester University Press, 2003); Delbourgo, "Whole History of the Cacao," 71–101; and Iannini, *Fatal Revolutions*, 35–75.

9. This phrase—which captures the detached, objective perspective characterizing scientific observation—comes from the title of a work by philosopher Thomas Nagel, *The View from Nowhere* (New York: Oxford University Press, 1986). In the field of natural history in particular, the elaboration of this epistemological stance has long been associated with the work of Michel Foucault. Of the practice of eighteenth-century natural historians, Foucault writes, "The documents of this new history are not other words, texts or records, but unencumbered spaces in which things are juxtaposed: herbariums, collections, gardens; the locus of this history is a non-temporal rectangle in which, stripped of all commentary, of all enveloping language, creatures present themselves one beside the other, their surfaces visible, grouped according to their common features, and thus already virtually analyzed, and bearers of nothing but their own individual names." Michel Foucault, *The Order of Things: An Archaeology of the Human Sciences* (New York: Vintage, 1994 [1966]), 131.

10. Dillon continues: "The story of the rise of freedom in the Atlantic world—the newfound authority of the commons within a politics of popular sovereignty—cannot be separated from its hidden dependence on the colonial relation." For Dillon, Atlantic theater performance, "where presence and absence appear in tandem," affords a privileged glimpse into the previously invisible relation. As I elaborate below, attending closely to Sloane's editorship can bring such relations into focus, further elaborating the varied rhetorical strategies used to efface them. Elizabeth Maddock Dillon, *New World Drama: The Performative Commons in the Atlantic World* (Durham: Duke University Press, 2014), 22–23.

11. Susan Scott Parrish has pointed out the importance of natural history discourse for creole authors, particularly the reliance upon non-Europeans for access to and knowledge about New World nature. Parrish emphasizes the ways "various peoples issuing from various parts of the Atlantic world, made facts about America in vexed chains of communication." Susan Scott Parrish, *American Curiosity: Cultures of Natural History in the Colonial British Atlantic World* (Chapel Hill: University of North Carolina Press, 2006), 23. James Delbourgo makes a similar argument about a distinctly "American" Enlightenment, characterized by experiments with electricity throughout the eighteenth century. Delbourgo traces a "science from below," routed through "non-elites," "fleshy bodies," and "experimental machines" rather than the disembodied rationalism of ideas and texts. James Delbourgo, *A Most Amazing Scene of Wonders: Electricity and Enlightenment in America* (Cambridge, MA: Harvard University Press, 2006), 7–9. A trio of recent studies highlights the complexity of these dynamics surrounding the production of medical knowledge in the West Indies in particular: James Sweet, *Domingos Álvares, African Healing, and*

the Intellectual History of the Atlantic World (Chapel Hill: University of North Carolina Press, 2011); Pablo Gómez, *The Experiential Caribbean: Creating Knowledge and Healing in the Early Modern Atlantic World* (Chapel Hill: University of North Carolina Press, 2017); and Schiebinger, *Secret Cures of Slaves*.

12. This number (24) does not include Sloane's paratextual contributions to the journal during his tenure as secretary. Additionally, in my accounting I distinguish those works authored or translated by Sloane from those that include a comment by him (e.g., "Part of a letter from Mr. T.M. in Salop, to Mr. William Baxter, concerning the Strange Effects from the eating of Dog-Mercury with Remarks thereon by Hans Sloan [sic] M.D. and S.R.S.") or those that label him as the recipient of a published correspondence (e.g., "Part of a Letter from Dr. Musgrave, Fellow of the Royal College of Physicians and R.S. to Dr. Sloane; Concerning a Piece of Antiquity Lately found in Somersetshire"). The latter category is by far the largest in number (63) and is the most difficult to discern the limits of. Many papers published in the journal during Sloane's tenure as secretary are likely drawn from his correspondence, but may not identify him as the correspondent. These contributions, as well as Sloane's rhetorical positioning in relationship to them, are discussed in more detail below.

13. Henry Oldenburg, "Epistle Dedicatory," *PT* 1 (1665–6): n.p.

14. Francis Bacon, *The Two Books of Francis Bacon. Of the Proficiencie and Advancement of Learning*, book I (Oxford: Leon Litchfield, 1605), 7.

15. Lorraine Daston and Katharine Park connect the emergence of the neutral, scientific fact to the domestication of wondrous phenomena by virtuosi of the late seventeenth century. Reports of monsters, strange lights in the sky, or sudden deaths—events previously enlisted as testimony of divine intervention in earthly affairs—were not so much rationally explained by natural philosophers as they were denied their status as evidence in purely religious disputes. Wonders, once ripe with meaning, instead served as the model for the stubborn, strange facts of modern science. See Lorraine Daston, "Marvelous Facts and Miraculous Evidence in Early Modern Europe," *Critical Inquiry* 18, no. 1 (1998), 93–124; and Daston and Park, *Wonders and the Order of Nature* (New York: Zone Books, 1998), 215–55.

16. Henry Oldenburg, "Introduction," *PT* 1 (1665–6): 1–2.

17. James Jurin, "The Preface," *PT* 17 (1693): 581.

18. Sloane, *A Voyage to the Islands of Madera, Barbadoes, Nieves, S. Christophers, and Jamaica*, vol. 1 (London: B.M, 1707), unpaginated preface. Julie Kim points out that Sloane identifies Amerindians and enslaved Africans on Jamaica as sources of knowledge about plants (native or transplanted) that can be made use of for "Food, Physic., &c." For Kim's analysis, see "Obeah and the Secret Sources of Atlantic Medicine," in *From Books to Bezoars*, 100–101.

19. Sloane, *A Voyage to . . . Jamaica*, cxxiv.

20. As Benjamin Breen's work in this volume attests, the accommodation of non-European healing practices into the European epistemological regime was frustrated when healing properties relied on ritualistic preparations or an individual's identity as a healer more than any isolable, and therefore commodifiable, compound (see chapter 7). Sloane's emphasis on the royal lineage of his African healer, a common trope in imperial discourse that aims to mitigate racial intimacy via an appeal to transracial status identity, nevertheless retains a sense of the wondrous authority of the healer herself. Breen's larger question—why there is a lack of African materia medica in official pharmacopoeias—is

key to understanding the conflicting epistemological regimes that jostle in the Atlantic medical complex. However, reading across the varied genres and formats that governed the res publica medica (e.g., natural histories, periodical publications, as well as official pharamacopeia) illuminates where non-European knowledge did inform European medical practice, thereby providing instructive understandings for how such knowledge was effaced. On the importance of self-experimentation for European physicians in the Caribbean, see Schiebinger, *Secret Cures of Slaves*, 10–14.

21. William Cockburn, *The Present Uncertainty in the Knowledge of Medicines* (London: R.J. for Benjamin Barker, 1703), 3.

22. Cinchona bark, or "quinquina," was introduced into the London *Pharamacopeia* for the first time in the 1667 edition. In the Anglo-American medical tradition, Thomas Sydenham, Benjamin Franklin, and Thomas Jefferson, among others, trumpeted the efficacy of "the bark." On the imperial and epistemological importance of cinchona bark in the wider Atlantic World, see Matthew Crawford, *The Andean Wonder Drug: Cinchona Bark and Imperial Science in the Spanish Atlantic, 1630–1800* (Pittsburgh: University of Pittsburgh Press, 2016). For an assessment of the debate over its efficacy, see Saul Jarcho, *Quinine's Predecessor: Francesco Torti and the Early History of Cinchona* (Baltimore: Johns Hopkins University Press, 1993).

23. Cockburn, *Present Uncertainty*, unpaginated prefatory material.

24. Hans Sloane, "The Preface," *PT*, 21 (1699): n.p.

25. Sloane, "Preface," *PT*, 21 (1699), n.p.

26. Julie Kim documents Sloane's interest in the corollary to New World materia medica: poisons, particularly those wielded by the enslaved as a form of resistance. See Kim, "Obeah," 100. This interest also plays out in a series of Sloane-related publications in the *Transactions* that recount experiments with rattlesnakes in Maryland, Virginia, and the Carolinas. Enslaved Africans appear regularly in these accounts. See Captain Hall, "An Account of Some Experiment on the Effects of the Poison of the Rattle Snake By Captain Hall. Communicated by Hans Sloane," *PT* 35 (1728): 309–15; John Ranby, "The Anatomy of the Poisonous Apparatus of a Rattle-Snake, Made by the Direction of Sir Hans Sloane, Bart. Praes. Soc. Reg. & Coll. Med. together with an Account of the Quick Effects of Its Poison," *PT* 35 (1728): 377–81; and Hans Sloane, "Conjectures on the Charming or Fascinating Power Attributed to the Rattle-Snake: Grounded on Credible Accounts, Experiments and Observations," *PT* 38 (1733): 321–31.

27. The letter recounts the sickening of the Matthews family (and subsequent death of one of their children) after consuming a meal of gathered "Herbs ... Fryed with Bacon." Among the herbs is a plant suspected to be dog-mercury, a commonly employed purgative and emetic. Its poisonous effects are well known today. Hans Sloane, "Part of a Letter from Mr. T.M. in Salop, to Mr. William Baxter, concerning the Strange Effects from the Eating of Dog-Mercury with Remarks thereon by Hans Sloan [sic] M.D. and S.R.S.," *PT* 17 (1693): 875–77.

28. Hans Sloane and George Dampier, "Part of a Letter from Mr. George Dampier, Dated Exmouth, November 10 1697 to Mr. William Dampier, his Brother, concerning the Care of Bitings of mad Creatures. With a Remark on the same by Hans Sloane, M.D.," *PT* 20 (1698): 49–52.

29. Sloane and Dampier, "Part of a Letter," 49.

30. The effectiveness of Dampier's model of observation (and his preparation of *Lichen Cinereus terrestris*) is attested to repeatedly in subsequent contributions to the

Transactions. For instance, Thomas Fuller dramatizes the ubiquity and easy identification of the fungus in a letter to Sloane, published in 1737. "Some years ago," Fuller writes, "a Mad-Dog or Cat ... had bit some Children and the Mother at Battle; ... and we all went out in a Snow, with a Broom, and found some of it, and mix'd it as the Account of Dampier directed ... [N]one of them had any bad effects from the bite." John Fuller, "A Letter from John Fuller, Esq; Jun. F.R.S. to Sir Hans Sloane, Bart. Pres. R.S. &c. concerning the Effects of Dampier's Powder, in curing the Bite of a Mad Dog," *PT* 40 (1738): 272.

31. [Anon.]. "Of the Use of the Root Ipecacuanha, for Loosenesses, Translated from a French Paper: With Some Notes on the Same, by Hans Sloane, M.D.," *PT* 20 (1698): 69–79.

32. Petiver was a conduit for such collections in the *Transactions* during Sloane's tenure. He also published similar catalogues from Maryland (1698) and a series of such catalogues from the East Indies (1701). Though beyond the scope of this particular essay, the latter is remarkable for the distinction Petiver draws between his own, learned observations (akin to the comparison and classification performed by Sloane elsewhere in the *Transactions*) and the firsthand observations of his colonial correspondent, Samuel Brown, a physician at Fort George. James Petiver, "A Catalogue of Some Guinea-Plants, with their Native Virtues; Sent to James Petiver, Apothecary, and Fellow of the Royal Society; with His Remarks on them. Communicated in a Letter to Dr. Hans Sloane. Secret. Reg. Soc," *Philosophical Transactions* 19 (1695): 677.

33. Sloane, *Philosophical Transactions*, 677.

34. Sloane, *Philosophical Transactions*, 677–79.

35. Sloane, *Philosophical Transactions*, 684.

36. For more on the double-sided nature of African healing knowledge, see Breen, chapter 7.

37. On Sloane's role in restoring the stature of the *Transactions*, see Delbourgo, *Collecting the World*, 160–68; and Marten Ultee, "Sir Hans Sloane, Scientist," *British Library Journal* (electronic version) (1988): 1–20.

38. For more on King's satires of Sloane, see Richard Coulton, "'The Darling of the *Temple-Coffee-House-Club*': Science, Sociability, and Satire in Early Eighteenth-Century London," *Journal for Eighteenth-Century Studies* 35, no. 1 (2012): 43–65; Barbara Benedict, "Collecting Trouble: Sir Hans Sloane's Literary Reputation in Eighteenth-Century Britain," *Eighteenth-Century Life* 36, no. 2 (2012): 111–42; and William J. Ryan, "'A New Strange Disease': The Feeling of Form in Hans Sloane's Case Studies of English Jamaica," *Eighteenth Century: Theory and Interpretation* 35, no. 1 (2018): 305–24.

39. William King, "The Transactioneer, With Some of his Philosophical Fancies: In Two Dialogues," in *The Original Works of William King*, vol. 2, ed. John Nichols (London, 1776), *unpaginated preface*.

40. King, "Transactioneer," 10.

41. King, "Transactioneer," 4.

42. King, "Transactioneer," 26 (emphasis original).

43. William King, *The Present State of Physick in the Island of Cajamai. To the Members of the R.S* (London, 1710), 2.

44. *Pharmacopoeia ... Londinensis* (London, 1721), 18. The pre-Linnaean botanical names are primarily derived from the work of Swedish botanists Gaspard and Johann Bauhin, as indicated by the abbreviations *C.B.* or *J.B.* (*Raii*, for John Ray, also appears, albeit quite rarely). The Bauhin system, a binomial precursor to the work of Linnaeus, was

highly influential on both Ray and Sloane. Sloane's own Catalogue of Jamaican Plants is referenced only once.

45. My quotations for the preface come from John Quincy's contemporaneously published English translation of the *Pharmacopoeia, The Dispensatory of the Royal College of Physicians in London* (London: W. Boyer, 1721), unpaginated preface.

46. Quincy, *Pharmacopoeia*, 24.

47. The lack of botanical classification in the 1721 edition of the *Pharmacopoeia* is not limited to New World or African simples. Other materia medica not catalogued in existing botanical guides (e.g., "ceterach," a derivative of the Rustyback fern (*Aspleneium ceterach*) native to central Europe) are also presented without botanical classification. However, the absence of taxonomical nomenclature specifically for Atlantic World materia medica highlights the complex and contested manner in which such plants came into European knowledge systems. For more on the inclusion (or lack thereof) of African materia medica in official pharmacopoeias, see Benjamin Breen's essay (chapter 7).

48. Edmund Halley, "The Preface," *Philosophical Transactions* 29 (1714): 3–4.

CHAPTER 7. The Flip Side of the Pharmacopoeia

1. Denis Diderot et al., *Encyclopédie, ou Dictionnaire raisonné des sciences, des arts et des métiers* (Paris, 1751), 164.

2. John Jacob Berlu, *The Treasury of Drugs Unlock'd* (London: Printed for John Harris and Thomas Hawkins, 1690), 34–35 (on *cranium humanum*) and 18 (for cannabis).

3. Berlu, *Treasury of Drugs*, 47–48. All three, Berlu wrote, were brought from "Guinea," or the coastal region stretching from present-day Senegal to Cameroon. Berlu recorded the locations of drugs mentioned in his book, using terms like "This is generally brought from Turkey and Alexandria" (57) or "the best cometh from Germany" (60). In tabulating the origins of drugs mentioned in his book, I have combined some of Berlu's labels, for instance Berlu's "Peru" and "Mexico" are both included under my "Spanish America."

4. Arquivo Nacional da Torre do Tombo (AN/TT), Lisbon, Portugal, Livros dos Feitos Findos no. 85, "Livro de Carregações de Productos de Botica de Manuel Ferreira de Castro," (1738–1750s, Lisbon), "Carregação que Manda ir Antonio de Lima Gomes, surgião... para o rio de Jan[eiro]," ["Cargo sent to Rio de Janeiro on behalf of of Antonio de Lima Gomes, surgeon"], fol. 3r–5r.

5. Stuart Schwartz, *Sugar Plantations in the Formation of Brazilian Society* (Cambridge: Cambridge University Press, 1985); Philip D. Morgan, *Slave Counterpoint Black Culture in the Eighteenth-Century Chesapeake and Lowcountry* (Chapel Hill: University of North Carolina, 1998).

6. For recent work on this topic see Pablo Gómez, *The Experiential Caribbean: Creating Knowledge and Healing in the Early Modern Atlantic* (Chapel Hill: University of North Carolina Press, 2017); Susan Scott Parrish, "Diasporic African Sources of Enlightenment Knowledge," in *Science and Empire in the Atlantic World*, ed. James Delbourgo and Nicholas Dew (New York: Routledge, 2008); and Susan Scott Parrish, *American Curiosity: Cultures of Natural History in the Colonial British Atlantic World* (Chapel Hill: University of North Carolina Press, 2006).

7. On "disturbance pharmacopoeias" see Robert Voeks, "Disturbance Pharmacopoeias: Medicine and Myth from the Humid Tropics," *Annals of the Association of American Geographers* 94 (2004): 868–88, and the discussion of the idea in Matthew Crawford's essay (chapter 3).

8. Judith Carney, *In the Shadow of Slavery: Africa's Botanical Legacy in the Atlantic World* (Berkeley: University of California Press, 2010); Londa Schiebinger, *Secret Cures of Slaves: People, Plants, and Medicine in the Eighteenth-Century Atlantic World* (Redwood City, CA: Stanford University Press, 2017); Gómez, *Experiential Caribbean*.

9. Willem Bosman, *A New and Accurate Description of the Coast of Guinea, Divided into the Gold, the Slave, and the Ivory Coasts* (London, 1705), 225.

10. Madagascar periwinkle, for instance, contains a powerful alkaloid, vincristine, in widespread use as a chemotherapy drug. See Abena Dove Osseo-Asare, *Bitter Roots: The Search for Healing Plants in Africa* (Chicago: University of Chicago Press, 2014).

11. Giovanni Antonio Cavazzi da Montecuccolo, *Istorica descrizione de' tre' regni Congo, Matamba et Angola* (Bologna, 1687).

12. King James I of England, *A Counterblaste to Tobacco* (London, 1604).

13. Despite the prominent success of certain novel Asian and American cures, early modern European pharmacopoeias continued to be dominated by remedies carried over from Greco-Roman and medieval Muslim pharmacy. On the translatability (or lack thereof) of New World medicines into European pharmacy, see Matthew Crawford's essay (chapter 3) and Emily Beck's essay (chapter 2) in this volume, as well as Timothy D. Walker, "The Medicines Trade in the Portuguese Atlantic World: Acquisition and Dissemination of Healing Knowledge from Brazil (c. 1580–1800)," *Social History of Medicine* 26, no. 3 (August 1, 2013): and M. Stephenson, "From Marvelous Antidote to the Poison of Idolatry: The Transatlantic Role of Andean Bezoar Stones during the Late Sixteenth and Early Seventeenth Centuries," *Hispanic American Historical Review* 90, no. 1 (2010): 3–39.

14. On the role of print culture and printing presses in the dissemination of pharmaceutical knowledge in Portuguese India and New Spain, see Charles R. Boxer, *Two Pioneers of Tropical Medicine: Garcia d'Orta and Nicolás Monardes* (London: Wellcome Library, 1963); and Simon Varey, Rafael Chabrán, and Dora B. Weiner, eds. *Searching for the Secrets of Nature: The Life and Works of Dr. Francisco Hernández* (Redwood City, CA: Stanford University Press, 2002).

15. Portuguese, Dutch, and British slave traders engaged in extensive commerce with West and West-Central African coastal polities, but this was trade that was highly limited both spatially and in terms of its unidirectional nature (with Atlantic World commodities like brandy, tobacco, cowries, or textiles being almost exclusively exchanged for enslaved Africans). For instance, Joseph Miller characterizes the "merchant communities" of slave traders in Africa as physically and culturally isolated and restricted to "small commercial enclaves" that failed to integrate with a larger "African political economy." Joseph Miller, *Way of Death: Merchant Capitalism and the Angolan Slave Trade, 1730–1830* (Madison: University of Wisconsin Press, 1997), 200. On the restricted nature of Portuguese trade in Angola see José C. Curto, *Enslaving Spirits: The Portuguese-Brazilian Alcohol Trade at Luanda and its Hinterland, c. 1550–1830* (Leiden: Brill, 2003).

16. Robert Boyle, *Of the reconcileableness of specifick medicines to the corpuscular philosophy* (London: Printed for Sam. Smith, 1685), 24–25.

17. Robert Hooke, "An Account of the Plant, call'd Bangue, before the Royal Society, Dec 18. 1689," reprinted in W. Derham, ed., *Philosophical experiments and observations of the late eminent Dr. Robert Hooke* (London: Printed by W. and J. Innys, 1726), 209.

18. Nehemiah Grew, *Musaeum Regalis Societatis* (London: Printed by W. Rawlins, for the author, 1681), 385.

19. Grew, *Musaeum*, 316
20. Grew, *Musaeum*, 181.
21. On the connections between European mercantile communities and the economies of the West-Central African interior, see Mariana P. Candido, *An African Slaving Port and the Atlantic World: Benguela and its Hinterland* (Cambridge: Cambridge University Press, 2013).
22. Pierre Pomet, *Traite General Des Drogues Simples Et Composees* (Paris, 1695) 220.
23. AN/TT, PT/TT/LFF/0085, Livro de Carregações de Productos de Botica de Manuel Ferreira de Castro, March 18, 1738.
24. James H. Sweet, *Domingos Álvares, African Healing, and the Intellectual History of the Atlantic World* (Chapel Hill: University of North Carolina Press, 2011).
25. For a new assessment of Makandal's supposed poisoning that argues that the actual deaths due to poison were in fact attributable to spoiled grain, see Trevor Burnard and John Garrigus, *The Plantation Machine: Atlantic Capitalism in French Saint-Domingue and British Jamaica* (Philadephia: University of Pennsylvania Press, 2016).
26. "An Act to Remedy the Evils Arising from Irregular Assemblies of Slaves, and to Prevent their Possessing Arms and Ammunition, and Going from Place to Place Without Tickets, and for Preventing the Practice of Obeah," January 1, 1761, in *Acts of Assembly Passed in the Island of Jamaica* (Assembly of Jamaica: St. Jago de la Vega, 1771).
27. Sweet, *Domingos Álvares*, 71.
28. Caetano de Santo António, *Pharmacopea Lusitana reformada: método prático de preparar os medicamentos na forma Galenica e chimica* (Lisbon: Real Mosteyro de São Vicente de Fóra, 1711), 2.
29. Sante Arduino, *De Venenis* (Basel, 1562), 1. For more on this book and on late medieval definitions of poison see Frederick W. Gibbs, "Specific Form and Poisonous Properties: Understanding Poison in the Fifteenth Century," *Preternature: Critical and Historical Studies on the Preternatural* 2, no. 1 (2016): 19–46.
30. Alisha Rankin, "On Anecdote and Antidotes: Poison Trials in Sixteenth-Century Europe," *Bulletin of the History of Medicine* 91, no. 2 (2017): 274–302.
31. João Curvo Semedo, *Polyanthea Medicinal: Noticias galenicas, e chymicas, repartidas em tres tratados* (Lisbon: Miguel Deslandes, 1697), 196.
32. Semedo, *Polyanthea*, 27.
33. Thomas Herbert, *Some yeares travels into Africa and Asia the Great* (London: Printed for Jacob Blome and Richard Bishop, 1638), 9.
34. Herbert, *Travels*, 9.
35. Luis Gomes Ferreira, *Erario Mineral* (Lisbon: Officina de Miguel Rodrigues, 1735), 336.
36. It is unclear if "Angolistas" in this context refers to Native Angolans or to Luso-African colonists in Angola. As Luiz Felipe de Alencastro notes, "Angolista" was an expression used in nineteenth-century Portuguese to refer to colonists. See Luiz Felipe de Alencastro, "Le versant brésilien de l'Atlantique-Sud: 1550–1850," *Annales. Histoire, Sciences Sociales* 2 (2006): ff. 7.
37. Ferreira, *Erario Mineral*, 338.
38. Ferreira, *Erario Mineral*, 478.
39. Mal de Loanda bears similarities to the "earth sickness" studied by Joyce Chaplin.

Early modern sufferers of scurvy on long sea voyages often conceptualized their illness not as a discrete disease but as a consequence of too long a time spent in unhealthy or poisoned landscapes. See Joyce Chaplin, "Terrestriality," in *Appendix: A Journal of Narrative and Experimental History* 2, no. 2 (April 2014), accessed November 12, 2018, http://the appendix.net/issues/2014/4/terrestriality; and Joyce Chaplin, "Earthsickness: Circumnavigation and the Terrestrial Human Body, 1520–1800," *Bulletin of the History of Medicine* 86, no. 4 (Winter 2012): 515–42.

40. The cure is inscribed in a commonplace book belonging to the Jordan family of Virginia, alongside a number of other household remedies. Virginia Historical Society (VHS), Mss 5: 5J7664:1. This 122-page text is listed in the VHS catalogue as a possession of Robert Jordan (1731–1810), but its initial entries are dated to the late 1680s, and it also appears to feature entries by Robert's son, Edmund, and several other relatives, serving a triple function as a record of medicinal cures, Jordan family births and deaths, and slaves owned by the family. Many recipes in the book appear to have been adapted from William Salmon, *Pharmacopoeia Londinensis, or, the New London Dispensatory* (London, 1685) and Nicolas Culpeper's *Complete Herbal* (London, 1653).

41. On Caesar see Mary Galvin, "Decoctions for Carolinians" in *Creolization in the Americas*, ed. David Buisseret and Steven G. Reinhardt (College Station: Texas A&M University Press, 2000), 76–77; and Claire Gherini, "Valuing Caesar's and Sampson's Cures," *The Recipes Project*, August 18, 2015, http://recipes.hypotheses.org/6419.

42. *South Carolina Gazette (SCG)*, May 9, 1750.

43. Caesar's owner, John Norman, also stood to testify that "Caesar had done many services in a Physical Way," including the curing of rattlesnake bites and pleurisy. J. H. Easterby, ed., *The Colonial Records of South Carolina: The Journal of the Commons Assembly* (Charleston, S.C, 1962), March 28, 1749–March 19, 1750.

44. Easterby, *Journal of the Commons Assembly*, April 28, 1750.

45. Robert A. Voeks, *Sacred Leaves of Candomblé: African Magic, Medicine, and Religion in Brazil* (Austin: University of Texas Press, 2010).

46. R. K. Gordon, *Anglo-Saxon Poetry* (London: Dent, 1962), 92–93.

47. My account of Kwasi's life is based on the discussion in Parrish, *American Curiosity*, 2–13.

48. Carolus Linnaeus, *Lignum qvassiae* (Uppsala, 1763), 5.

49. John Stedman, *Narrative, of a Five Years Expedition, against the revolted Negroes of Surinam, in Guiana, on the wild coast of South America* (London: J. Johnson and T. Payne, 1806-1813), 359.

50. Stedman, *Narrative*, 360.

51. On the role of these objects in diasporic African healing, see Gómez, *Experiential Caribbean*, 42–43.

52. Society for the Propagation of the Gospel Archives, London, Letter Books, Series B, vol. 6, 62. Letter from Arthur Holt to Henry Newman, February 18, 1729, cited in Jerome Handler, "Slave Medicine and Obeah in Barbados, circa 1650 to 1834," *Nieuwe West-Indische Gids–New West Indian Guide* 74 (2000): 74–75.

53. Quoted in Parrish, *American Curiosity*, 275. In 1765, the Assembly of Georgia enacted legislation with almost the exact same wording. See Allen D. Cander et al., eds., *The Colonial Records of the State of Georgia*, vol. 18 (Athens: University of Georgia Press, 1974), 662.

54. On Cartagena see Gómez, *Experiential Caribbean*; and on Jamaica see William Ryan's essay on Hans Sloane (chapter 6).

55. Biblioteca Nacional de Portugal (BNP), F.R. 437 (Microfilm), codigo 13114, Francisco de Buytrago, "Arvore da Vida e Thesouro descuberto" (the manuscript was apparently written in Lisbon in 1731, but describes events in Angola in the 1710s and 1720s: Buytrago describes it as the work of "Sargento Mor Francisco de Buytrago, knight of the Order of Christ, in the space of twenty years in this kingdom").

56. BNP, Buytrago, "Arvore da Vida," fol. 5v.

57. Willem Bosman, *A New and Accurate Description of the Coast of Guinea, Divided Into the Gold, the Slave, and the Ivory Coasts* (London: Printed for James Knapton and Daniel Midwinter, 1705), 224.

58. Bosman, 225.

59. See for instance Arquivo Historico Ultramarino (AHU), Lisbon, Portugal, Angola, Cx. 8, doc. 32, August 11, 1664, petition of surgeon Pedro da Silva for funds to purchase "purgas para a botica."

60. AHU, Cx 9, doc. 33, July 8, 1666, testimony of Luiz Gonçalves de Andrade.

61. AN/TT, Ministerio do Reino, Informações das governadores, No. 604, "Das Preciosidades de Africa" undated (eighteenth century), 6–9.

62. Caetano de Santo Antonio, *Pharmacopea Lusitana Reformada* (Lisbon: Real Mosteyro de São Vincente de Fóra, 1711), 286. According to Santo Antonio (citing the German chemical physician Frederic Hoffman) solimão was made via a chemical reaction involving quicksilver mixed with "Caparroza de Ungria" (Hungarian copper sulfite) and vinegar.

63. Thomas Short, *Medicina Britannica: or, A treatise on such physical plants, as are generally to be found in Great-Britain* (London, 1747).

64. Short, *Medicina Britannica*, 148, 187, 213.

65. Carney, *In the Shadow of Slavery*.

66. Grains of Paradise (*Aframomum melegueta*) were by far the most important African medicine in the early modern Atlantic world. Cited by Bosman as among the most prized medicines of Gold Coast shaman/healers, *Aframomum melegueta* today figures prominently in the healing rituals of Haitian Voodou. Donald C. Simmons, "Efik Divination, Ordeals, and Omens," *Southwestern Journal of Anthropology* 12, no. 2 (July 1, 1956): 223–28, doi:10.1086/soutjanth.12.2.3629116.223f.

67. On Semedo's links to African naturalia see Benjamin Breen, "Curvo's Sixteen Secrets: Tracing Pharmaceutical Networks in the Portuguese Tropics," in *Empires of Knowledge: Scientific Networks in the Early Modern World*, ed. Paula Findlen (London: Routledge, 2018).

68. Vincent Brown, *The Reaper's Garden: Death and Power in the World of Atlantic Slavery* (Cambridge, MA: Harvard University Press, 2008), 10–12.

69. On the rise of "scientific" and professionalized pharmacy in the nineteenth century, see Gregory J. Higby, *In Service to American Pharmacy: The Professional Life of William Procter Jr.* (Tuscaloosa: University of Alabama Press, 1991).

CHAPTER 8: Consuming Canada

1. Boughton Cobb, Elizabeth Farnsworth, and Cheryl Lowe, *A Field Guide to Ferns and their Related Families: Northeastern and Central North America*, 2nd ed. (New York: Houghton Mifflin Company, 2005), 56.

2. For the best overview of Gaultier's life and medical practice, see Stéphanie Tésio, "De La Croix-Avranchin à Québec, Jean-François Gaultier, médecin du roi, de 1742 à 1756," *Annales de Normandie* 55, no. 5 (2005): 403–26. To understand the broader scientific context in which Gaultier worked and collected, see Roland Lamontagne, "L'influence de Maurepas sur les sciences: Le botaniste Jean Prat à La Nouvelle-Orléans, 1735–1746," *Revue d'histoire des sciences* 49, no. 1 (1996): 113–24.

3. Jean-François Gaultier, "Description de plusieurs plantes du Canada par M Gaultier," Cote: P91,D3, Fonds Jean-François Gaultier, Bibliothèque et Archives nationales du Québec, Québec, 61.

4. Gaultier, "Description," 73, 323, 423.

5. Rénald Lessard, "Aux XVIIe et XVIIIe siècles: L'exportation de plantes médicinales canadiennes en Europe," *Cap-aux-Diamants* no. 46 (1996): 21.

6. Stéphanie Tésio, *Histoire de la pharmacie en France et en Nouvelle-France au XVIIIe siècle* (Sainte-Foy, QC: Les Presses de l'Université Laval, 2009), 158.

7. Catherine Desbarats and Allan Greer, "Où est la Nouvelle-France?" *Revue d'histoire de l'Amérique française* 64, no. 3–4 (2011): 31–62.

8. Desbarats and Greer, "Où est la Nouvelle-France?"; Gilles Havard, *Empire et métissages. Indiens et Français dans le Pays d'en Haut, 1660–1715* (Sillery, QC: Septentrion, 2003); Paul W. Mapp, *The Elusive West and the Contest for Empire, 1713–1763* (Chapel Hill: University of North Carolina Press, 2011).

9. Desbarats and Greer, "Où est la Nouvelle-France?" See also Gervais Carpin, *Histoire d'un mot: l'ethnonyme "Canadien" de 1535–1691* (Sillery, QC: Septentrion, 1995), 25–66.

10. For an overview of this history, see Gilles Havard and Cécile Vidal, *Histoire de l'Amérique française* (Paris: Flammarion, 2004).

11. For an overview of the evolutionary history of *Adiantum pedatum* and *A. capillus-veneris*, see Jin-Mei Lu et al., "Biogeographic Disjunction between Eastern Asia and North America in the Adiantum Pedatum Complex (Pteridaceae)," *American Journal of Botany* 98, no. 10 (2011): 1680–93.

12. Northern maidenhair is also mentioned in popular accounts such as L. Blair, "Martyrdom by toads' tongues. Early Canadian doctors and their victims," *Canadian Family Physician* 45 (1999): 2293–304.

13. Tésio, *Histoire de la pharmacie*, 157–58. On ginseng, see Andreas Motsch, "Le ginseng d'Amérique: Un lien entre les deux Indes, entre curiosité et science," *Études Epistémè: Revue de littérature et de civilisation (XVIe-XVIIIe siècles)* 26 (2014), http://episteme.revues.org/331; Christopher M. Parsons, "The Natural History of Colonial Science: Joseph-François Lafitau's Discovery of Ginseng and Its Afterlives," *William and Mary Quarterly* 73, no. 1 (2016): 37–72.

14. Allan Greer, "Commons and Enclosure in the Colonization of North America," *American Historical Review* 117, no. 2 (2012): 365–86.

15. Colin Coates, *The Metamorphoses of Landscape and Community in Early Quebec* (Montreal, QC: McGill-Queen's University Press, 2000); idem., "The Colonial Landscape of the Early Town," in *Metropolitan Natures: Environmental Histories of Montreal*, ed. Stéphane Castonguay and Michèle Dagenais (Pittsburgh: University of Pittsburgh Press, 2011), 19–36. See also Christopher M. Parsons, *A Not-So-New World: Empire and Environment in French Colonial North America* (Philadelphia: University of Pennsylvania Press, 2018).

16. See, for example, Saliha Belmessous, "Etre français en Nouvelle-France: Identité

française et identité coloniale aux dix-septième et dix-huitième siècles," *French Historical Studies* 27, no. 3 (2004): 507–40; Gilles Havard, "Les forcer à devenir Cytoyens," *Annales. Histoire, Sciences Sociales* 64, no. 5 (2009): 985–1018; Sophie White, *Wild Frenchmen and Frenchified Indians: Material Culture and Race in Colonial Louisiana* (Philadelphia: University of Pennsylvania Press, 2012).

17. Ina Baghdiantz McCabe, *Orientalism in Early Modern France: Eurasian Trade, Exoticism and the Ancien Régime* (Oxford: Berg, 2008), ch. 6. On the troubling familiarity of African materia medica, see Benjamin Breen, "The Flip Side of the Pharmacopoeia? Poisons in the Atlantic World" (chapter 7).

18. The debate about the putative value of indigenous and exotic plants in this period is best discussed in Alix Cooper, *Inventing the Indigenous: Local Knowledge and Natural History in Early Modern Europe* (Cambridge: Cambridge University Press, 2007).

19. See, for example, Daniela Bleichmar, "Seeing the World in a Room: Looking at Exotica in Early Modern Collections," in *Collecting Across Cultures: Material Exchanges in the Early Modern Atlantic World*, ed. Daniela Bleichmar and Peter C. Mancall (Philadelphia: University of Pennsylvania Press, 2011), 19–20.

20. On Guillaume Delisle, see Nelson-Martin Dawson, *L'atelier Delisle: l'Amérique du Nord sur la table à dessin* (Sillery, QC: Septentrion, 2000).

21. Mapp, *Elusive West and the Contest for Empire, 1713–1763*, 173. See also Dawson, *L'atelier Delisle*; Raymonde Litalien, Denis Vaugeois, and Jean-François Palomino, *La mesure d'un continent: Atlas historique de l'Amérique du Nord, 1492–1814* (Sillery, QC: Septentrion, 2007).

22. On Franquelin, see Litalien, Vaugeois, and Palomino, *La mesure d'un continent*, 104–7; On Coronelli's globes see Litalien, Vaugeois, and Palomino, *La mesure d'un continent*, 136–39, and Christian Jacob and Edward H. Dahl, *The Sovereign Map: Theoretical Approaches in Cartography throughout History* (Chicago: University of Chicago Press, 2006), 165–72.

23. James S. Pritchard, "Early French Hydrographic Surveys in the Saint Lawrence River," *International Hydrographic Review* 56, no. 1 (1979): 126–27.

24. Jean-François Palomino, "Pratiques cartographiques en Nouvelle-France: La prise en charge de l'État dans la description de son espace colonial à l'orée du XVIIIe siècle," *Lumen* 31 (2012): 22.

25. Chandra Mukerji, *Impossible Engineering: Technology and Territoriality on the Canal du Midi* (Princeton, NJ: Princeton University Press, 2009), 15–35. On the Atlantic dimensions (and difficulty) of this project, see Nicholas Dew, "Scientific Travel in the Atlantic World: The French Expedition to Gorée and the Antilles, 1681–1683," *British Journal for the History of Science* 43, no. 1 (2010): 1–17. On the wider European dimensions of this project, see David Turnbull, "Cartography and Science in Early Modern Europe: Mapping the Construction of Knowledge Spaces," *Imago Mundi* 48, no. 1 (1996): 5–24.

26. Desbarats and Greer, "Où est la Nouvelle-France?" 37.

27. See Helen Dewar, "'Y establir nostre auctorité': Assertions of Imperial Sovereignty through Proprietorships and Chartered Companies in New France, 1598–1663" (PhD diss., University of Toronto, 2012).

28. See Carpin, *Histoire d'un mot: l'ethnonyme "Canadien" de 1535–1691*.

29. Pierre Margry, *Découvertes et établissements des Français dans l'ouest et dans le sud de l'Amérique Septentrionale, 1614–1754* (Paris: Maisonneuve et cie, 1879–1888), 1: 438.

30. Brian Brazeau, *Writing a New France, 1604–1632: Empire and Early Modern French Identity* (Farnham, UK: Ashgate, 2009), 12. See also Brian Slattery, "French Claims in North America, 1500–59," *Canadian Historical Review* 59, no. 2 (1978): 139–69.

31. André Thevet, *André Thevet's North America: A Sixteenth-Century View* (Montreal, QC: McGill-Queen's University Press, 1986), 48.

32. See, for example, James S. Pringle, "How 'Canadian' is Cornut's *Canadensium Plantarum Historia*? A Phytogeographic and Historic Analysis," *Canadian Horticultural History: An Interdisciplinary Journal* 1, no. 4 (1988): 190–209.

33. François-Marc Gagnon, "Louis Nicolas's Depiction of the New World in Figures and Text," in *The Codex Canadensis and the Writings of Louis Nicolas*, ed. François-Marc Gagnon (Montreal, QC: McGill-Queen's University Press, 2011), 31.

34. Jorge Cañizares-Esguerra, *Puritan Conquistadors: Iberianizing the Atlantic, 1550–1700* (Palo Alto: Stanford University Press, 2006), 147.

35. Louis Nicolas, *The Codex Canadensis and the Writings of Louis Nicolas*, ed. François-Marc Gagnon (Montreal, QC: McGill-Queen's University Press, 2011), 148.

36. Cañizares-Esguerra, *Puritan Conquistadors*, 149.

37. See, for example, François-Marc Gagnon, "Louis Nicolas's Depiction of the New World in Figures and Text"; Germaine Warkentin, "Aristotle in New France: Louis Nicolas and the Making of the Codex Canadensis," *French Colonial History* 11, no. 1 (2010): 71–107.

38. Nicolas, *Codex Canadensis*, 280.

39. Lynn Berry, "The Delights of Nature in this New World: A Seventeenth-Century Canadian View of the Environment," in *Decentring the Renaissance: Canada and Europe in Multidisciplinary Perspective, 1500–1700*, ed. Germaine Warkentin and Carolyn Podruchny (Toronto, ON: University of Toronto Press, 2001).

40. Pierre Boucher, *Histoire véritable et naturelle des moeurs et productions du pays de la Nouvelle-France, vulgairement dite le Canada* (Boucherville, QC: Société historique de Boucherville, 1964), 86.

41. Alain Asselin, Jacques Cayouette, and Jacques Mathieu, *Curieuses histoires de plantes du Canada*, vol. 1 (Québec: Septentrion, 2014), 118.

42. See, for example, Jean Talon, "Mémoire de Talon au Roi sur le Canada (2 Novembre 1671)," *Rapport de l'Archiviste de la province de Québec 1930–1931* (1931): 156–62.

43. Jean Talon, "Lettre de Talon au ministre Colbert (27 Octobre 1667)," *Rapport de l'Archiviste de la Province de Québec 1930–1931* (1931): 80.

44. As Stéphanie Tésio has demonstrated, this remained true throughout the seventeenth and eighteenth centuries. Tésio, *Histoire de la pharmacie*, 133–74.

45. François Gendron, *Quelques particularitez du pays des Hurons en la Nouvelle France* (Paris: Denys Bechet et Louis Billaine, 1660).

46. Rénald Lessard, "Pratique et praticiens en contexte colonial: le corps médical canadien aux XVIIe et XVIIIe siécles" (PhD diss., Université Laval, 1994), 194.

47. Nicolas Lémery, *Traité universel des drogues simples mises en ordre alphabétique* (Paris: L. d'Houry, 1698), 164.

48. Chrestien Leclercq, *Nouvelle relation de la Gaspésie* (Montreal, QC: Les Presses de l'Université de Montréal, 1999), 528–29.

49. Tésio, *Histoire de la pharmacie*, 157.

50. Lessard, "Pratique et praticiens en contexte colonial," 265.

51. Tésio, *Histoire de la pharmacie*, 157.
52. Reuben Gold Thwaites, ed., *The Jesuit Relations and Allied Documents*, 73 vols. (Cleveland: Burrows Brothers Company, 1896–1901), vol. 49, 205 [hereafter JR].
53. See, for more information, "FORESTIER, MARIE, dite de Saint-Bonaventure-de-Jésus," *Dictionary of Canadian Biography*, http://biographi.ca/fr/bio/forestier_marie_1F.html.
54. JR 49: 199.
55. JR 49: 205–11.
56. The list is not dissimilar from those analyzed in Tésio, *Histoire de la pharmacie*, 133–74.
57. François Rousseau, *L'Oeuvre de chère en Nouvelle-France: Le régime des malades à l'Hôtel-Dieu de Québec* (Sainte-Foy, QC: Les Presses de l'Université Laval, 1983), 57.
58. Lessard, "Pratique et praticiens," 184.
59. On Cadillac, see "LAUMET, dit de Lamothe Cadillac, ANTOINE," *Dictionary of Canadian Biography*, http://www.biographi.ca/en/bio/laumet_antoine_2E.html.
60. M. de Lamothe Cadillac, "Lettre de M. de Lamothe Cadillac (28 Septembre 1694)," *Rapport de l'Archiviste de la province de Québec 1923–1924* (1924): 92.
61. See "DURET DE CHEVRY DE LA BOULAYE, CHARLES," *Dictionary of Canadian Biography*, http://www.biographi.ca/en/bio/duret_de_chevry_de_la_boulaye_charles_1F.html.
62. Lahontan, *Oeuvres complètes* (Montreal, QC: Les Presses de l'Université de Montréal, 2000), 1: 166.
63. Lessard, "Pratique et praticiens," 173.
64. Jean-François Lozier, "In Each Other's Arms: France and the St. Lawrence Mission Villages in War and Peace, 1630–1730" (PhD diss., University of Toronto, 2012), 314.
65. JR 66: 155.
66. Gédéon de Catalogne, "Report on the Seigniories and Settlements in the Districts of Quebec, Three Rivers, and Montreal, by Gedeon de Catalogne/ Engineer, November 7, 1712," in *Documents Relating to the Seigniorial Tenure in Canada, 1598–1854*, ed. William Bennett Munro (Toronto: Champlain Society, 1908), 109.
67. Rousseau, *L'Oeuvre de chère en Nouvelle-France*, 188.
68. Simon Lorène, "Intérêt pharmaceutique des lettres adressées à l'apothicaire dieppois Féret par les religieuses de l'Hôtel-Dieu de Québec" (PhD diss., Université de Rouen, 2014), 125.
69. Pringle, "How 'Canadian' is Cornut's *Canadensium Plantarum Historia?*"
70. Pierre-François-Xavier de Charlevoix, "Description des plantes principales de l'Amérique septentrionnale," in *Histoire et description generale de la Nouvelle France avec le journal historique d'un Voyage fait par ordre du Roi dans l'Amérique Septentrionnale* (Paris, 1744), vol. 2, 3.
71. Charlevoix, "Description des plantes principales," 3.
72. Charlevoix, "Description des plantes principales," vol. 5, 241.
73. See Boucher, *Histoire véritable et naturelle*, 86; capillaire was one of the few American plants integrated into colonial medicine and that acquired a reputation in the French medical marketplace. For a discussion of the medical use of and commerce in capillaire see Rénald Lessard, "Pratique et praticiens," 238–42.
74. Lémery, *Traité universel*, 12.

75. Pierre Jean Baptiste Chomel, *Abrégé de l'histoire des plantes usuelles* (Paris: Chez Charles Osmont, 1712), 66–67.

76. Lémery, *Traité universel*, 11–12.

77. Joseph Pitton de Tournefort, *Traité de la matière médicale; ou l'histoire et l'usage des médicamens, et leur analyse chymique* (Paris: Laurent d'Houry, 1717), 3: 185.

78. Jacques Savary des Brûlons and Philémon-Louis Savary, *Dictionnaire universel de commerce contenant tout ce qui concerne le commerce qui se fait dans les quatre parties du monde* (Amsterdam: Chez les Jansons à Waesberge, 1726), 1: 540.

79. Nicolas Lémery, *Pharmacopée universelle, contenant toutes les compositions de pharmacie qui sont en usage dans la medecine, tant en France que par toute l'Europe* (Paris: Laurent d'Houry, 1716), 165.

80. Lémery, *Traité universel*, 625.

81. Lorène, "Intérêt pharmaceutique des lettres," 148–49.

82. CAOM, C11A vol. 13, 178r-191v; The medical historian Rénald Lessard also shows that the superior of the Hôtel-Dieu shipped syrup made from the plant regularly in the early eighteenth century; "Pratique et praticiens," 238–39.

83. Lahontan, *Oeuvres complètes*, 1: 76.

84. Philbert Guybert, *Toutes les oeuvres charitables de Philebert Guibert* (Rouen: François Vaultier, 1667), 167.

85. Chomel, *Abrégé de l'histoire des plantes usuelles*, 83.

86. This is little discussed in the sources, but could have been a major difference in the experience of capillaire du Canada. This also suggests that capillaire may have been a vehicle for introducing that most Canadian of plants—the sugar maple—to France and French markets. John Evelyn, *Silva, or A discourse of forest-trees, and the propagation of timber in His Majesty's dominions* (London, 1706), 120.

87. Pehr Kalm, *Travels into North America containing its natural history, and a circumstantial account of its plantations and agriculture in general: With the civil, ecclesiastical and commercial state of the country, the manners of the inhabitants, and several curious and important remarks on various subjects* (London: T. Lowndes, 1771), 118–19.

88. Louis-Antoine de Bougainville, *Écrits sur le Canada: mémoires, journal, lettres* (Sillery, QC: Septentrion, 2003), 89.

89. Helen Dewar, "Canada or Guadeloupe?: French and British Perceptions of Empire, 1760–1763," *Canadian Historical Review* 91, no. 4 (2010): 637–60.

CHAPTER 9: Rethinking Pharmacopoeic Forms

Thanks to *Early American Studies* for permission to reprint this essay. For support with that original article, I thank then editor Elaine Crane and the anonymous reviewers for *Early American Studies* for their insightful comments. I'm very grateful to Katy Chiles and Cassander L. Smith for reading early versions of the article and to Ivy Schweitzer and Christopher Parsons for conversations about Occom's herbal. Peter Carini and the staff at Rauner Library, Dartmouth College, provided invaluable support during my research trip and, later, equally invaluable assistance with images. Tricia Royston at the New London County Historical Society graciously provided access to Occom's herbal. Research for this article was supported by a University of North Texas Research Initiation Grant.

For their work assembling this volume and suggestions for the chapter version of this piece, I thank Matthew Crawford and Joseph Gabriel.

This chapter was originally published as Kelly Wisecup, "Medicine, Communication, and Authority in Samson Occom's Herbal," *Early American Studies* 10, no. 3 (Fall 2012): 540–65.
Reprinted with permission of the University of Pennsylvania Press.

1. It is unclear whether the booklets were once in one piece, or whether Occom initially wrote two books. Nonetheless, they are undoubtedly related, for Occom numbered each recipe, and the first booklet ends with cure number 5; the second booklet begins with number 6.

2. One booklet is held at the New London County Historical Society and the second at Rauner Library, Dartmouth College's special collections library.

3. Samson Occom, *The Collected Writings of Samson Occom, Mohegan: Leadership and Literature in Eighteenth-Century America*, ed. Joanna Brooks (Oxford: Oxford University Press, 2006); Philip Rabito-Wyppensenwah and Robert Abiuso, "The Montaukett Use of Herbs: A Review of the Recorded Material," *The History and Archaeology of the Montauk*, vol. 3 of *Readings in Long Island Archeology and Ethnohistory*, 2nd ed., ed. Gaynell Stone (Stony Brook: Suffolk County Archeological Association, 1993), 585; Edward Connery Lathem, *Ten Indian Remedies: From Manuscript Notes on Herbs and Roots* (N.p.: n.p. 1954).

4. Rabito-Wyppensenwah and Abiuso have found a likely reference to Ocus on a Suffolk County muster list from 1760, which notes that he is about 5′ 9″ in height and that he enlisted at age forty-eight. See Rabito-Wyppensenwah and Abiuso, 585.

5. On Henry Quaquaquid, see Brooks, 13, 18, and 423.

6. For references to Occom's herbal, see Joanna Brooks, "'This Indian World: An Introduction to the Writings of Samson Occom," in Occom, *Collected Writings*, 4 and 42. John J. Kucich examines Occom's herbal as evidence of the way in which Occom used the environment to feed his family. Kucich argues that Occom distanced himself from the environment and from "native religious rituals" and transfigured Native and European relationships to the environment. By contrast, I argue that the herbal documents Occom's relationship to the environment, a relationship that he configured in terms of Native medical knowledge and that was not separate from Christian religious practices. See John J. Kucich, "Sons of the Forest: Environment and Transculturation in Jonathan Edwards, Samson Occom and William Apess," in *Assimilation and Subversion in Earlier American Literature*, ed. Robin de Rosa (Newcastle: Cambridge Scholars Press, 2006), 13.

7. Occom, "Saturday, Octr 13," *Collected Writings*, 380–81; and Monday, July 16," *Collected Writings*, 373.

8. Occom, "'In Christ, He is a New Creature,' 2 Corinthians 5:17 (July 13, 1766)," *Collected Writings*, 174.

9. Occom's writing has been at the center of the growing field of early Native studies and efforts to foreground writing by—rather than colonial representations of—Native people and to consider Native writing within tribally specific contexts. See Lisa Brooks, *The Common Pot: The Recovery of Native Space in the Northeast* (Minneapolis: University of Minnesota Press, 2008), especially chapter 2; Kristina Bross and Hilary E. Wyss, eds. *Early Native Literacies in New England: A Documentary and Critical Anthology* (Amherst: University of Massachusetts Press, 2008); Drew Lopenzina, *Red Ink: Native Americans Picking Up the Pen in the Colonial Period* (Albany: State University of New York Press, 2012); Phillip H. Round, *Removable Type: Histories of the Book in Indian Country, 1663–1880* (Chapel Hill: University of North Carolina Press, 2010); and Hilary E. Wyss, *Writing*

Indians: Literacy, Christianity, and Native Community in Early America (Amherst: University of Massachusetts Press, 2003) and Wyss, *English Letters and Indian Literacies: Reading, Writing, and New England Missionary Schools, 1750–1830* (Philadelphia: University of Pennsylvania Press, 2012).

10. English pharmacological texts were reprinted in American editions starting in the early eighteenth century; colonists also brought copies of English-printed medical texts with them to North America. See, for example, Nicholas Culpeper, *The English Physician, containing admirable and approved remedies for several of the most usual diseases* (Boston, 1708); Culpeper, *Pharmacopoeia Londinensis; or, The London dispensatory further adorned by the studies and collections of the fellows now living, of the said college* (Boston, 1720); and Thomas Short, *Medicina Britannica: Or a Treatise on such Physical Plants as are Generally to be found in the Fields or Gardens in Great-Britain: Containing a Particular Account of their Nature, Virtues, and Uses* (London, 1747).

11. Thomas Short, *Medicina Britannica: Or a Treatise on such Physical Plants as are Generally to be found in the Fields or Gardens in Great-Britain: Containing a Particular Account of their Nature, Virtues, and Uses* (Philadelphia, 1751), xviii–xix.

12. Short, *Medicina Britannica*, xviii.

13. On such practices in the context of natural historical illustrations, see Daniela Bleichmar's study of collation, a process in which plants were placed among "illustrations and textual descriptions in published works in order to classify new specimens or rectify mistakes, in this way staking claims about the novelty and significance of their observations." See Daniela Bleichmar, *Visible Empire: Botanical Expeditions and Visual Culture in the Hispanic Enlightenment* (Chicago: University of Chicago Press, 2012), 8.

14. Gladys Tantaquidgeon, *Folk Medicine of the Delaware and Related Algonkian Indians* (Harrisburg: Pennsylvania Historical and Museum Commission, 2001), 12.

15. See Tantaquidgeon, *Folk Medicine*, 12.

16. Tantaquidgeon, *Folk Medicine*, 12.

17. Occom, "Herbs & Roots," 1754–1756, Rauner DC History, Rauner Library, Dartmouth College. Occom did not number the pages of "Herbs & Roots," so I will refer to recipes by providing the numbers Occom gave them.

18. Frank G. Speck, "Medicine Practices of the Northeastern Algonquians," *Proceedings of the 19th International Congress of Americanists* (Washington, DC: 1917), 305.

19. Tantaquidgeon, "Mohegan Medicinal Practices, Weather-Lore and Superstition," *Annual Report of the Bureau of American Ethnology to the Secretary of the Smithsonian Institution, 1925–1926* (Washington DC: 1926), 264 and 265.

20. Occom, "Herbs & Roots," Rauner DC History.

21. Speck, "Medicine Practices," 306.

22. Lloyd G. Carr and Carlos Westey, "Surviving Folktales and Herbal Lore among the Shinnecock Indians of Long Island," *Journal of American Folklore* 58, no. 228 (April–June 1945): 114. See also Daniel E. Moerman, *Native American Ethnobotany* (Portland, OR: Timber Press, 1998), especially 29.

23. Speck, "Medicine Practices," 305.

24. Occom, "Herbs & Roots," Rauner DC History.

25. For one study of such taxonomies, see Michel Foucault, *The Order of Things: An Archaeology of the Human Sciences* (New York: Vintage Books, 1994).

26. Tantaquidgeon and Jayne G. Fawcett, "Symbolic Motifs on Painted Baskets on the Mohegan-Pequot," *A Key Into the Language of Woodsplint Baskets*, ed. Ann McMullen and

Russell G. Handsman (Washington, CT: American Indian Archaeological Institute, 1987), 98. See also Stephanie Fitzgerald, "The Cultural Work of a Mohegan Painted Basket," in *Early Native Literacies in New England: A Documentary and Critical Anthology*, ed. Kristina Bross and Hilary E. Wyss (Amherst: University of Massachusetts Press, 2008), 52–56.

27. Occom, "Herbs & Roots," Rauner DC History.
28. Tantaquidgeon and Fawcett, "Symbolic Motifs," 98.
29. Tantaquidgeon and Fawcett, "Symbolic Motifs," 99.
30. Occom, "Herbs & Roots," Rauner DC History.
31. William S. Simmons, *Spirit of the New England Tribes; Indian History and Folklore, 1620–1984* (Hanover: University Press of New England, 1986), 109; and Tantaquidgeon, *Folk Medicine*, 15.

32. Tantaquidgeon's field notes include information on "projects" by which one could determine who one's lover would be. For example, girls would break an egg in water and examine the shape it took; or a girl would throw a ball of cord or yarn into a pit or cellar and wind it up, with the expectation that her lover would appear at the other end of the string. See Tantaquidgeon qtd. in Simmons, 108–10. She also explains that Delaware practitioners employed two herbs as love medicine: one named "flower hangs down" and one named "pull-up." See Tantaquidgeon, *Folk Medicine*, 15.

33. See David D. Hall, *Worlds of Wonder, Days of Judgment: Popular Religious Belief in Early New England* (Cambridge, MA: Harvard University Press, 1990) and Karen Ordahl Kupperman, *Indians and English: Facing Off in Early America* (Ithaca: Cornell University Press, 2000), 128–37.

34. On the professionalization of colonial medicine, see the articles in *Medicine in Colonial Massachusetts 1620–1820: A Conference Held 25 & 26 May 1978 by the Colonial Society of Massachusetts* (Boston: Colonial Society of Massachusetts, 1980).

35. Susan Scott Parrish, *American Curiosity: Cultures of Natural History in the Colonial British Atlantic World* (Chapel Hill: University of North Carolina Press, 2006), 216. For an argument that this shift occurred earlier, in the seventeenth century, see Joyce E. Chaplin, *Subject Matter: Technology, the Body, and Science on the Anglo-American Frontier, 1500–1676* (Cambridge, MA: Harvard University Press, 2003), chapter eight. On colonists' attempts to counter European biases against their knowledge, see Ralph Bauer, *The Cultural Geography of Colonial American Literatures: Empire, Travel, Modernity* (Cambridge: Cambridge University Press, 2003).

36. Parrish, 217. On occult virtues see also Keith Hutchison, "What Happened to Occult Qualities in the Scientific Revolution?" in *The Scientific Enterprise in Early Modern Europe: Readings from Isis*, ed. Peter Dear (Chicago: University of Chicago Press, 1997), 86–106.

37. Tantaquidgeon, *Folk Medicine*, 12.

38. A. Irving Hallowell, "Ojibwa Ontology, Behavior, and World View," in *Culture in History: Essays in Honor of Paul Radin*, ed. Stanley Diamond (New York: GP Putnam's Sons, 1960), 23. For more recent commentary on Hallowell's essay, see Kenneth M. Morrison, "The Cosmos as Intersubjective: Native American Other-Than-Human Persons," *Indigenous Religions: A Companion*, ed. Graham Harvey (London: Cassell, 2000), 23–36.

39. Tantaquidgeon, *Folk Medicine*, 15.
40. Occom, "Account of the Montauk Indians, on Long Island (1761)," *Collected Writings*, 49.
41. Occom, "Account," 49.

42. Occom, "Account," 49.

43. Joshua David Bellin describes such processes of drawing upon knowledge from multiple origins as constituting a "medicine bundle": "the bringing together of diverse medicine acts, all of which derive their form and power through contact with their others." See Bellin, *Medicine Bundle: Indian Sacred Performance and American Literature, 1824–1932* (Philadelphia: University of Pennsylvania Press, 2008), 9.

44. Jeffrey Glover, "Early American Archives and the Evidence of History," *Early American Literature* 46, no. 1 (2011): 177.

45. Glover, "Early American Archives," 177.

46. Robert Allen Warrior, *Tribal Secrets: Recovering American Indian Intellectual Traditions* (Minneapolis: University of Minnesota Press, 1995), 124.

47. Warrior, *Tribal Secrets*, 124.

48. Occom, "Herbs & Roots," 1754–56, New London County Historical Society.

49. Nicholas Thomas, *Entangled Objects: Exchange, Material Culture and Colonialism in the Pacific* (Cambridge, MA: Harvard University Press, 1991), 50.

50. Thomas, *Entangled Objects*, 51.

51. See Tantaquidgeon, *Folk Medicine*, 67.

52. Gladys Tantaquidgeon, "Notes on the Gay Head Indians of Massachusetts," *Indian Notes* 7, no. 1 (Jan. 1930): 17. See also Melissa Jayne Fawcett, *Medicine Trail: The Life and Lessons of Gladys Tantaquidgeon* (Tucson: University of Arizona Press, 2000), 38.

53. Thomas, *Entangled Objects*, 51.

54. Thomas, *Entangled Objects*, 51.

55. Fawcett, 37. Tantaquidgeon published some herbal remedies in *Folk Medicine of the Delaware and Related Algonkian Indians*.

56. Occom, "Herbs & Roots," New London County Historical Society.

57. Laura M. Stevens, *The Poor Indians: British Missionaries, Native Americans, and Colonial Sensibility* (Philadelphia: University of Pennsylvania Press, 2004), 18.

58. Stevens, *Poor Indians*, 16.

59. Stevens, *Poor Indians*, 21. See also 173–78.

60. Occom, "Herbs & Roots," Rauner DC History.

61. Occom, "Herbs & Roots," New London County Historical Society. This recipe is unnumbered in the manuscript.

62. Occom, "Herbs & Roots," Rauner DC History.

63. Occom, "'In Christ, He is a New Creature,' 2 Corinthians 5:17 (July 13, 1766)," *Collected Writings*, 174.

64. Occom was in England between 1765 and 1768.

65. Occom to Nathaniel Shaw, December 8, 1767, New London County Historical Society. Letters at Rauner Library show that Shaw was not always a reliable contact: in 1766, Bezaleel Woodward informed Wheelock that Mary Occom was uncertain regarding how much money she should repay Shaw, since he had earlier refused to loan her money. See Bezaleel Woodward to Eleazar Wheelock, June 13, 1766, in the Rauner Library Special Collections, Dartmouth College Library.

66. For this suggestion, I thank Tricia Royston, librarian at the New London County Historical Society.

67. On Occom's separation from Wheelock, see Brooks, "This Indian World," 19–22.

68. Walter Ong, *Orality and Literacy: The Technologizing of the Word* (New York: Methuen, 1988), 99.

NOTES TO PAGES 194-202

69. Ong, *Orality and Literacy*, 99.
70. Ong, *Orality and Literacy*, 8.
71. Tantaquidgeon and Fawcett, 98. See also Matt Cohen, *The Networked Wilderness: Communicating in Early New England* (Minneapolis: University of Minnesota Press, 2009), 5–6, for a consideration of how Native American studies complicates Ong's theory of literacy.
72. Londa Schiebinger, *Plants and Empire: Colonial Bioprospecting in the Atlantic World* (Cambridge, MA: Harvard University Press, 2004), 198.
73. In this sense, his lists align with women's and familial medicinal recipe books that were created through acts of collaborative authorship and offered spaces for intellectual and medical experimentation. See Elaine Leong, "Collecting Knowledge for the Family: Recipes, Gender and Practical Knowledge in the Early Modern English Household," *Centaurus* 55 (2013): 81–103; Leong, "Making Medicines in the Early Modern Household," *Bulletin of the History of Medicine* 82, no. 1 (2008): 145–68; and Wendy Wall, *Recipes for Thought: Knowledge and Taste in the Early Modern English Kitchen* (Philadelphia: University of Pennsylvania Press, 2016).
74. Robert Warrior, *The People and the Word: Reading Native Nonfiction* (Minneapolis: University of Minnesota Press, 2005), xxix.
75. Warrior, *People and the Word*, xxix.

CHAPTER 10: National Identities, Medical Politics, and Local Traditions

1. David L. Cowen, "The Edinburgh Pharmacopoeia," in *The Early Years of the Edinburgh Medical School*, ed. R.G.W. Anderson and A.D.C. Simpson (Edinburgh: Royal Scottish Museum, 1976), 1–20.
2. George Urdang, *Pharmacopoeia Londinensis of 1618, with a Historical Introduction by G. Urdang* (Madison: State Historical Society of Wisconsin, 1944), 1.
3. David L. Cowen, "The Edinburgh Pharmacopoeia," in Anderson and Simpson, *Early Years of the Edinburgh Medical School*, 1–20.
4. T. P. Kirkpatrick, *The Dublin Pharmacopoeias* (Dublin: Royal College of Physicians of Ireland, 1921), 1–20.
5. George Urdang, "Pharmacopoeias as Witnesses of World History," *Journal of the History of Medicine* 1 (1946): 46–70.
6. Urdang, "Pharmacopoeias as Witnesses of World History," 58–61.
7. Penelope Hunting, *A History of the Society of Apothecaries* (London: The Society of Apothecaries, 1998), 23.
8. Margaret Connolly, "Evidence for the Continued use of Medieval Medical Prescriptions in the Sixteenth Century: A Fifteenth Century Remedy Book and its Later Owner," *Medical History* 60 (2016): 133–54.
9. Maria Amalia D'Aronco, "The Old English Pharmacopoeias," *AVISTA Forum Journal* 13, no. 2 (2003): 9–18.
10. F. Hoeniger, and J. Hoeniger, *The Development of Natural History in Tudor England* (Charlottesville: University of Virginia Press, 1973).
11. George Clark, *A History of the Royal College of Physicians of London*, vol. 1 *1518–1689* (Oxford: Clarendon Press, 1964), 55.
12. Clark, *History of the Royal College of Physicians of London*, 58.
13. Clark, *History of the Royal College of Physicians of London*, 40.

14. W. Monk, *Roll of the College of Physicians*, vol. 1, 2nd ed. (London: Royal College of Physicians, 1878), 2.
15. Clark, *History of the Royal College of Physicians of London*, 68.
16. Clark, *History of the Royal College of Physicians of London*, 69, 72.
17. Clark, *History of the Royal College of Physicians of London*, 158.
18. Urdang, *Preface to Pharmacopoeia Londinensis 1618*, 82.
19. Clark, *History of the Royal College of Physicians of London*, 159.
20. Clark, *History of the Royal College of Physicians of London*, 159.
21. David Jacques, *Essential to the Pracktick Part of Physic: The London Apothecaries 1540–1617* (London: Society of Apothecaries, 1992), 47.
22. Hunting, *History of the Society of Apothecaries*, 24.
23. A. C. Wootton, *Chronicles of Pharmacy*, vol. 2 (London: Macmillan and Co, 1910), 60.
24. The full list is printed in Monk, *Roll of the College of Physicians*, vol. 3, 373. Cited in Clark, *History of the Royal College of Physicians of London*, 161.
25. Wootton, *Chronicles of Pharmacy*, vol. 2, 60.
26. Clark, *History of the Royal College of Physicians of London*, 161.
27. Clark, *History of the Royal College of Physicians of London*, 165.
28. Clark, *History of the Royal College of Physicians of London*, 199.
29. Clark, *History of the Royal College of Physicians of London*, 218.
30. Clark, *History of the Royal College of Physicians of London*, 219.
31. Hunting, *History of the Society of Apothecaries*, 32.
32. Clark, *History of the Royal College of Physicians of London*, 219.
33. Clark, *History of the Royal College of Physicians of London*, 220.
34. Clark, *History of the Royal College of Physicians of London*, 159.
35. Clark, *History of the Royal College of Physicians of London*, 221.
36. Hunting, *History of the Society of Apothecaries*, 34.
37. Clark, *History of the Royal College of Physicians of London*, 220.
38. James Grier, *A History of Pharmacy* (London: Pharmaceutical Press, 1937), 141.
39. A. C. Wootton, *Chronicles of Pharmacy*, vol. 1 (London: Macmillan and Co, 1910) 256.
40. Wootton, *Chronicles of Pharmacy* vol. 2, 64.
41. Hunting, *History of the Society of Apothecaries*, 48.
42. Clark, *History of the Royal College of Physicians of London*, 228.
43. *Pharmacopoea Londinensis, in Qva Medicamenta Antiqva et Nova, Londinensis, Opera Medicorum Collegij, 1618*. First issue of the first edition of the *London Pharmacopoeia*.
44. Glenn Sonnedecker, *Kremers and Urdang's History of Pharmacy* (Madison: American Institute of the History of Pharmacy, 1976), 117.
45. Clark, *History of the Royal College of Physicians of London*, 228. See also Hunting, *History of the Society of Apothecaries*, reference 43, 268.
46. Monk, *History of the Royal College of Physicians of London*, vol. 1 170.
47. Clark, *History of the Royal College of Physicians of London*, 228.
48. *Pharmacopoea Londinensis, in Qva Medicamenta Antiqva et Nova, Londinensis, Opera Medicorum Collegij 1618*. Second issue of the first edition of the *London Pharmacopoeia*.
49. Sonnedecker, *Kremers and Urdang's History*, 430.

50. W. Brockbank, "Sovereign Remedies: A Critical Depreciation of the 17th Century London Pharmacopoeia," *Medical History* 8, no. 1 (1964): 1–14.

51. Urdang, *Pharmacopoeia Londinensis of 1618*, 24, 77–81.

52. W. Ferguson, *Scotland's Relations with England: A Survey to 1707* (Edinburgh: John Donald Publishers, 1977), 201.

53. K. O. Morgan, *The Oxford History of Britain*, vol. 3, *The Tudors and Stuarts* (Oxford: Oxford University Press, 1989), 34.

54. Morrice McCrae, *Physicians and Society: A Social History of the Royal College of Physicians of Edinburgh* (Edinburgh: John Donald, 2007), 17.

55. J. Geyer-Kordesch and F. MacDonald, *Physicians and Surgeons in Glasgow: The History of the Royal College of Physicians and Surgeons of Glasgow, 1599–1858* (London: Hambledon Press, 1999).

56. Morgan, *Oxford History of Britain*, vol. 3, 170.

57. McCrae, *Physicians and Society*, 15.

58. McCrae, *Physicians and Society*, 15.

59. J. F. McHarg, "Dr John Maklure and the 1630 Attempt to Establish the College," *Proceedings of the Royal College of Physicians of Edinburgh Tercentenary Congress* (1981): 49.

60. Clark, *History of the Royal College of Physicians of London*, 284.

61. Clark, *History of the Royal College of Physicians of London*, 284.

62. McCrae, *Physicians and Society*, 15.

63. McCrae, *Physicians and Society*, 18.

64. McCrae, *Physicians and Society*, 7.

65. McCrae, *Physicians and Society*, 12.

66. McCrae, *Physicians and Society*, 21.

67. T. M. Devine, *Scotland's Empire 1600–1815* (London: Penguin Books, 2003), 37.

68. Clark, *History of the Royal College of Physicians of London*, 336.

69. McCrae, *Physicians and Society*, 27.

70. McCrae, *Physicians and Society*, 37.

71. Cowen, *Edinburgh Pharmacopoeia*, 3.

72. F. P. Hett, ed., *The Memoirs of Sir Robert Sibbald 1641–1722* (London: Oxford University Press, 1932), 93.

73. Cowen, *Edinburgh Pharmacopoeia*, 3.

74. McCrae, *Physicians and Society*, 34.

75. McCrae, *Physicians and Society*, 35.

76. McCrae, *Physicians and Society*, 35.

77. McCrae, *Physicians and Society*, 36.

78. W. B. Howie, "Sir Archibald Stevenson, His Ancestry and the Riot in the College of Physicians at Edinburgh," *Medical History* 11, no. 3 (1967): 269–84.

79. McCrae, *Physicians and Society*, 38.

80. H. L. Coulter, *Divided Legacy: A History of Modern Western Medicine, J. B. Helmont to Claude Bernard* (Berkeley: North Atlantic Books, 2000), 110.

81. *Pharmacopoea Collegii Regii Medicorum Edimburgensium, Edimburgi, MDCXCIX*. First Edition of the *Edinburgh Pharmacopoeia*.

82. K. Kenny, *Ireland and the British Empire* (Oxford: Oxford University Press, 2004), 6.

83. J.D.H. Widdess, *A History of the Royal College of Physicians of Ireland 1654–1963* (Edinburgh: E. & S. Livingstone, 1963), 3.

84. Clark, *History of the Royal College of Physicians of London*, 250.

85. Widdess, *History of the Royal College of Physicians of Ireland*, 7.
86. Widdess, *History of the Royal College of Physicians of Ireland*, 68.
87. T. Bartlett, "Ireland, Empire and Union 1690–1801," in *Ireland and the British Empire*, ed. K. Kenny (Oxford: Oxford University Press, 2004), 69.
88. Kirkpatrick, *Dublin Pharmacopoeias*, 3.
89. Kirkpatrick, *Dublin Pharmacopoeias*, 4.
90. Kirkpatrick, *Dublin Pharmacopoeias*, 4.
91. N. C. Cooper, "Development of a Pharmaceutical Profession in Ireland," *Pharmacy in History* 29, no. 4 (1987): 165–76.
92. Kirkpatrick, *Dublin Pharmacopoeias*, 5.
93. Kirkpatrick, *Dublin Pharmacopoeias*, 5.
94. Kirkpatrick, *Dublin Pharmacopoeias*, 6.
95. Kirkpatrick, *Dublin Pharmacopoeias*, 7.
96. Hunting, *History of the Society of Apothecaries*, 224.
97. Kirkpatrick, *Dublin Pharmacopoeias*, 7.
98. Kirkpatrick, *Dublin Pharmacopoeias*, 7.
99. Kirkpatrick, *Dublin Pharmacopoeias*, 8.
100. Widdess, *History of the Royal College of Physicians of Ireland*, 75.
101. Kirkpatrick, *Dublin Pharmacopoeias*, 9–10.
102. Kirkpatrick, *Dublin Pharmacopoeias*, 10.
103. Kirkpatrick, *Dublin Pharmacopoeias*, 10.
104. Kirkpatrick, *Dublin Pharmacopoeias*, 12.
105. *Pharmacopoeia Collegii Medicorum Regis et Retinae in Hibernia, Dublinii, 1807*. First edition of the *Dublin Pharmacopoeia*.
106. Cowen, *Edinburgh Pharmacopoeia*, 2.
107. *Edinburgh Pharmacopoeia, Second Edition* (Edinburgh: Royal College of Physicians of Edinburgh, 1722), i–ii.
108. Cowen, *Edinburgh Pharmacopoeia*, 7.
109. J. V. Pickstone, *Ways of Knowing: A New History of Science, Technology and Medicine* (Chicago: Chicago University Press, 2000), 2.
110. Pickstone, *Ways of Knowing*, 5.
111. *British Pharmacopoeia, published under the direction of the General Council of Medical Education and Registration of the United Kingdom pursuant to the Medical Act 1858*, London, 1864. First edition of the *British Pharmacopoeia*.

CHAPTER 11: The Codex Nationalized

1. Michel de Certeau, Dominique Julia, and Jacques Revel, *Une politique de la langue* (Paris: Gallimard, 1975); David Bell, "Tearing down the Tower of Babel: Grégoire and French Multilingualism," in *The Abbé Grégoire and his World*, ed. Jeremy Popkin and Richard Popkin (Dordrecht: Kluwer Academic Publishers, 2000).

2. In this rich historiography, France features prominently. See Ben Kafka, *The Demon of Writing : Powers and Failures of Paperwork* (New York: Zone Books, 2012); Gérard Noiriel, *État, nation et immigration. Vers une histoire du pouvoir* (Paris: Belin, 2001), chapters 10–14; Jacob Soll, *The Information Master* (Ann Arbor: University of Michigan Press, 2009).

3. Michael Stainpeis was a professor at Vienna's medical school at the turn of the sixteenth century. Quoted in Christoph Friedrich and Wolf-Dieter Müller-Jahncke, *Geschichte der Pharmazie II: von der frühen Neuzeit bis zur Gegenwart* (Eschborn: Govi, 2005),

190–91. I abbreviated Stainpeis's enumeration, trusting his point would come across in the quoted excerpt.

4. See also Paula De Vos, "European Materia Medica in Historical Texts: Longevity of a Tradition and Implications for Future Use," *Journal of Ethnopharmacology* 132 (2010): 28–47, and "The 'Prince of Medicine': Yuhanna ibn Masawayh and the Foundations of the Western Pharmaceutical Tradition," *Isis* 104 (2013): 667–712.

5. Glenn Sonnedecker, "The Founding Period of the US Pharmacopeia: I. European Antecedents," *Pharmacy in History* 35, no. 4 (1993): 151–62, provides a useful overview on the history of, and historiography on, early modern European pharmacopoeias. On the hopes inspired by the technology of printing for the purification of both texts and substances, as well as the limits thereof, see Adrian Johns, *Piracy: The Intellectual Property Wars from Gutenberg to Gates* (Chicago: University of Chicago Press, 2009), 100–101.

6. "Lectori Benevolo," *Codex medicamentarius* (1638), n.p. The dispensatory of Valerius Cordus is the Nuremberg *Dispensatorium* of 1546. On the making of the first edition of the Parisian Codex, see Jean Bergounioux, "Les éditions du *Codex Medicamentarius*," *Bulletin de la société française d'histoire de la pharmacie* 15 (1927): 382–89.

7. Wolf-Dieter Müller-Jahncke, "Platon im Arzneibuch und der Heller am Tresen: Pharmazie im 16. Jahrhundert zwischen Humanismus, Stadtgesellschaft und Ökonomie," *Berichte zur Wissenschaftsgeschichte* 23 (2000): 1–15.

8. Bergounioux, "Les éditions du *Codex Medicamentarius*," 389.

9. For a broad portrait of the inclusive world of early modern French health care, see also Laurence Brockliss and Colin Jones, *The Medical World of Early Modern France* (Oxford: Clarendon Press, 1997); and on pharmacy specifically, see Matthew Ramsey, "Traditional Medicine and Medical Enlightenment: The Regulation of Secret Remedies in the Ancien Régime," *Historical Reflections/Réflexions historiques* 9, nos. 1–2 (1982): 215–32.; and Bénédicte Dehillerin and Jean-Pierre Goubert, "À la conquête du monopole pharmaceutique: Le Collège de Pharmacie de Paris (1777–1796)," *Historical Reflections/Réflexions historiques* 9, nos. 1–2 (1982): 233–48.

10. Nicolas Lémery, *Pharmacopée universelle* (Paris: Houry, 1698), v.

11. Friedrich and Müller-Jahncke, *Geschichte der Pharmazie*, 405–7.

12. Antoine François de Fourcroy, *Discours prononcé à la Société libre des pharmaciens de Paris* (Paris: Quillau, 1797)4, 7. On Fourcroy's career, see Jonathan Simon, *Chemistry, Pharmacy and Revolution in France, 1777–1809* (Aldershot: Ashgate, 2005), chapter 4.

13. Dehillerin and Goubert, "À la conquête du monopole pharmaceutique," 245–48.

14. This shift is the subject of Matthew Ramsey, *Professional and Popular Medicine in France, 1770–1830: The Social World of Medical Practice* (Cambridge: Cambridge University Press, 1988).

15. From the French *officine*, i.e., pharmacy laboratory.

16. *Codex medicamentarius* (1758), v–vi, and *Codex medicamentarius* (1818), iii. The last Parisian Codex listed 617 formulas, the first French Codex 558.

17. "Ordonnance du Roi sur la publication d'un nouveau code pharmaceutique," 115–17.

18. Charles-Louis Cadet, "Examen du Codex Medicamentarius," *Journal de pharmacie et des sciences accessoires* 5 (1819): 121–37, 203–29. especially 121, 203, 212–20.

19. See for example the unforgiving review by Richard Phillips of the Royal College of Physicians in London: "Remarks on the '*Code des médicamens ou pharmacopée française*,'" *Quarterly Journal of Science, Literature and the Arts* 9, no. 18 (1820): 239–50.

20. The trial, held in Paris's Tribunal Correctionnel on October 7, 1819, is well covered in France's leading pharmacy journal of the time, whose chief editor was none other than Virey himself. See "Affaire du Codex Medicamentarius," cited and discussed in Jean Flahaut, "La vie difficile du premier Codex national français," *Revue d'histoire de la pharmacie* 327, no. 3 (2000): 341–42. For a comparison of the contents of Virey's handbooks and the Codex, see Philotime, *Lettre à M. le Docteur Virey* (Paris: J. Gratiot, 1819).

21. James C. Scott, *Seeing Like a State: How Certain Schemes to Improve the Human Condition Have Failed* (New Haven, CT: Yale University Press, 1998), 64–71.

22. See Valentin Groebner, *Who Are You? Identification, Deception, and Surveillance in Early Modern Europe* (New York: Zone Books, 2007).

23. Gérard Noiriel, "L'identification des citoyens. Naissance de l'état civil républicain," *Genèses. Sciences Sociales et Histoire* 13 (1993): 3–28.; and "Surveiller les déplacements ou identifier les personnes? Contribution à l'histoire du passeport en France de la Ier à la IIIe République," *Genèses. Sciences Sociales et Histoire* 30 (1998): 77–100.

24. Noiriel, "L'identification des Citoyens," 26–27.

25. P. Grand, "Metz, 11 février 1857," *Journal du Palais: Jurisprudence française* (1857): 449–53. 450.

26. Jones, "The Great Chain of Buying: Medical Advertisement, the Bourgeois Public Sphere, and the Origins of the French Revolution," *American Historical Review* 101, no. 1 (2009): 13–40.; and Matthew Ramsey, "Property Rights and the Right to Health: the Regulation of Secret Remedies in France, 1789–1815," in *Medical Fringe and Medical Orthodoxy, 1750–1850*, ed. William F. Bynum and Roy Porter (London: Croom Helm, 1987).

27. Société Royale de Médecine, *Nouveau plan de constitution pour la médecine* (Paris: 1790), 127.

28. Fourcroy, *Discours*, 36.

29. *Codex Medicamentarius* (1818), v & sq.

30. Noiriel, "L'identification des Citoyens," especially pages 7–16.

31. Victor Duruy and Adolphe Laugier, *Pandectes pharmaceutiques* (Paris: Colas, 1837), 281.

32. The secret remedies commission was transferred from the Ministry of the Interior to the Royal Academy of Medicine in 1823. For a statement of how its members understood their mission, see: "Rapport à s. Exc. Mgr. le Ministre secrétaire d'État de l'Intérieur, par l'Académie Royale de Médecine," May 16, 1823, in Archives de Paris, DM5 Box #10. On the work of the secret remedies commissions during and after the revolutionary period: Ramsey, "Property Rights and the Right to Health"; and "Academic Medicine and Medical Industrialism: The Regulation of Secret Remedies in Nineteenth-Century France," in *French Medical Culture in the Nineteenth Century*, ed. Ann La Berge and Mordechai Feingold (Amsterdam: Rodopi, 1994). On the persistence of the trade in illicit proprietary drugs in Paris in the late 1820s, see Préfecture de Police, "Ordonnance du 21 juin 1828," in Archives de Paris, DM5 Box #10.

33. Maurice Cassier, "Brevets pharmaceutiques et santé publique en France: opposition et dispositifs spécifiques d'appropriation des médicaments en France entre 1791 et 2004," *Entreprises et histoire* 36 (2004): 29–47. 30–33.

34. André Narodetzki, *Le remède secret. Législation et jurisprudence de la Loi du 21 Germinal an XI au Decret du 13 juillet 1926* (Paris: Delrieu, 1928), 21 and sq.

35. Grand, "Metz, 11 février 1857," 449.

36. This is best seen in reports of the chemical and pharmaceutical sections of the

world fairs in Paris in 1867 and 1900: Charles Louis Barreswil and Victor Fumouze, "Section VIII. Produits pharmaceutiques," in *Exposition universelle de 1867 a Paris. Rapports du jury international publiés sous la direction de M. Michel Chevalier,* vol. 7 (Paris: Imprimerie impériale, 1868), 285–89 and 306–7; and Maurice Leprince, *1er congrès international de l'industrie et du commerce des spécialités pharmaceutiques* (Paris: Jourdan, 1900), 43–44.

CHAPTER 12: Indian Secrets, Indian Cures, and the *Pharmacopoeia of the United States of America*

1. Jonathan Redman Coxe, *The American Dispensatory* (1806), v–vi.
2. Duncan, *Edinburgh New Dispensatory* (1804), 271, 296, 325.
3. Benjamin Smith Barton, *Collections for an Essay Towards a Materia Medica of the United States, Part First* (1801), vi. Other sources used by Coxe include the *Philadelphia Medical Museum* and the *Philadelphia Medical and Physical Journal*, two recently introduced medical journals, and dissertations written by medical students graduating from the University of Pennsylvania. Coxe, *American Dispensatory*, 209, 356, 549, 364, 419, 328, 241, 399, 585, 527, 290, 390, 354. Quote on 354. Biographical details are from "Obituary. John Redman Coxe, M.D.," *American Medical Times*, May 7, 1864, 226.
4. M. I. Wilbert, "The Pharmacist and Pharmacopoeia," *Pharmaceutical Era*, May 19, 1904, 481.
5. Coxe, *American Dispensatory*, v.
6. Coxe, *American Dispensatory*, v–vii, 2.
7. *The Pharmacopeia of the United States of America* (1820), 17–18. On the founding and early history of the USP, see Edward Kremers and George Urdang, *History of Pharmacy*, rev. Glenn Sonnedecker (Philadelphia: Lippincott, 1976), 255–63; Gregory J. Higby, "The Early History of the USP," in Lee Anderson and Gregory J. Higby, *The Spirit of Volunteerism: A Legacy of Commitment and Contribution: The United States Pharmacopeia 1820–1995* (Rockville, MD: United States Pharmacopeial Convention, Inc., 1995), 3–39; Glenn Sonnedecker, "The Founding Period of the U.S. Pharmacopeia I. European Antecedents," *Pharmacy in History* 35, no. 4 (1993): 151–62; Glen Sonnedecker, "The Founding Period of the U.S. Pharmacopeia II. A National Movement Emerges," *Pharmacy in History* 36, no. 1 (1994): 3–25; Glen Sonnedecker, "The Founding Period of the U.S. Pharmacopeia III. The First Edition," *Pharmacy in History* 36, no. 3 (1994): 103–21.
8. George B. Wood and Franklin Bache, *The Dispensatory of the United States of America* (1833).
9. "Miscellaneous Medical Facts," *Philadelphia Medical and Physical Journal* (February 1, 1806): 87–88.
10. John Tennent, *Every Man his Own Doctor: or, The Poor Planter's Physician* (1734).
11. John Tennent, *An Essay on the Pleurisy* (1736), 46.
12. John Tennent, "A Memorial, humbly addressed to the learned, impartial, and judicious world," *Virginia Gazette*, September 22, 1738, 1; "Continuation of the Memorial, began in our last," *Virginia Gazette*, September 29, 1738, 1; John Tennent, "To the Northern Colonies. The Memorial and Remonstrance of John Tennent, Practitioner in Physick," *Pennsylvania Gazette*, July 2, 1739, 1.
13. John Tennent, "Remainder of Dr. Tennent's Memorial," *Pennsylvania Gazette*, July 26, 1739, 1–2.
14. Richard Saunders, *Poor Richard, 1737. An Almanack For the Year of Christ 1737* (1736), n.p.

15. Wood and Bache, *Dispensatory* (1833), 581–82.

16. *The Pharmacopoeia of the Royal College of Physicians at Edinburgh*, 4th edition (1744), trans. William Lewis (London, 1748), 63.

17. *Pharmacopoeia simpliciorum et efficaciorum* (Philadelphia, 1778).

18. Duncan, *Edinburgh New Dispensatory* 315; Coxe, *American Dispensatory*, 528.

19. Virgil J. Vogel, *American Indian Medicine* (Norman: University of Oklahoma Press, 1970); Cristobal Silva, *Miraculous Plagues: An Epidemiology of Early New England Narrative* (New York: Oxford University Press, 2011); Paul Kelton, *Cherokee Medicine, Colonial Germs: An Indigenous Nation's Fight against Smallpox, 1518–1824* (Norman: University of Oklahoma Press, 2015).

20. Quoted in Martha Robinson, "New Worlds, New Medicines," *Early American Studies* 3, no. 1 (Spring 2005): 93.

21. Colin G. Calloway, *New Worlds for All: Indians, Europeans, and the Remaking of Early America* (Cambridge, MA: Harvard University Press), 31.

22. Robinson, "New Worlds, New Medicines," 97.

23. Philip Pauly, *Biologists and the Promise of American Life: From Meriwether Lewis to Alfred Kinsey* (Princeton, NJ: Princeton University Press, 2002), 18.

24. For another early example, see William Bartram, *Travels through North and South Carolina, Georgia, East and West Florida, the Cherokee Country, the Extensive Territories of the Muscogulges, or Creek Confederacy, and the Country of the Chactaws* (Philadelphia, 1791).

25. Jacob Bigelow, *American Medical Botany*, vol. 1 (1817), vii.

26. *Pharmacopoeia simpliciorum et efficaciorum* (Philadelphia, 1778).

27. *Pharmacopoeia of the Massachusetts Medical Society* (1808); James Thacher, *The American New Dispensatory* (1810), 198.

28. Andrew J. Lewis, "Gathering for the Republic: Botany in Early Republic America," in *Colonial Botany: Science, Commerce, and Politics in the Early Modern World*, ed. Londa Schiebinger and Claudia Swan (Philadelphia: University of Pennsylvania Press, 2005), 68.

29. Examples include John Thobald, *Every Man His Own Physician* (1767); Thomas Johnson, *Every man his own Doctor, or the Poor Man's Family Physician* (1798), Samuel Thomson, *New guide to Health; or, Botanic Family Physician* (1822).

30. Samuel Henry, *A New and Complete American Medical Family Herbal* (New York, 1814).

31. "Review of Henry's American Herbal," *Analytic Magazine*, March 1816, 249–65.

32. Anonymous, "Review of Henry's American Herbal," *Analytic Magazine*, March 1816, 249.

33. *Pharmacopoeia nosocomii New-Eboracensis; or the Pharmacopoeia of the New-York Hospital* (1816), vii–viii.

34. Mitchill founded the *Medical Repository* in 1797 with Elihu Smith and Edward Miller. It was the only medical journal in the country until 1802, and ceased publication in 1824. Anton Sebastian, "American Medical Journals," in *A Dictionary of the History of Medicine* (Pearl River, NY: Pantheon, 1999), 39.

35. Quoted in Kremers and Urdang, *History of Pharmacy*, 262.

36. *Pharmacopeia of the United States of America* (1820), 17–18.

37. *Pharmacopeia of the United States of America* (1820), 19.

38. Quoted in Kremers and Urdang, *History of Pharmacy*, 263.

39. *Pharmacopoeia of the United States of America* (1820), 117, 35, 56, 43, 53, 155, 151.

40. *Pharmacopeia of the United States of America* (1820), 21.
41. "Art. XIII The Pharmacopoeia of the United States of America, 1820," *American Medical Recorder* (1821).
42. Coxe, *New American Dispensatory* (1831), vi.
43. George B. Wood, "Article XV. Observations on the Pharmacopoeia of the United States," *Journal of the Philadelphia College of Pharmacy* (July 1832): 94–107.
44. Wood and Bache, *Dispensatory* (1833).
45. Wood and Bache, *Dispensatory* (1833), 581–82.
46. Richard Slotkin, *Regeneration through Violence: The Mythology of the American Frontier, 1600–1860* (Norman: University of Oklahoma Press, 2000).
47. Barton, *Collections*, part ii, 75; Wood and Bache, *Dispensatory* (1836), 617.
48. Wood and Bache, *Dispensatory* (1845), 681.
49. Alexander Coventry, "Observations on Endemic Fever," *New York Medical and Physical Journal* (January–March 1825): 9–10.
50. Joseph M. Gabriel, *Medical Monopoly: Intellectual Property Rights and the Origins of the Modern Pharmaceutical Industry* (Chicago: University of Chicago Press, 2014), especially chapter 2.
51. For example, see John D. Hunter, "Remarks on the Diseases of the Females of Several Indian Tribes West of the Mississippi," *New York Medical and Physical Journal* (July–September, 1822): 404.
52. Rich Comp, "A Cure for a Wen. An Indian Prescription," *American Farmer*, September 17, 1819, 200.
53. Joshua David Bellin, "Taking the Indian Cure: Thoreau, Indian Medicine, and the Performance of American Culture," *New England Quarterly* (March 2006): 3–36; Philip J. Deloria, *Playing Indian* (New Haven, CT: Yale University Press, 1999). On representations of Indians in popular culture, see also Joshua David Bellin, *Medicine Bundle: Indian Sacred Performance and American Literature, 1824–1932* (Philadelphia: University of Pennsylvania Press, 2007).
54. Peter Smith, *The Indian Doctor's Dispensatory* (1813).
55. James Harvey Young, *The Toadstool Millionaires: A Social History of Patent Medicines in America before Federal Regulation* (Princeton, NJ: Princeton University Press, 1961), 9. As is well known, these products were usually not patented. I discuss patent medicines extensively in Gabriel, *Medical Monopoly*.
56. Gabriel, *Medical Monopoly*, especially chapters 1 and 2.
57. "Cure of Venereal Complaints," *Raleigh Register*, December 9, 1800.
58. "Romlindorf Digestive Lozenges—An Indian Specific," *Charleston Courier*, April 18, 1812, 4.
59. "Correspondence," *New-England Journal of Medicine* (1823): 437.
60. "Tolerated Quackery," *New-England Galaxy and United States Literary Advertiser*, June 25, 1824, 2.
61. Owen Whooley, *Knowledge in the Time of Cholera: The Struggle over American Medicine in the Nineteenth Century* (Chicago: University of Chicago Press, 2013).
62. Stephen W. Williams, "Indigenous Medical Botany, No. 3," *New York Journal of Medicine and Collateral Sciences* (Sept. 1846): 179.
63. Frederick C. Brightly, *A Digest of the Laws of Pennsylvania, from the Year One Thousand Seven Hundred to the Twenty-First Day of May, One Thousand Eight Hundred and Sixty-One* (1862), 782.

64. Henry Row Schoolcraft, *Historical and Statistical Information Respecting the History, Condition and Prospects of the Indian Tribes of the United States, Part 1* (1851), 518.

65. R. P. Lukens to Wilson Popenoe (October 21, 1947), Records of B. A. Krukoff, Metz Library, New York Botanical Society, New York, b. 30, f. 8

66. R. P. Lukens to Randolph Major (Nov. 26, 1947), Records of B. A. Krukoff, Metz Library, New York Botanical Society, New York, b. 30, f. 8.

AFTERWORD

1. Here I use the term "pharmacopoeia" under the capacious definition advanced by the editors of this volume ("Introduction," 3–4). See also, Robert E. Voeks, "Disturbance Pharmacopoeias: Medicine and Myth from the Humid Tropics," *Annals of the Association of American Geographers* 94 (2004): 868–88.

2. There's a large literature on the subject of how the West creates its "others." See, for instance, Ashis Nandy, "History's Forgotten Doubles," *History and Theory* 34 (1995): 44–66; Michel-Rolph Trouillot, *Silencing the Past: Power and the Production of History* (Boston: Beacon Press, 2015); Dipesh Chakrabarty, "The Muddle of Modernity," *American Historical Review*, 116 (2011): 663–75; Bruno Latour, *On the Modern Cult of the Factish Gods* (Durham: Duke University Press, 2011); or the classic Johannes Fabian, *Time and the Other: How Anthropology Makes Its Object* (New York: Columbia University Press, 2014).

3. Among others, see, Marcy Norton, "Subaltern Technologies and Early Modernity in the Atlantic World," *Colonial Latin American Review* 26, no. 1 (2017): 18–38; Benjamin Breen, "Drugs and Early Modernity," *History Compass*, 2017; Timothy D. Walker, "The Medicines Trade in the Portuguese Atlantic World: Acquisition and Dissemination of Healing Knowledge from Brazil (c. 1580–1800)," *Social History of Medicine* 26, no. 3 (August 1, 2013): 403–31; Matthew James Crawford, *The Andean Wonder Drug: Cinchona Bark and Imperial Science in the Spanish Atlantic World, 1630–1800* (Pittsburgh: University of Pittsburgh Press, 2016); Neil Safier, "Global Knowledge on the Move: Itineraries, Amerindian Narratives, and Deep Histories of Science," *Isis* 101 (2010): 133–45; Pratik Chakrabarti, *Materials and Medicine: Trade, Conquest and Therapeutics in the Eighteenth Century* (Manchester: Manchester University Press, 2010).

4. See, Pablo F. Gómez, *The Experiential Caribbean: Creating Knowledge and Healing in the Early Modern Atlantic* (Chapel Hill: University of North Carolina Press, 2017), 118–44.

5. Philippe Descola, *Beyond Nature and Culture* (Chicago: University of Chicago Press, 2013). For the classic pieces of the "ontological turn" in anthropology see, Eduardo Viveiros de Castro, "Cosmological Deixis and Amerindian Perspectivism," *Journal of the Royal Anthropological Institute* 4, no. 3 (1998): 469–88. For assemblages of people and things see, among others, Amiria Henare, Martin Holbraad, and Sari Wastell, eds. *Thinking through Things: Theorising Artefacts Ethnographically* (London: Routledge, 2007).

6. For a materialistic answer to the ontological turn, see Pablo F. Gómez, "Caribbean Stones and the Creation of Early-modern Worlds," *History and Technology* (34): 11–20.

7. Henare, Holbraad, and Wastell, *Thinking*.

8. Antoine Lentacker, "The Codex Nationalized: Naming People and Things in the Wake of a Revolution" (chapter 11).

9. Stuart Anderson, "National Identities, Medical Politics and Local Traditions: The Origins of the *London, Edinburgh*, and *Dublin Pharmacopoeias* 1618 to 1807" (chapter 10).

10. Emily Beck, "Authority, Authorship, and Copying: The *Ricettario Fiorentino* and Sixteenth-Century Florentine Manuscript Recipe Culture" (chapter 2).

11. Joseph M. Gabriel, "Indian Secrets, Indian Cures, and the Pharmacopoeia of the United States of America" (chapter 12).

12. This type of biomedically framed analysis has been more fruitful in histories of modernity. For instance, Abena Dove Osseo-Asare, *Bitter Roots: The Search for Healing Plants in Africa* (Chicago: University of Chicago Press, 2014); Kristin Peterson, *Speculative Markets: Drug Circuits and Derivative Life in Nigeria* (Durham: Duke University Press, 2014); Gabriela Soto Laveaga, *Jungle Laboratories: Mexican Peasants, National Projects and the Making of the Pill* (Durham: Duke University Press, 2009).

13. Paula De Vos, "Pharmacopoeias and the Textual Tradition in Galenic Pharmacy" (chapter 1).

14. William Ryan, "Imperfect Knowledge: Medicine, Slavery, and Silence in Hans Sloane's *Philosophical Transactions* and the 1721 London Pharmacopoeia" (chapter 6).

15. Christopher Parsons, "Consuming Canada: *Capillaire du Canada* in the French Atlantic World" (chapter 8).

16. Matthew J. Crawford, "An Imperial Pharmacopoeia? Pharmacopoeias as Witnesses and Agents of World History in the Eighteenth-Century Spanish Atlantic World" (chapter 3); Timothy D. Walker, "Crown Authorities, Colonial Physicians, and the Exigencies of Empire: The Codification of Indigenous Therapeutic Knowledge in India and Brazil during the Enlightenment Era" (chapter 5).

17. Justin Rivest, "Beyond the Pharmacopoeia? Secret Remedies, Exclusive Privileges, and Trademarks in Early Modern France" (chapter 4).

18. Walker, "Crown Authorities."

19. Harold J. Cook and Timothy D. Walker, "Circulation of Medicine in the Early Modern Atlantic World," *Social History of Medicine* 26 (2013): 337–51. Daniela Bleichmar, *Visible Empire: Botanical Expeditions and Visual Culture in the Hispanic Enlightenment* (Chicago: University of Chicago Press, 2012); Neil Safier, *Measuring the New World: Enlightenment Science and South America* (Chicago: University of Chicago Press, 2008); Londa Schiebinger, *Plants and Empire: Colonial Bioprospecting in the Atlantic World* (Cambridge: Harvard University Press, 2004); James Delbourgo and Nicholas Dews, eds., *Science and Empire in the Atlantic World* (New York: Routledge, 2007).

20. I borrow the concept of suturing from Nancy Hunt, *Suturing New Medical Histories of Africa* (Munster: LIT Verlag, 2013).

21. Kelly Wisecup, "Rethinking Pharmacopoeic Forms: Samson Occom and Mohegan Medicine" (chapter 9)

22. Michael Taussig, *Defacement: Public Secrecy and the Labor of the Negative* (Stanford: Stanford University Press, 1999).

23. Benjamin Breen, "The Flip Side of the Pharmacopoeia? Sub-Saharan African Medicines and Poisons in the Atlantic World" (chapter 7); Gómez, *Experiential Caribbean*, ch. 5, 7.

24. Taussig, *Defacement*, 134.

25. Gómez, *Experiential Caribbean*, 70–94, 118–44.

26. As Lorraine Daston and Katharine Park have observed, "To count as one of the 'curious' was to combine a thirst to know with an appetite for marvels." Lorraine Daston and Katharine Park, *Wonders and the Order of Nature, 1150–1750* (New York: Zone Books, 1998), 218.

BIBLIOGRAPHY

Archival Sources

Alexandre Rodrigues Ferreira Collection. Manuscripts Division. Biblioteca Nacional do Rio de Janeiro, Rio de Janeiro, Brazil.

Alfandega Collection. Arquivo Histórico do São Tomé e Príncipe, São Tomé e Principe.

Autos de Concurso de Acreedores a Bienes de Don Jacinto Herrera Dueno de Botica en esta Ciudad. Civil, V. 72, Exp. 9, 1774. Archivo General de la Nación, Mexico City, México.

Buytrago, Francisco de. "Arvore da Vida e Thesouro descuberto." F.R. 437 (microfilm), codigo 13114. Biblioteca Nacional de Portugal, Lisbon, Portugal.

Chancelarias de Dom José I Collection. Arquivo Nacional da Torre do Tombo, Lisbon, Portugal.

Conseil privé, Minutes d'arrêt (May 25, 1657). AN V⁶ 346, Archives Nationales, Paris, France.

Conselho Ultramarino Collection. Arquivo Histórico Ultramarino, Lisbon, Portugal.

Correspondência Collection. Biblioteca da Academia das Ciências de Lisboa, Lisbon, Portugal.

Fonds des colonies, Série C11A, Correspondance générale Canada (1540–1784). Archives nationales d'outre-mer, Aix-en-Provence, France.

Fonds Jean-François Gaultier. Bibliothèque et Archives nationales du Québec, Québec, Canada.

Goncalves de Andrade, Luiz. "Testimony." 8 July 1666. Luanda, Angola. Cx. 9, Doc. 33. Arquivo Histórico Ultramarino, Lisbon, Portugal.

Jordan, Robert. "Commonplace Book, 1736-1958," MSS 5:5J7664:1. Virginia Historical Society, Richmond, VA, United States.

Letters patent of Louis-Anne Contugi (December 27, 1686). AN O¹ 30. Fol. 397. Archives Nationales, Paris, France.

Livros dos Feitos Findos. Arquivo Nacional da Torre do Tombo, Lisbon, Portugal.

Livros dos Monções do Reino Collection. Historical Archive of Goa, Panaji, Goa, India.

Manuscrítos Collection. Biblioteca Nacional do Portugal, Lisbon, Portugal.

Manuscrítos (Série Azul) Collection. Biblioteca da Academia das Ciências de Lisboa, Lisbon, Portugal.

Ministério do Reino Collection. Arquivo Nacional da Torre do Tombo, Lisbon, Portugal.

Ministério dos Negócios Estrangeiros Collection. Arquivo Nacional da Torre do Tombo, Lisbon, Portugal.

Occom, Samson. "Herbs & Roots," 1754–1756. New London County Historical Society, New London, CT, United States.

Occom, Samson. "Herbs & Roots," 1754–1756. Rauner DC History. Rauner Library. Dartmouth College, Darmouth, NH, United States.

Opera Nostrorum Collection. Archivum Romanum Societatis Iesu, Rome, Italy.

Passport of Louis-Anne Countugi (February 6, 1686). AN V^5 1246. Fol. 209, Archives Nationales, Paris, France.

Records of B. A. Krukoff. Metz Library, New York Botanical Society, New York, NY, United States.

Recueil de pièces relatives à la Faculté de médecine et à l'Académie de médecine de Paris, aux chirurgiens et aux apothicaires, à diverses institutions médicales de France, à la charge de Premier médecin du Roi, etc. Fol. 310. Mémoire pour l'analyse des remèdes. Signé: Hérault. (October 16, 1729). Bibliothèque interuniversitaire de la santé, Paris, France.

Ricettario (ms. 2376). Manuscript. c. sixteenth century. Biblioteca Riccardiana, Florence, Italy.

Ricettario (ms. 3044). Manuscript. c. fifteenth–sixteenth century. Biblioteca Riccardiana, Florence, Italy.

Ricettario (ms. 3049). Manuscript. c. fifteenth century. Biblioteca Riccardiana, Florence, Italy.

Ricettario (ms. 3059). Manuscript. c. fifteenth century. Biblioteca Riccardiana, Florence, Italy.

Rimino, Bernardino da. Ms. 691. Wellcome Library, London, England.

São Tomé e Príncipe Collection. Arquivo Histórico Ultramarino, Lisbon, Portugal.

Segni delle Febbri and *Ricettario.* (ms. 3057). Manuscript. c. sixteenth century. Biblioteca Riccardiana, Florence, Italy.

Série E Collection. Biblioteca do Museu do Jardim Botânico, Lisbon, Portugal.

Series DM5 (Hygiene and Public Health). Archives de Paris, Paris, France.

Silva, Pedro da. "Petition." 11 August 1664. Angola, Cx. 8, Doc. 32. Arquivo Histórico Ultramarino, Lisbon, Portugal.

"Transaction faite entre le Sieur et veuve Contugi et les apothiquaires privilégiés sur l'instance à l'occasion de la saisie faite à le Sieur Boulogne," (August 14, 1685). Communauté des apothicaires-épiciers de Paris. Requêtes-transactions,

BIBLIOGRAPHY

1501–1699. Fol. 272–75. Pharmacie. Archives. Reg. 31. Bibliothèque interuniversitaire de la santé, Paris, France.

Unpublished Manuscripts Collection. Manuscripts Division, Biblioteca Nacional do Rio de Janeiro, Rio de Janeiro, Brazil.

Printed Primary Sources

"An Act to Remedy the Evils Arising from Irregular Assemblies of Slaves, and to Prevent their Possessing Arms and Ammunition, and Going from Place to Place Without Tickets, and for Preventing the Practice of Obeah," January 1, 1761, in *Acts of Assembly Passed in the Island of Jamaica*. Assembly of Jamaica: St. Jago de la Vega, 1771.

"Affaire du Codex Medicamentarius, ou de sa prétendue contrefaçon dans le Traité de Pharmacie de J.-J. Virey." *Journal de Pharmacie et des sciences accessoires* 6, no. 2 (1820): 102–4.

Aguilera, Antonio. *Exposicion sobre las preparaciones de Mesue*. Alcala: Casa de Juan de Villanueva, 1569.

Al-Zahrāwī/Abulcasis. *Servidor de Abulcasis*. Translated by Rodrigues de Tudela. Valladolid, 1515.

[Anonymous]. "Of the Use of the Root Ipecacuanha, for Loosenesses, Translated from a French Paper: With Some Notes on the Same, by Hans Sloane, M.D." *Philosophical Transactions* 20 (1698): 69–79. doi:10.1098/rstl.1698.0016.

Antonil, André João. *Cultura e Opulência do Brasil por suas Drogas e Minas . . .* Lisbon: Officina Real Deslandesiana, 1711.

Arduino, Sante. *De Venenis*. Basel, 1562.

Arrêt du conseil d'Etat du Roy, qui défend à toutes personnes de distribuer des remèdes spécifiques et autres sans en avoir obtenu de nouvelles permissions. Du 3 Juillet 1728. Paris, 1728.

Arrêt du Conseil d'Etat du Roy, qui defend à toutes sortes de personnes de distribuer des Remedes sans en avoir obtenu de nouvelles Permissions. Extrait des registres du Conseil d'Etat du Roy. Du 25 octobre 1728. Paris, 1728.

Ascoli, Saladino da. *Compendio de los boticarios*. Translated by Alonso Rodríguez de Tudela. Valladolid, 1515. Facsimile edition. Seville: Extramuros Edición, 2008.

Authority of the Medical Societies and Colleges. *The Pharmacopeia of the United States of America*. Boston: Printed by Wells and Lilly, 1820.

Bacon, Francis. *The Two Books of Francis Bacon. Of the Proficiencie and Advancement of Learning*, Bk 1. Oxford: Leon Litchfield, 1605.

Barreswil, Charles Louis, and Victor Fumouze. "Section VIII. Produits pharmaceutiques." In *Exposition universelle de 1867 à Paris. Rapports du jury international publiés sous la direction de M. Michel Chevalier* Vol. 7. Paris: Imprimerie impériale, 1868.

BIBLIOGRAPHY

Barton, Benjamin Smith. *Collections for an Essay Towards a Materia Medica of the United States, Part First*. Philadelphia: Printed for the Author, 1801.

Bate, George. *Pharmacopoeia Bateana*. London: Impensis Sam Smith, 1691.

Berlu, John Jacob. *The Treasury of Drugs Unlock'd*. London: Printed for John Harris and Thomas Hawkins, 1690.

Bosman, Willem. *A New and Accurate Description of the Coast of Guinea, Divided Into the Gold, the Slave, and the Ivory Coasts*. London, 1705.

Boucher, Pierre. *Histoire véritable et naturelle des moeurs et productions du pays de la Nouvelle-France, vulgairement dite le Canada*. Boucherville, QC: Société historique de Boucherville, 1964.

Bougainville, Louis-Antoine de. *Écrits sur le Canada: mémoires, journal, lettres*. Sillery, QC: Septentrion, 2003.

Boyle, Robert. *Of the reconcileableness of specifick medicines to the corpuscular philosophy*. London: Printed for Sam Smith, 1685.

Brightly, Frederick C. *A Digest of the Laws of Pennsylvania, from the Year One Thousand Seven Hundred to the Twenty-First Day of May, One Thousand Eight Hundred and Sixty-One*. Philadelphia: Kay & Brother, 19 South Sixth Street, 1862.

Brihuega, Francisco de. *Examen Pharmaceutico Galeno-Chimico Teorico-Practico*. Madrid: En la Imprenta de los Reynos, 1776.

Cadet, Charles-Louis. "Examen du Codex Medicamentarius." *Journal de pharmacie et des sciences accessoires* 5 (1819): 121–37, 203–29.

Candler, Allen D., et al., eds. *The Colonial Records of the State of Georgia*. 20 vols. Athens: University of Georgia Press, 1974.

Castels, Juan Antonio. *Theorica y pratica de boticarios en que se trata de la arte y forma como se han de componer las confectiones ansi interiores como Exteriores*. Barcelona: En casa de Sebastian de Cormellas, 1592.

Catalogne, Gédéon de. "Report on the Seigniories and Settlements in the Districts of Quebec, Three Rivers, and Montreal, by Gedeon de Catalogne/ Engineer, November 7, 1712." In *Documents Relating to the Seigniorial Tenure in Canada, 1598–1854*, ed. William Bennett Munro, 109. Toronto: Champlain Society, 1908.

Charas, Moyse. *Pharmacopée royale galénique et chymique*. Paris, 1676.

Charlevoix, Pierre-François-Xavier de. "Description des plantes principales de l'Amérique septentrionnale." In *Histoire et description generale de la Nouvelle France avec le journal historique d'un Voyage fait par ordre du Roi dans l'Amérique Septentrionnale*. Paris: Didot, 1744.

Chollegio degli Eximii Doctorii della Arte et Medicina della Inclita Cipta di Firenze. *Il Nuovo Ricettario Fiorentino (1498): Testo e Lingua*. Edited and transcribed by Olimpia Fittipaldi. Pluteus: 2011. Accessed February 1, 2018. http://www.pluteus.it/wp-content/uploads/2014/01/nuovo%20ricettario.pdf.

BIBLIOGRAPHY

Chomel, Pierre Jean Baptiste. *Abrégé de l'histoire des plantes usuelles*. Paris: Chez Charles Osmont, 1712.

Cockburn, William. *The Present Uncertainty in the Knowledge of Medicines*. London: R.J. for Benjamin Barker, 1703.

Codex medicamentarius seu Pharmacopoea Parisiensis. Paris: Varennes, 1638.

Codex medicamentarius seu Pharmacopoea Parisiensis. Paris: Cavelier, 1758.

Codex medicamentarius sive Pharmacopoea Gallica. Paris: Hacquart, 1818.

Collegio de Medici di Firenze. *Ricettario Fiorentino*. Florence: Heredi di Bernardo Giunti, 1567.

Collegio de Medici. *Ricettario Fiorentino, Di Nuovo Illustrato*. Florence: Giorgio Marescotti, 1597.

Collegium Medicum. *Pharmacopoeia Augustana*. Augustae Vindelicorum: Ex Off. A. Apergeri : Prostant apud J. Krugerum, 1622.

Comp, Rich. "A Cure for a Wen. An Indian Prescription." *American Farmer* 1 (1819): 200.

Concordie Apothecariorum Barchinone 1511, la primera farmacopea española. Tarragona: Colegio Oficial de Farmacéuticos de Tarragona, 1941.

Cortesão, Armando. *Suma Oriental of Tomé Pires: An Account of the East, from the Red Sea to China, Written in Malacca and India in 1512–1515*. New Delhi: Asia Educational Services, 2005.

Coxe, John Redman. *The American Dispensatory, containing the operations of pharmacy; together with the natural, chemical, pharmaceutical and medical history of the different substances employed in medicine*. Philadelphia: Printed by A. Bartram, 1806.

Culpeper, Nicholas. *The English Physician*. Edited by Michael Flannery. Huntsville: University of Alabama Press, 2007.

Culpeper, Nicholas. *The English Physician, containing admirable and approved remedies for several of the most usual diseases*. Boston, 1708.

Culpeper, Nicholas. *The English Physician, or An astrological Discourse on the vulgar herbs of this Nation being a compleat method of physick, whereby a man may preserve his body in health, or cure himself being sick for three pence charge, with such things only as grow in England, they being most fit for English bodies*. London: Printed by William Bentley, 1652.

Culpeper, Nicholas. *Pharmacopoeia Londinensis; or, The London dispensatory further adorned by the studies and collections of the fellows now living, of the said college*. Boston, 1720.

"Cure of Venereal Complaints." *Raleigh Register*, December 9, 1800.

Curvo Semedo, João. *Polyanthea Medicinal: Noticias galenicas, e chymicas, em tres tratados*. Lisbon: Miguel Deslandes, 1697.

Diderot, Denis. *Encyclopédie, ou Dictionnaire raisonné des sciences, des arts et des métiers*. Paris: Chez Briasson, 1751.

Duncan, Andrew. *The Edinburgh New Dispensatory*. Edinburgh: Printed for Bell & Bradfute, 1804.

Duruy, Victor, and Adolphe Laugier. *Pandectes pharmaceutiques*. Paris: Colas, 1837.

Easterby, J. H., ed. *The Colonial Records of South Carolina: The Journal of the Commons Assembly. March 28, 1749–March 19, 1750*. Charleston, SC, 1962.

Estatutos del Real Colegio de Professores Boticarios de Madrid, aprobados y confirmados por su Magestad, que Dios guarde. Madrid: Imprenta Real, 1737.

Evelyn, John. *Silva, or A discourse of forest-trees, and the propagation of timber in His Majesty's dominion*. London, 1706.

A Facsimile Edition of the First Edition of the Pharmacopoeia Augustana with Introductory Essays by Theodor Husemann. Madison: State Historical Society of Wisconsin, 1927.

Ferreira, Alexandre Rodrigues. "Diario da Viagem Philosophica pela Capitania de São José do Rio Negro . . ." *Revista Trimestral do Instituto Histórico e Geográfico Brasileiro*, vols. 48–51. Rio de Janeiro, Brazil: 1885–1888.

Fontenelle, Bernard Le Bovier de. "Éloge de Fagon." *Histoire de l'Académie royale des sciences* (1718): 94–101.

Fourcroy, Antoine François de. *Discours prononcé à la Société libre des pharmaciens de Paris*. Paris: Quillau, 1797.

Fuller, John. "A Letter from John Fuller, Esq; Jun. F.R.S. to Sir Hans Sloane, Bart. Pres. R.S. &c. concerning the Effects of Dampier's Powder, in curing the Bite of a Mad Dog." *Philosophical Transactions* 40 (1738): 272–73. Accessed August 17, 2018. http://dx.doi:10.1098/rstl.1737.0043.

Gendron, François. *Quelques particularitez du pays des Hurons en la Nouvelle France*. Paris: Denys Bechet et Louis Billaine, 1660.

General Medical Council. *British Pharmacopoeia, published under the direction of the General Council of Medical Education and Registration of the United Kingdom pursuant to the Medical Act 1858*. London, 1864.

Gomes Ferreira, Luis. *Erario Mineral*. Lisbon: Officina de Miguel Rodrigues, 1735.

Grand, P. "Metz, 11 février 1857," *Journal du Palais: Jurisprudence française* (1857): 449–53.

Grew, Nehemiah. *Musaeum Regalis Societatis*. London: Printed by W. Rawlins, for the author, 1681.

Guybert, Philbert. *Toutes les oeuvres charitables de Philebert Guibert*. Rouen: Chez François Vaultier, 1667.

Hall, Captain. "An Account of Some Experiment on the Effects of the Poison of the Rattle Snake. By Captain Hall. Communicated by Hans Sloane." *Philosophical*

Transactions 35 (1728): 309–15. Accessed August 17, 2018. http://dx.doi:10.1098/rstl.1727.0008.

Halley, Edmund. "The Preface." *Philosophical Transactions* 29 (1714): 3–4. Accessed August 17, 2018. http//doi:10.1098/rstl.1714.0001.

Henry, Samuel. *A New and Complete American Medical Family Herbal.* New York: Published by Samuel Henry, 1814.

Herbert, Thomas. *Some yeares travels into Africa and Asia the Great.* London: Printed for Jacob Blome and Richard Bishop, 1638.

Hunter, John. "Remarks on the Diseases of the Females of Several Indian Tribes West of the Mississippi." *New York Medical and Physical Journal* 1 (1822): 404.

Ibn-at-Tilmīd, Hibatallāh Ibn-Ṣāʿid, and Oliver Kahl. *The Dispensatory of Ibn at-Tilmid: Arabic Text, English Translation, Study and Glossaries.* Leiden: Brill, 2007.

James I, King of England. *A Counter-blaste to Tobacco.* London, 1604.

Johnson, Thomas. *Every Man his own Doctor; or the Poor Man's Family Physician.* Salisbury: Printed for the Author, 1798.

Jubera, Alonso de. *Dechado y reformacion de todas las medicinas compuestas usuales, con declaracion de todas las dudas enellas cōtenidas, assi de los simples que enellos entrā y succedaneos que por los dudosos se hayā de poner, como en el mondo de las hazer.* Valladolid: Por Diego Fernandez de Cordova, 1578.

Jurin, James. "The Preface." *Philosophical Transactions* 17 (1693): 581–82. Accessed August 17, 2018. http://dx.doi:10.1098/rstl.1693.0001.

Kahl, Oliver. *The Dispensatory of Ibn at-Tilmīd.* Leiden: Brill, 2007.

Kahl, Oliver. *Sābūr ibn Sahl, The Small Dispensatory.* Leiden: Brill, 2003.

Kalm, Pehr. *Travels into North America containing its natural history, and a circumstantial account of its plantations and agriculture in general: With the civil, ecclesiastical and commercial state of the country, the manners of the inhabitants, and several curious and important remarks on various subjects.* London: T. Lowndes, 1771.

King, William. *The Present State of Physick in the Island of Cajamai. To the Members of the R.S.* London, 1710.

King, William. "The Transactioneer, With Some of his Philosophical Fancies: In Two Dialogues." *Original Works of William King,* vol 2., edited by John Nichols. London, 1776.

La Martinière, Pierre-Martin de. *Traitté des compositions du mitridat, du thériaque, de l'orviétan, et des confections d'alkermes et d'hyacinthe et autres compositions antidotoires.* Paris, 1665.

Lahontan. *Oeuvres completes.* Montreal, QC: Les Presses de l'Université de Montréal, 2000.

Lamothe Cadillac, Antoine de. "Lettre de M. de Lamothe Cadillac (28 Septembre 1694)." *Rapport de l'Archiviste de la province de Québec 1923–1924* (1924): 92.

Laredo, Bernardino de. *Modus faciendi cum ordine medicando*. Seville: por Jacobo Cromberger, 1527. Facsimile edition, edited by Milagro Laín and Doris Ruiz Otín. Madrid: Ediciones Doce Calles, 2001.

Lavoisien, Jean François. *Dictionnaire portatif de médecine*. 2 vols. Paris, 1771.

Leclercq, Chrestien. *Nouvelle relation de la Gaspésie*. Montreal, QC: Les Presses de l'Université de Montréal, 1999 [1691].

Lémery, Nicolas. *Cours de Chymie*. Paris: N. Lémery, 1675.

Lémery, Nicolas. *Pharmacopée universelle, contenant toutes les compositions de pharmacie qui sont en usage dans la médicine, tant en France que par toute l'Europe; Avec plusieurs remarques & raisonnemens sur chaque operation*. Paris: Houry, 1698.

Lémery, Nicolas. *Pharmacopée universelle, contenant toutes les compositions de pharmacie qui sont en usage dans la medecine, tant en France que par toute l'Europe*. Paris: Laurent d'Houry, 1716.

Lémery, Nicolas. *Traité universel des drogues simples mises en ordre alphabétique*. Paris: L. d'Houry, 1698.

Leprince, Maurice. *Ministère du Commerce, de l'industrie, des postes et des télégraphes. Exposition universelle internationale de 1900. Direction générale de l'exploitation. 1er congrès international de l'industrie et du commerce des spécialités pharmaceutiques, tenu à Paris les 3 et 4 septembre 1900. Compte rendu*. Paris: Jourdan, 1900.

Levey, Martin, and Noury al-Khaledy. *The medical formulary of al-Samarqandī and the relation of early Arabic simples to those found in the indigenous medicine of the Near East and India*. Philadelphia: University of Pennsylvania Press, 1967.

Lewis, William. *The Pharmacopoeia of the Royal College of Physicians at Edinburgh, Faithfully translated from the Fourth Edition*. London: Printed for John Nourse, 1748.

Linnaeus, Carolus. *Lignum qvassiae*. Uppsala, 1763.

Margry, Pierre. *Découvertes et établissements des Français dans l'ouest et dans le sud de l'Amérique Septentrionale, 1614–1754*. Paris: Maisonneuve et cie, 1879–1888.

Martinez de Leache, Miguel. *Discurso Pharmaceutico sobre los Canones de Mesue*. Pamplona: Por Martín de Labayen y Diego de Zabala, 1652.

Mateo, Pedro Benedicto. *Loculentissimi viri . . . Liber in examen apothecariorum . . .* Barcelona: Joan Rosembach, 1521.

"Miscellaneous Medical Facts." *Philadelphia Medical and Physical Journal* 2 (1806): 87–88.

Munro, William Bennett. *Documents Relating to the Seigniorial Tenure in Canada, 1598–1854*. Toronto: Champlain Society, 1908.

Narodetzki, André. *Le remède secret. Législation et jurisprudence de la Loi du 21 Germinal an XI au Décret du 13 juillet 1926*. Paris: Delrieu, 1928.

Nicolas, Louis. *The Codex Canadensis and the Writings of Louis Nicolas*. Edited by François-Marc Gagnon. Montreal, QC: McGill-Queen's University Press, 2011.

BIBLIOGRAPHY

"Obituary. John Redman Coxe, M.D." *American Medical Times* 8 (May 7, 1864): 226.

Occom, Samson. *The Collected Writings of Samson Occom, Mohegan: Leadership and Literature in Eighteenth-Century America*. Edited by Joanna Brooks. Oxford: Oxford University Press, 2006.

Oldenburg, Henry. "Epistle Dedicatory." *Philosophical Transactions* 1 (1665–1666): unpaginated. doi:10.1098/rstl.1665.0001.

Oldenburg, Henry. "The Introduction." *Philosophical Transactions* 1 (1665–1666): 1–2. doi:10.1098/rstl.1665.0002.

"Ordonnance du Roi sur la publication d'un nouveau code pharmaceutique." *Bulletin des lois du Royaume de France* 7, no. 3 (1817): 115–17.

Oviedo, Luis de. *Método de la colección y reposición de las medicinas simples y de su corrección y preparación*. Madrid: Luis Sánchez, 1609.

Paiva, Manoel Joaquim Henriques de. *Farmacopéa Lisbonense, ou Collecçao dos simplices, preparaçoes e composiçoes mais efficazes e de major uso*. Lisbon: Filippe da Silva e Azevedo, 1785 and 1802.

Palacios, Félix. *Palestra Pharmaceutica Chymico-Galenica*. Madrid: Juan García Infanzón, 1706.

Palacios, Félix. *Palestra Pharmaceutica Chymico-Galenica*. Madrid: Juan de Sierra, 1730.

Paul of Aegina. *The Seven Books of Paulus Aegineta*. Translated by Francis Adams. London: Printed for the Sydenham Society, 1846.

Petiver, James. "A Catalogue of Some Guinea-Plants, with their Native Virtues; Sent to James Petiver, Apothecary, and Fellow of the Royal Society; with His Remarks on them. Communicated in a Letter to Dr. Hans Sloane. Secret. Reg. Soc." *Philosophical Ttransactions* 19 (1695): 677–86. Accessed August 17, 2018. http://dx.doi:10.1098/rstl.1695.0124.

Pharmacopeia Collegii Regalis Medicorum Londinensis. London: G. Bowyer, 1721.

Pharmacopea Hispana. Madrid: Typographia Ibarriana, 1794.

Pharmacopoeia Matritensis. Madrid: Typis Antonii Perez de Soto, 1762.

Pharmacopoeia nosocomii New-Eboracensis; or the Pharmacopoeia of the New-York Hospital. New York: Published by Collins & Co. 189 Pearl Street, 1816.

Pharmacopoeia of the Massachusetts Medical Society. Boston: Published by E. & J. Larkin, No. 47, Cornhill.

Pharmacopoeia simpliciorum et efficaciorum. Philadelphia: Ex Officina Styner & Cist, 1778.

Phillips, Richard. "Remarks on the '*Code des médicamens ou pharmacopée française*.'" *Quarterly Journal of Science, Literature and the Arts* 9, no. 18 (1820): 239–50.

Philotime [pseud.]. *Lettre à M. le Docteur Virey sur la saisie de la seconde édition de son Traité de pharmacie théorique et pratique*. Paris: J. Gratiot, 1819.

Pomet, Pierre. *Traite General des Drogues Simples et Composees*. Paris, 1695.

BIBLIOGRAPHY

"Préparation célèbre de la Thériaque nouvellement faite à Paris par M. de Rouvière." *Journal des Sçavans* 13 (1685): 228–31.

Quincy, John, trans. *The Dispensatory of the Royal College of Physicians in London*. London: W. Boyer, 1721.

Ranby, John. "The Anatomy of the Poisonous Apparatus of a Rattle-Snake, Made by the Direction of Sir Hans Sloane, Bart. Praes. Soc. Reg. & Coll. Med. together with an Account of the Quick Effects of Its Poison." *Philosophical Transactions* 35 (1728): 377–81. Accessed August 17, 2018. http://dx.doi:10.1098/rstl.1727.0025.

Raudot, Antoine-Denis. *Relations par lettres de l'Amérique septentrionalle (Années 1709 et 1710)*. Paris: Letouzey et ané, 1904.

Requeste servant de Factum pour Antoine Boulogne Ayde-Apoticaire du Corps du Roy, Défendeur. Contre Roberte Richard, veuve Contugi, dit l'Orvietan, et son fils, Demandeurs et Défendeurs. Paris: s.n., 1684.

"Review of Henry's American Herbal." *Analytic Magazine* 7 (1816): 249–65.

Riollet, Jean-Thomas. *Remarques curieuses sur la thériaque, avec un excellent traité sur l'orvietan*. Bourdeaux, 1665.

"Romlindorf Digestive Lozenges—An Indian Specific." *Charleston Courier*, April 18, 1812.

Royal College of Physicians of Dublin. *Pharmacopoeia Collegii Medicorum Regis et Retinae in Hibernia, Dublinii*. 1st ed. Dublin, 1807.

Royal College of Physicians of Edinburgh. *Pharmacopoea Collegii Regii Medicorum Edimburgensium*. 1st ed. Edinburgh, 1699.

Royal College of Physicians of Edinburgh. *Pharmacopoea Collegii Regii Medicorum Edimburgensium*. 2nd ed. Edinburgh, 1722.

Royal College of Physicians of London. *Pharmacopoea Londinensis, in Qva Medicamenta Antiqva et Nova, Londinensis, Opera Medicorum Collegij*. First issue of the 1st ed. London: Printed by Edwardus Griffin for Iohannis Marriot, 1618 (May).

Royal College of Physicians of London. *Pharmacopoea Londinensis, in Qva Medicamenta Antiqva et Nova, Londinensis, Opera Medicorum Collegij*. Second issue of the 1st ed. London: Printed by E. Griffin for Iohn Marriot, 1618 (December).

Royal College of Physicians of London. *Pharmacopoeia Londinensis: of 1618*, reproduced in facsimile with a historical introduction by George Urdang. Madison: State Historical Society of Wisconsin, 1944 [1618].

Royal College of Physicians of London. *Pharmacopoeia Londinensis*. London: Printed for John Marriott, 1639.

Sābūr ibn Sahl. *Dispensatorium parvum (al-Aqrabadhin al-saghir)*. Edited and translated by Oliver Kahl. Leiden: Brill, 1994.

Sābūr ibn Sahl and Oliver Kahl. *Sābūr Ibn Sahl's Dispensatory in the Recension of the 'Aḍudī Hospital*. Leiden: Brill, 2009.

BIBLIOGRAPHY

Sampaio, Francisco António de. *História dos Reinos Vegital, Animal e Mineral.* 2 vols. *Anais da Biblioteca Nacional,* vol. 89. Rio de Janeiro, Brazil, 1969.

Saunders, Richard. *Poor Richard, 1737. An Almanack For the Year of Christ 1737.* Philadelphia: Printed and sold by B. Franklin, at the New Printing-Office near the Market, 1736.

Savary des Brûlons, Jacques, and Philémon-Louis Savary. *Dictionnaire universel de commerce contenant tout ce qui concerne le commerce qui se fait dans les quatre parties du monde.* Amsterdam: Chez les Jansons à Waesberge, 1726.

Schoolcraft, Henry Row. *Historical and Statistical Information Respecting the History, Condition and Prospects of the Indian Tribes of the United States, Part 1.* Philadelphia: Lippincott, Grambo & Co., 1851.

Schröder, Johann. *Pharmacopoeia medico-chymica sive Thesaurus pharmacologicus.* 4th ed. Ulm, 1655.

Short, Thomas. *Medicina Britannica: Or a Treatise on such Physical Plants as are Generally to be found in the Fields or Gardens in Great-Britain: Containing a Particular Account of their Nature, Virtues, and Uses.* London, 1747.

Sloane, Hans. "Conjectures on the Charming or Fascinating Power Attributed to the Rattle-Snake: Grounded on Credible Accounts, Experiments and Observations." *Philosophical Transactions* 38 (1733): 321–31. Accessed August 17, 2018. http://dx.doi:10.1098/rstl.1733.0050.

Sloane, Hans. "A Description of the Pimienta or Jamaica-Pepper Tree." *Philosophical Transactions* 17 (1686): 462–68. Accessed August 17, 2018. http://dx.doi:10.1098/rstl.1686.0083.

Sloane, Hans. *Magic and Mental Disorder: Sir Hans Sloane's Memoir of John Beaumont.* Edited by Michael Hunter. London: Robert Boyle Project, 2011.

Sloane, Hans. "Preface." *Philosophical Ttransactions* 21 (1699): unpaginated. Accessed August 17, 2018. http://dx.doi:10.1098/rstl.1699.0081.

Sloane, Hans. *A Voyage to the Islands of Madera, Barbadoes, Nieves, S. Christophers, and Jamaica.* Vol. 1. London: B.M., 1707.

Sloane, Hans, and George Dampier. "Part of a Letter from Mr. George Dampier, Dated, Exmouth, November 10 1697 to Mr. William Dampier, his Brother, concerning the Care of Bitings of mad Creatures. With a Remark on the same by Hans Sloane, M.D." *Philosophical Transactions* 20 (1698): 49–52. Accessed August 17, 2018. http://dx.doi:10.1098/rstl.1698.0009.

Smith, Peter. *The Indian Doctor's Dispensatory; being Father Smith's advice respecting diseases and their cure.* Cincinnati: Printed by Browne and Looker, for the author, 1813.

Société Royale de Médecine. *Nouveau plan de constitution pour la médecine.* Paris, 1790.

South Carolina Gazette. May 9, 1750.
Stedman, John. *Narrative, of a Five Years Expedition, against the revolted Negroes of Surinam, in Guiana, on the wild coast of South America.* London: J. Johnson and T. Payne, 1806–1813.
Suarez de Figueroa, Cristobal. *Plaza universal de todas ciencias y artes.* Madrid: Por Luis Sanchez, 1615.
Talon, Jean. "Lettre de Talon au ministre Colbert (27 Octobre 1667)." *Rapport de l'Archiviste de la province de Québec* 1930–1931 (1931): 80.
Talon, Jean. "Mémoire de Talon au Roi sur le Canada (2 Novembre 1671)." *Rapport de l'Archiviste de la province de Québec* 1930–1931 (1931): 156–62.
Tennent, John. "Continuation of the Memorial, began in our last." *Virginia Gazette,* September 29, 1738.
Tennent, John. *An Essay on the Pleurisy.* Williamsburg: Printed and Sold by William Parks, 1736.
Tennent, John. *Every Man his Own Doctor: or, The Poor Planter's Physician.* Philadelphia: Re-printed and Sold by B. Franklin, near the Market, 1734.
Tennent, John. "A Memorial, humbly addressed to the learned, impartial, and judicious world." *Virginia Gazette,* September 22, 1738.
Tennent, John. "Remainder of Dr. Tennent's Memorial." *Pennsylvania Gazette,* July 26, 1739.
Tennent, John. "To the Northern Colonies. The Memorial and Remonstrance of John Tennent, Practitioner in Physick." *Pennsylvania Gazette,* July 19, 1739.
Thacher, James. *The American New Dispensatory.* Boston: Printed and Published by T.B. Wait and Co., 1810.
Thevet, André. *André Thevet's North America: A Sixteenth-Century View.* Montreal, QC: McGill-Queen's University Press, 1986.
Thobald, John. *Every Man His Own Physician.* Boston: Printed for Cox and Berry, 1767.
Thomson, Samuel. *New Guide to Health, or, Botanic Family Physician.* Boston: Printed for the author, by E.G. House, 1822.
Thwaites, Reuben Gold, ed. *The Jesuit Relations and Allied Documents.* 73 vols. Cleveland: Burrows Brothers Company, 1896–1901.
"Tolerated Quackery." *New-England Galaxy and United States Literary Advertiser* 7 (1824): 2.
Tournefort, Joseph Pitton de. *Traité de la matière médicale; ou l'histoire et l'usage des médicamens, et leur analyse chymique.* Paris: Laurent d'Houry, 1717.
Tribunal del Real Protomedicato. *Pharmacopoeia Matritensis Regii ac Supremi Hispaniarum.* Madrid: Typographia Regia Michaelis Rodriguez, 1739.
Université de Paris, Faculté de médecine. *Codex Medicamentarius seu Pharmacopoea Parisiensis.* Lutetiae Parisiorum: Sumptibus Olivarii de Varennes, 1645.

Vandelli, Domingos. *Diccionario dos termos technicos de Historia Natural, extrahidos das obras de Linneo, e a memoria sobre a utilidade dos Jardins Botanicos.* Coimbra: Real Officina da Universidade, 1788.

Vandelli, Domingos. *Florae lusitanicae et brasiliensis specimen* [. . .] *et epistolae ab eruditsviris.* Coimbra: Typografia Academico-Regia, 1788.

Venel, Gabriel-François. "Magistral, remède (Thérapeut.)." *Encyclopédie ou Dictionnaire raisonné des sciences, des arts et des métiers* 9 (1751–1772): 855.

Venel, Gabriel-François. "Officinal, adj. (Pharmacie.)." *Encyclopédie ou Dictionnaire raisonné des sciences, des arts et des métiers* 9 (1751–1772): 855.

Verdier, Jean. *La jurisprudence de la medicine en France.* 2 vols. Alençon, 1762.

Villa, Esteban de. *Examen de boticarios.* Burgos: por Pedro de Huydobro, 1632.

Vinaburu, Pedro de. *Cartilla Pharmaceutica Chimico-Galenica.* Pamplona: por Joseph Joachin Martinez, 1778.

Wilbert, M. I. "The Pharmacist and Pharmacopoeia." *Pharmaceutical Era* 31 (1904): 481.

Williams, Stephen W. "Indigenous Medical Botany, No. 3." *New York Journal of Medicine and Collateral Sciences* 7 (1846): 179.

Wood, George B. "Article XV. Observations on the Pharmacopoeia of the United States." *Journal of the Philadelphia College of Pharmacy* 4 (1832): 94–107.

Wood, George B. and Franklin Bache. *The Dispensatory of the United States of America.* Philadelphia: Published by Grigg & Elliot, 1833.

Worm, Ole. *Museum Wormianum: seu historia rerum rariorum, tam naturalium, quam artificialium, tam domesticarum, quam exoticarum, quae hasniae danorum in aedibus authoris fervantur.* Lugduni Batavorum: Apud Iohannem Elsevirium, 1655.

Ximénez, Francisco. *Quatro libros. De la naturaleza y virtudes de las plantas y animales que estan recividos en el use de Medicina en la Nueva España.* Mexico: Casa de la Viuda de Diego Lopez Davalos, 1615.

Yaʾkūb ibn Ishāk ibn Subbāh, Abū Yūsuf, al-Kindī, and Martin Levey. *The Medical Formulary or Aqrabadhin of Al-Kindi.* Madison: University of Wisconsin Press, 1966.

Zwelfer, Johann. *Pharmacopoeia regia.* Noribergae: Typis & Sumptibus Balthasaris Joachimi Endteri, 1693.

Secondary Sources

Abernethy, David. *The Dynamics of Global Dominance: European Overseas Empires, 1415–1980.* New Haven, CT: Yale University Press, 2002.

Agrimi, Jole, and Chiara Crisciani. *Edocere medicos: medicina scolastica nei secoli XIII-XV.* Naples: Guerini e associati, 1988.

Alden, Dauril. "The Gang of Four and the Campaign against the Jesuits in Eighteenth-Century Brazil." In *The Jesuits II: Cultures, Sciences, and the Arts,*

BIBLIOGRAPHY

1540–1773. Edited by John W. O'Malley, S.J., et al. Toronto: University of Toronto Press, 2006.

Alencastro, Luiz Felipe de. "Le versant brésilien de l'Atlantique-Sud: 1550–1850." *Annales. Histoire, Sciences Sociales* 2 (2006): 339–82.

Alvarez Peláez, Raquel. *La conquista de la naturaleza americana*. Madrid: CSIC, 1993.

Amaro, A. M. *Introdução da Medicina ocidental em Macau e as receitas de segredo da botica do Colégio de São Paulo*. Macau: Instituto Cultural de Macau, 1992.

Anderson, Lee, and Gregory J. Higby. *The Spirit of Volunteerism: A Legacy of Commitment and Contribution: The United States Pharmacopeia 1820–1995*. Rockville, MD: United States Pharmacopeial Convention, Inc., 1995.

Anderson, Stuart. "Pharmacy and Empire: The 'British Pharmacopoeia' as an Instrument of Imperialism, 1864–1932." *Pharmacy in History* 52 (2010): 112–21.

Arber, Agnes. *Herbals: Their Origin and Evolution*. 3rd ed. Cambridge: Cambridge University Press, 1987 [1912].

Armitage, David. "Three Concepts of Atlantic History." In *The British Atlantic World, 1500–1800*, edited by David Armitage and Michael J. Braddick, 11–27. New York: Palgrave Macmillan, 2002.

Asselin, Alain, Jacques Cayouette, and Jacques Mathieu. *Curieuses histoires de plantes du Canada*. Vol. 1. Quebec: Septentrion, 2014.

Azzolini, Monica. "In Praise of Art: Text and Context of Leonardo's *Paragone* and Its Critique of the Arts and Sciences." *Renaissance Studies* 19 (2005): 487–508.

Baghdiantz McCabe, Ina. *Orientalism in Early Modern France: Eurasian Trade, Exoticism and the Ancien Régime*. Oxford: Berg, 2008.

Barrera-Osorio, Antonio. *Experiencing Nature: The Spanish American Empire and the Early Scientific Revolution*. Austin: University of Texas Press, 2006.

Bartlett, T. "Ireland, Empire and Union 1690–1801." In *Ireland and the British Empire*, edited by K. Kenny, 69. Oxford: Oxford University Press, 2004.

Bauer, Ralph. *The Cultural Geography of Colonial American Literatures: Empire, Travel, Modernity*. Cambridge: Cambridge University Press, 2003.

Bec, Christian. "I mercanti scrittori." In *Letteratura italiana, Volume secondo. Produzione e consumo*, edited by Alberto Asor Rosa, 269–97. Turin: Giulio Einaudi, 1983.

Bec, Christian. "Lo statuto socio-professionale degli scrittori (Trecento e Cinquecento)." In *Letteratura italiana, Volume secondo. Produzione e consumo*, edited by Alberto Asor Rosa, 229–67. Turin: Giulio Einaudi, 1983.

Beckert, Sven, and Seth Rockman. *Slavery's Capitalism: A New History of American Economic Development*. Philadelphia: University of Pennsylvania Press, 2016.

Bell, David. "Tearing down the Tower of Babel: Grégoire and French Multilingualism." In *The Abbé Grégoire and his World*, edited by Jeremy Popkin and Richard Popkin, 109–29. Dordrecht: Kluwer Academic Publishers, 2000.

Bellin, Joshua David. *Medicine Bundle: Indian Sacred Performance and American Literature, 1824–1932*. Philadelphia: University of Pennsylvania Press, 2008.
Bellin, Joshua David. "Taking the Indian Cure: Thoreau, Indian Medicine, and the Performance of American Culture." *New England Quarterly* (March 2006): 3–36.
Bellorini, Cristina. *The World of Plants in Renaissance Tuscany: Medicine and Botany*. New York: Routledge, 2016.
Belmessous, Saliha. "Etre français en Nouvelle-France: Identité française et identité coloniale aux dix-septième et dix-huitième siècles." *French Historical Studies* 27, no. 3 (2004): 507–40.
Benedict, Barbara. "Collecting Trouble: Sir Hans Sloane's Literary Reputation in Eighteenth-Century Britain." *Eighteenth-Century Life* 36, no. 2 (2012): 111–42.
Bennett, Bradley C., and Ghilean T. Prance. "Introduce Plants in the Indigenous Pharmacopoeia of Northern South America." *Economic Botany* 54 (2000): 90–120.
Bennett, J. A. "Practical Geometry and Operative Knowledge." *Configurations* 6, no. 2 (1998): 195–222.
Bergounioux, Jean. "Les éditions du *Codex Medicamentarius* de l'ancienne Faculté de Médecine de Paris." *Bulletin de la société française d'histoire de la pharmacie* 15 (1927): 376–89.
Berry, Lynn. "The Delights of Nature in this New World: A Seventeenth-Century Canadian View of the Environment." In *Decentering the Renaissance: Canada and Europe in Multidisciplinary Perspective, 1500–1700*, edited by Germaine Warkentin and Carolyn Podruchny, 223–35. Toronto, ON: University of Toronto Press, 2001.
Biagioli, Mario. "Patent Republic: Representing Inventions, Constructing Rights and Authors." *Social Research* 73, no. 4 (2006): 1150–72.
Blair, L. "Martyrdom by toads' tongues. Early Canadian Doctors and their Victims." *Canadian Family Physician* 45 (1999): 2293–304.
Bleichmar, Daniela. "Books, Bodies, and Fields: Sixteenth-Century Transatlantic Encounters with New World Materia Medica." In *Colonial Botany: Science, Commerce, and Politics in the Early Modern World*, edited by Londa L. Schiebinger and Claudia Swan, 83–99. Philadelphia: University of Pennsylvania Press, 2005.
Bleichmar, Daniela. "Seeing the World in a Room: Looking at Exotica in Early Modern Collections." In *Collecting across Cultures: Material Exchanges in the Early Modern Atlantic World*, edited by Daniela Bleichmar and Peter C. Mancall, 11–30. Philadelphia: University of Pennsylvania Press, 2011.
Bleichmar, Daniela. *Visible Empire: Botanical Expeditions and Visual Culture in the Hispanic Enlightenment*. Chicago: University of Chicago Press, 2012.
Bleichmar, Daniela, Paula De Vos, Kristine Huffine, and Kevin Sheehan, eds. *Sci-

ence in the Spanish and Portuguese Empires. Stanford: Stanford University Press, 2009.

Borges, Charles J. *The Economics of the Goa Jesuits, 1542–1759*. New Delhi: Concept Publishing, 1994.

Bouvet, Maurice. "Histoire sommaire du remède secret." *Revue d'histoire de la pharmacie* 45, nos. 153–54 (1957): 57–63, 109–18.

Bouvet, Maurice. "Les apothicaires royaux." *Bulletin de la sociétié de l'histoire de la pharmacie* 5, nos. 58–66 (1929–1928): 58–66, 104–9, 149–55, 187–90, 230–32, 263–71, 308–20, 408–15.

Bouvet, Maurice. "Sur l'historique du conditionnement de la spécialité pharmaceutique." *Revue des spécialités* (1928): 101–43, 213–23, 297–315.

Boxer, Charles R. *Two Pioneers of Tropical Medicine: Garcia d'Orta and Nicolás Monardes*. Wellcome Library: London, 1963.

Bracht, Fabiano. "Condicionantes sociais e políticas nos processos de produção de conhecimento: o caso da Índia portuguesa do século XVIII." In *História & Ciência: Ciência e Poder Na Primeira Idade Global*, edited by Amélia Polónia, Fabiano Bracht, Gisele C. Conceição, and Monique Palma. Porto: Biblioteca da Faculdade de Letras da Universidade do Porto, 2016.

Brazeau, Brian. *Writing a New France, 1604–1632: Empire and Early Modern French Identity*. Farnham, UK: Ashgate, 2009.

Breen, Benjamin. "Curvo's Sixteen Secrets: Tracing Pharmaceutical Networks in the Portuguese Tropics." In *Empires of Knowledge: Scientific Networks in the Early Modern World*, edited by Paula Findlen. New York: Routledge, forthcoming.

Breen, Benjamin. "Drugs and Early Modernity." *History Compass* 15 (2017). Accessed August 17, 2018. http://dx.doi.org/10.1111/hic3.12376.

Brockbank, W. "Sovereign Remedies: A Critical Depreciation of the 17th Century London Pharmacopoeia." *Medical History* 8, no. 1 (1964): 1–14.

Brockliss, Laurence, and Colin Jones. *The Medical World of Early Modern France*. Oxford: Clarendon Press, 1997.

Brooks, Joanna. "'This Indian World: An Introduction to the Writings of Samson Occom." In *The Collected Writings of Samson Occom, Mohegan: Leadership and Literature in Eighteenth-Century America*, edited by Joanna Brooks, 3–39. Oxford: Oxford University Press, 2006.

Brooks, Lisa. *The Common Pot: The Recovery of Native Space in the Northeast*. Minneapolis: University of Minnesota Press, 2008.

Bross, Kristina, and Hilary E. Wyss, eds. *Early Native Literacies in New England: A Documentary and Critical Anthology*. Amherst: University of Massachusetts Press, 2008.

Brown, Vincent. *The Reaper's Garden: Death and Power in the World of Atlantic Slavery*. Cambridge, MA: Harvard University Press, 2008.

BIBLIOGRAPHY

Burnard, Trevor, and John Garrigus. *The Plantation Machine: Atlantic Capitalism in French Saint-Domingue and British Jamaica*. Philadelphia: University of Pennsylvania Press, 2016.

Calloway, Colin G. *New Worlds for All: Indians, Europeans, and the Remaking of Early America*. Baltimore: Johns Hopkins University Press, 1998.

Calmon, Pedro. *História do Brasil*. 3rd ed. 7 vols. São Paulo: José Olympio, 1959.

Calvet de Magalhães, José. *História das Relações Diplomáticas entre Portugal e os Estados Unidos da América (1776–1911)*. Lisbon: Publicações Europa-América, 1991.

Campos Díez, María Soledad. *El Real Tribunal del protomedicato castellano, siglos XIV–XIX*. Cuenca: University de Castilla La Mancha, 1999.

Candido, Mariana P. *An African Slaving Port and the Atlantic World: Benguela and Its Hinterland*. Cambridge: Cambridge University Press, 2013.

Cañizares-Esguerra, Jorge. "'Enlightened Reform' in the Spanish Empire: An Overview." In *Enlightened Reform in Southern Europe and its Colonies*, edited by Gabriel Paquette, 33–37. Burlington: Ashgate, 2009.

Cañizares-Esguerra, Jorge. *How to Write the History of the New World: Histories, Epistemologies and Identities in the Eighteenth-Century Spanish Atlantic*. Stanford: Stanford University Press, 2001.

Cañizares-Esguerra, Jorge. *Puritan Conquistadors: Iberianizing the Atlantic, 1550–1700*. Stanford: Stanford University Press, 2006.

Carney, Judith. *In the Shadow of Slavery: Africa's Botanical Legacy in the Atlantic World*. Berkeley: University of California Press, 2010.

Carpin, Gervais. *Histoire d'un mot: l'ethnonyme "Canadien" de 1535–1691*. Sillery, QC: Septentrion, 1995.

Carr, Lloyd G., and Carlos Westey. "Surviving Folktales and Herbal Lore among the Shinnecock Indians of Long Island." *Journal of American Folklore* 58, no. 228 (April–June 1945): 113–23.

Carrillo-Linares, Maria Jose. "Antidotarium Nicolai," *Oxford Dictionary of the Middle Ages*, edited by Robert E. Bjork. Oxford: Oxford University Press, 2010. Accessed November 9, 2018. http://www.oxfordreference.com/view/10.1093/acref/9780198662624.001.0001/acref-9780198662624-e-0382?rskey=s69olU&result=1

Cash, Philip, Eric H. Christianson, and J. Worth Estes, eds. *Medicine in Colonial Massachusetts 1620–1820: A Conference Held 25 & 26 May 1978 by the Colonial Society of Massachusetts*. Boston: Colonial Society of Massachusetts, 1980.

Cassier, Maurice. "Brevets pharmaceutiques et santé publique en France: opposition et dispositifs spécifiques d'appropriation des médicaments en France entre 1791 et 2004." *Entreprises et histoire* 36 (2004): 29–47.

Castro e Almeida, Eduardo de. *Inventário dos Documentos Relativos ao Brasil Existantes no Archivo de Marinha e Ultramar de Lisboa.* Vol. I, tome 1, Bahia, 1613–1762. Rio de Janeiro: Officinas Graphicas da Biblioteca Nacional, 1913.

Catellani, Patrizia, and Renzo Console. *L'Orvietano.* Pisa: ETS, 2004.

Cavallo, Sandra, and Tessa Storey. *Healthy Living in Late Renaissance Italy.* Oxford: Oxford University Press, 2014.

Certeau, Michel de, Dominique Julia, and Jacques Revel. *Une politique de la langue.* Paris: Gallimard, 1975.

Chakrabarti, Pratik. *Materials and Medicine: Trade, Conquest and Therapeutics in the Eighteenth Century.* Manchester: Manchester University Press, 2010.

Chakrabarty, Dipesh. "The Muddle of Modernity." *American Historical Review*, 116 (2011): 663–75.

Chaplin, Joyce. "Earthsickness: Circumnavigation and the Terrestrial Human Body, 1520–1800." *Bulletin of the History of Medicine* 86, no. 4 (Winter 2012): 515–42.

Chaplin, Joyce E. *Subject Matter: Technology, the Body, and Science on the Anglo-American Frontier, 1500–1676.* Cambridge, MA: Harvard University Press, 2003.

Chaplin, Joyce. "Terrestriality." *The Appendix: A Journal of Narrative and Experimental History* 2, no. 2 (April 2014). Accessed November 12, 2018. http://the appendix.net/issues/2014/4/terrestriality.

Chipman, Leigh. *The World of Pharmacy and Pharmacists in Mamlūk Cairo.* Leiden: Brill, 2010.

Clarence-Smith, William Gervase. *Cocoa and Chocolate, 1765–1914.* London: Routledge, 2002.

Clark, G. *A History of the Royal College of Physicians of London.* Volume 1, 1518–1689. Oxford: Clarendon Press, 1964.

Clouse, Michele L. *Medicine, Government and Public Health in Philip II's Spain: Shared Interests, Competing Authorities.* New York: Routledge, 2011.

Clucas, Stephen. "Galileo, Bruno and the Rhetoric of Dialogue in Seventeenth-Century Natural Philosophy." *History of Science* 46(2008): 405–29.

Coates, Colin. "The Colonial Landscape of the Early Town." In *Metropolitan Natures: Environmental Histories of Montreal*, edited by Stéphane Castonguay and Michèle Dagenais, 19–36. Pittsburgh: University of Pittsburgh Press, 2011.

Coates, Colin. *The Metamorphoses of Landscape and Community in Early Québec.* Montreal, QC: McGill-Queen's University Press, 2000.

Cobb, Boughton, Elizabeth Farnsworth, and Cheryl Lowe. *A Field Guide to Ferns and their Related Families: Northeastern and Central North America.* 2nd ed. New York: Houghton Mifflin Company, 2005.

Cohen, Matt. *The Networked Wilderness: Communicating in Early New England.* Minneapolis: University of Minnesota Press, 2009.

Conceição, Gisele C. "Estudos De Filosofia Natural No Brasil Ao Longo Do Século XVIII." In *História & Ciência: Ciência e Poder Na Primeira Idade Global*, edited by Amélia Polónia, Fabiano Bracht, Gisele C. Conceição, and Monique Palma. Porto: Biblioteca da Faculdade de Letras da Universidade do Porto, 2016.

Connolly, Margaret. "Evidence for the Continued Use of Medieval Medical Prescriptions in the Sixteenth Century: A Fifteenth Century Remedy Book and Its Later Owner." *Medical History* 60 (2016): 133–54.

Cook, Harold J. "Markets and Cultures: Medical Specifics and the Reconfiguration of the Body in Early Modern Europe." *Transactions of the Royal Historical Society* 21 (2011): 123–45.

Cook, Harold J., and Timothy D. Walker. "Circulation of Medicine in the Early Modern Atlantic World." *Social History of Medicine* 26 (2013): 337–51.

Cooper, Alix. *Inventing the Indigenous: Local Knowledge and Natural History in Early Modern Europe*. Cambridge: Cambridge University Press, 2007.

Cooper, N. C. "Development of a Pharmaceutical Profession in Ireland." *Pharmacy in History* 29, no. 4 (1987): 165–76.

Cordonnier, E. "Sur le Liber Servitoris d' Aboulcasis." *Janus* 9 (1904): 425–87.

Coste, Joël. "La medicine pratique et ses genre littéraires en France a l'epoque moderne." Bibliothèque interuniversitair de Santé. Accessed June 18, 2018. http://www.biusante.parisdescartes.fr/histoire/medica/medecine-pratique-en.php.

Coulter, H. L. *Divided Legacy: A History of Modern Western Medicine, J.B. Helmont to Claude Bernard*. Berkeley: North Atlantic Books, 2000.

Coulton, Richard. "'The Darling of the *Temple-Coffee-House-Club*': Science, Sociability, and Satire in Early Eighteenth-Century London." *Journal for Eighteenth-Century Studies* 35, no. 1 (2012): 43–65.

Courtwright, A. C. *The British Pharmacopoeia, 1864–2014: Medicines, International Standards, and the State*. Farnham: Ashgate, 2015.

Cowen, David L. "The Boston Editions of Nicholas Culpeper." *Journal of the History of Medicine and Allied Sciences* 11 (1956): 156–65.

Cowen, David L. "The Edinburgh Pharmacopoeia." In *The Early Years of the Edinburgh Medical School*, edited by R.G.W. Anderson and A.D.C. Simpson, 1–20. Edinburgh: Royal Scottish Museum, 1976.

Cowen, David L. *Pharmacopoeias and Related in Literature in Britain and America, 1618–1847*. Burlington: Ashgate, 2000.

Crawford, Matthew James. *The Andean Wonder Drug: Cinchona Bark and Imperial Science in the Spanish Atlantic, 1630–1800*. Pittsburgh: University of Pittsburgh Press, 2016.

Crisciani, Chiara. "Alchemy and Medieval Universities: Some Proposals for Research." *Universitas* 10 (1997). Accessed August 17, 2018. http://www.cis.unibo.it/universitas/10_1997/crisciani.html.

Crisciani, Chiara. "History, Novelty, and Progress in Scholastic Medicine." *Osiris* 6 (1990): 118–39.
Curtin, Phillip. *The Rise and Fall of the Plantation Complex*. Cambridge: Cambridge University Press, 1990.
Curto, José C. *Enslaving Spirits: The Portuguese Brazilian Alcohol Trade at Luanda and its Hinterlands, c. 1550–1830*. Leiden: Brill, 2003.
D'Aronco, M. A. "The Old English Pharmacopoeias." *AVISTA Forum Journal* 13, no. 2 (2003): 9–18.
Dames, Mansel Longworth, trans. *The Book of Duarte Barbosa*. 2 vols. New Delhi: Asia Educational Services, 2002. Reprint of the 1918–1921 London edition.
Daston, Lorraine. "Marvelous Facts and Miraculous Evidence in Early Modern Europe." *Critical Inquiry* 18, no. 1 (1991): 93–124.
Daston, Lorraine, and Katharine Park. *Wonders and the Order of Nature, 1150–1750*. New York: Zone Books, 1998.
Davis, Charles, and María Luz López Terrada. "Protomedicato y farmacia en Castilla a finales del siglo XVI: Edición crítica del *Catálogo de las cosas que los boticarios han de tener en sus boticas*, de Andrés de Zamudio de Alfaro, Protomédico General (1592–1599)." *Asclepio* 42 (2010): 579–626.
Dawson, Nelson-Martin. *L'atelier Delisle: l'Amérique du Nord sur la table à dessin*. Sillery, QC: Septentrion, 2000.
De Vos, Paula. "The Art of Pharmacy in Seventeenth- and Eighteenth-Century Mexico." PhD diss., University of California, Berkeley, 2001.
De Vos, Paula. "European Materia Medica in Historical Texts: Longevity of a Tradition and Implications for Future Use." *Journal of Ethnopharmacology* 132 (2010): 28–47.
De Vos, Paula. "From Herbs to Alchemy: The Introduction of Chemical Medicine to Mexican Pharmacy in the Seventeenth and Eighteenth Centuries." *Journal of Spanish Cultural Studies* 8 (2007): 135–68.
De Vos, Paula. "The 'Prince of Medicine': Yuhanna ibn Masawayh and the Foundations of the Western Pharmaceutical Tradition." *Isis* 104 (2013): 667–712.
De Vos, Paula. "Rosewater and Philosophers' Oil: Thermo-Chemical Processing in Medieval and Early Modern Spanish Pharmacy." *Centaurus* (forthcoming).
Debus, Allen G. *Alchemy and Chemistry in the Seventeenth Century*. Los Angeles: William Andrews Clark Memorial Library, University of California, 1966.
Debus, Allen G. *The Chemical Philosophy: Paraclesian Science and Medicine in the Sixteenth and Seventeenth Centuries*. Revised edition. New York: Dover Publications, 2002 [1977].
Debus, Allen G. *The English Paracelsians*. Chicago: University of Chicago Press, 1968.

Debus, Allen G. *The French Paracelsians: The Chemical Challenge to Medical and Scientific Tradition in Early Modern France*. Cambridge: Cambridge University Press, 1977.

Dehillerin, Bénédicte, and Jean-Pierre Goubert. "À la conquête du monopole pharmaceutique: Le Collège de Pharmacie de Paris (1777–1796)." *Historical Reflections/Réflexions historiques* 9, nos. 1–2 (1982): 233–48.

Delbourgo, James. *Collecting the World: Hans Sloane and the Origins of the British Museum*. Cambridge, MA: Harvard University Press, 2017.

Delbourgo, James. *A Most Amazing Scene of Wonders: Electricity and Enlightenment in America*. Cambridge, MA, Harvard University Press, 2006.

Delbourgo, James. "Sir Hans Sloane's Milk Chocolate and the Whole History of the Cacao." *Social Text* 29, no. 1 (2011): 71–101.

Delbourgo, James, and Nicholas Dew, eds. *Science and Empire in the Atlantic World*. New York: Routledge, 2007.

Deloria, Philip J. *Playing Indian*. New Haven, CT: Yale University Press, 1999.

Desbarats, Catherine, and Allan Greer. "Où est la Nouvelle-France?" *Revue d'histoire de l'Amérique française* 64, nos. 3–4 (2011): 31–62.

Descola, Philippe. *Beyond Nature and Culture*. Chicago: University of Chicago Press, 2013.

Devine, T. M. *Scotland's Empire 1600–1815*. London: Penguin Books, 2003.

Dew, Nicholas. "Scientific Travel in the Atlantic World: The French Expedition to Gorée and the Antilles, 1681–1683." *British Journal for the History of Science* 43, no. 1 (2010): 1–17.

Dewar, Helen. "Canada or Guadeloupe? French and British Perceptions of Empire, 1760–1763." *Canadian Historical Review* 91, no. 4 (2010): 637–60.

Dewar, Helen. "'Y establir nostre auctorité': Assertions of Imperial Sovereignty through Proprietorships and Chartered Companies in New France, 1598–1663." PhD diss., University of Toronto, 2012.

Dillon, Elizabeth Maddock. *New World Drama: The Performative Commons in the Atlantic World*. Durham: Duke University Press, 2014.

Disney, Anthony R. *A History of Portugal and the Portuguese Empire*. 2 vols. Cambridge University Press, 2009.

Duhr, B. *Pombal: Sein Charakter und Sein Politik: Ein Beitrag zur Geschichte des Absolutism*. Freiburg in Briesgau: Herder Verlagshandlung, 1891.

Eamon, William. *Science and the Secrets of Nature: Books of Secrets in Medieval and Early Modern Culture*. Princeton: Princeton University Press, 1994.

Elsheikh, Mahmoud Salem. *Medicina e farmacologia nei manoscritti della Biblioteca Riccardiana di Firenze*. Rome: Vecchiarelli Editore, 1990.

Eltis, David. *The Rise of African Slavery in the Americas*. Cambridge: Cambridge University Press, 1999.

BIBLIOGRAPHY

Fabian, Johannes. *Time and the Other: How Anthropology Makes Its Object*. New York: Columbia University Press, 2014.

Fabricius, Cajus. *Galens Exzerpte aus älteren Pharmacologen*. Berlin: Walter de Gruyter, 1972.

Fawcett, Melissa Jayne. *Medicine Trail: The Life and Lessons of Gladys Tantaquidgeon*. Tucson: University of Arizona Press, 2000.

Ferguson, W. *Scotland's Relations with England: A Survey to 1707*. Edinburgh: John Donald Publishers, 1977.

Ferrão, José E. Mendes. *A Aventura das Plantas e os Descubrimentos Portugueses*. Lisbon: Chaves Ferreira Publicações, 2005.

Ferrão, José E. Mendes, et al. *Plantas do Brasil: Flora económica do Brasil no Século XVIII; Plantas do Maranhão-Piauí*. Lisbon: Chaves Ferreira Publicações, 2002.

Figueroa Saavedra, Miguel, and Guadalupe Melgarejo Rodríguez. "La *Materia Medicinal de la Nueva España* de Fray Francisco Ximénez. Reapropiación y resignificación del conocimiento médico novohispano." *Dynamis* 38 (2018): 219–41.

Findlen, Paula. *Athanasius Kircher: The Last Man Who Knew Everything*. New York: Routledge, 2004.

Findlen, Paula. *Possessing Nature: Museums, Collecting, and Scientific Culture in Early Modern Italy*. Los Angeles: University of California Press, 1994.

Fitzgerald, Stephanie. "The Cultural Work of a Mohegan Painted Basket." In *Early Native Literacies in New England: A Documentary and Critical Anthology*, edited by Kristina Bross and Hilary E. Wyss, 52–56. Amherst: University of Massachusetts Press, 2008.

Flahaut, Jean. "La vie difficile du premier Codex national français." *Revue d'histoire de la pharmacie* 327, no. 3 (2000): 338–44.

Folch Andreu, Rafael. "La Concordia o Farmacopea de Zaragoza de 1553." *Farmacia Nueva* 13 (1948): 132.

Folch Andreu, Rafael. "Las Farmacopeas Nacionales españolas." *Actas del XV Congreso Internacional de Historia de la Medicina* 1 (1956): 247–67.

Fondazione Memofonte. "Cataloghi de Inventari della Biblioteca." Accessed June 11, 2018. http://www.memofonte.it/ricerche/cataloghi-ed-inventari-della-biblioteca.html.

Foucault, Michel. *The Order of Things: An Archaeology of the Human Sciences*. New York: Vintage, 1994 [1969].

Foust, Clifford. *Rhubarb: The Wonder Drug*. Princeton: Princeton University Press, 1992.

Francis, Susan, and Anne Stobart, eds. *Critical Approaches to the History of Western Herbal Medicine: From Classical Antiquity to the Early Modern Period*. London: Bloosmbury Publishing, 2014.

Freedberg, David. *The Eye of the Lynx: Galileo, His Friends, and the Beginnings of Modern Natural History.* Chicago: University of Chicago Press, 2002.

French, Roger. *Dissection and Vivisection in the European Renaissance.* Aldershot: Ashgate, 1999.

Friedrich, Christoph, and Wolf-Dieter Müller-Jahncke. *Geschichte der Pharmazie, II: von der frühen Neuzeit bis zur Gegenwart.* Eschborn: Govi, 2005.

Gabriel, Joseph M. *Medical Monopoly: Intellectual Property Rights and the Origins of the Modern Pharmaceutical Industry.* Chicago: University of Chicago Press, 2014.

Gagnon, François-Marc. "Louis Nicolas's Depiction of the New World in Figures and Text." In *The Codex Canadensis and the Writings of Louis Nicolas,* edited by François-Marc Gagnon, 3–92. Montreal, QC: McGill-Queen's University Press, 2011.

Galvin, Mary. "Decoctions for Carolinians." In *Creolization in the Americas,* edited by David Buisseret and Steven G. Reinhardt, 76–77. College Station, TX: Texas A&M University Consortium Press, 2000.

Games, Alison. "Atlantic History: Definitions, Challenges, and Opportunities." *American Historical Review* 111 (2006): 741–57.

Garcia Ballester, Luis. *Galeno en la sociedad y en la ciencia de su tiempo.* Madrid: Ediciones Guadarrama, 1972.

Garcia Ballester, Luis, and Jon Arrizbalaga, et al. *Galen and Galenism: Theory and Medical Practice from Antiquity to the European Renaissance.* Aldershot: Ashgate, 2002.

Gentilcore, David. "Apothecaries, 'Charlatans,' and the Medical Marketplace in Italy, 1400–1750." *Pharmacy in History* 45 (2003): 91–94.

Gentilcore, David. *Healers and Healing in Early Modern Italy.* Manchester: Manchester University Press, 1998.

Gentilcore, David. *Medical Charlatanism in Early Modern Italy.* Oxford: Oxford University Press, 2006.

Gerbi, Antonello. *The Dispute of the New World: From Christopher Columbus to Gonzalo Fernandez de Oviedo.* Translated by Jeremy Moyle. Pittsburgh: University of Pittsburgh Press, 1985.

Geyer-Kordesch, J., F. and MacDonald. *Physicians and Surgeons in Glasgow: The History of the Royal College of Physicians and Surgeons of Glasgow, 1599–1858.* London: The Hambledon Press, 1999.

Gherini, Claire. "Valuing Caesar's and Sampsons' Cures." *Recipes Project.* August 18, 2015. Accessed November 11, 2018. http://recipes.hypotheses.org/6419.

Gibbs, Frederick W. "Medical Understandings of Poison circa 1250–1600." PhD diss., University of Wisconsin–Madison, 2009.

Gibbs, Frederick W. "Specific Form and Poisonous Properties: Understanding Poison in the Fifteenth Century." *Preternature: Critical and Historical Studies on the Preternatural* 2 (2016): 19–46.

Glover, Jeffrey. "Early American Archives and the Evidence of History." *Early American Literature* 46, no. 1 (2011): 165–84.

Gómez, Pablo F. *The Experiential Caribbean: Creating Knowledge and Healing in the Early Modern Atlantic*. Chapel Hill: University North Carolina Press, 2017.

González Bueno, Antonio. "An Account [of] the History of the Spanish Pharmacopoeias." History of Pharmacopoeias Working Group, International Society for the History of Pharmacy. Accessed May 30, 2017. http://www.histpharm.org/ISHPWG%20Spain.pdf.

Gordon, R. K. *Anglo-Saxon Poetry*. London: Dent, 1962.

Graeber, David. "Fetishism as Social Creativity: Or, Fetishes Are Gods in the Process of Construction." *Anthropological Theory* 5 (2005): 407–38.

Graeber, David. "Radical Alterity Is Just Another Way of Saying 'Reality': A Reply to Eduardo Viveiros de Castro." *HAU: Journal of Ethnographic Theory* 5 (2015): 1–41.

Greene, Jeremy A. *Prescribing by the Numbers: Drugs and the Definition of Disease*. Baltimore: Johns Hopkins University Press, 2008.

Greer, Allan. "Commons and Enclosure in the Colonization of North America." *American Historical Review* 117, no. 2 (2012): 365–86.

Grier, J. *A History of Pharmacy*. London: Pharmaceutical Press, 1937.

Grivetti, Louis E. "Medicinal Chocolate in New Spain, Western Europe, and North America." In *Chocolate: History, Culture and Heritage*, edited by Louis Grivetti and Howard Shapiro. Malden: Wiley, 2009.

Groebner, Valentin. *Who Are You? Identification, Deception, and Surveillance in Early Modern Europe*. New York: Zone Books, 2007.

Gutas, Dimitri. *Greek Thought, Arabic Culture: The Graeco-Arabic Translation Movement in Baghdad and Early 'Abbasaid Society (2nd-4th/5th-10th c.)*. London: Routledge, 1998.

Hall, David D. *Worlds of Wonder, Days of Judgment: Popular Religious Belief in Early New England*. Cambridge, MA: Harvard University Press, 1990.

Hallowell, A. Irving. "Ojibwa Ontology, Behavior, and World View." In *Culture in History: Essays in Honor of Paul Radin*, edited by Stanley Diamond, 19–52. New York: G. P. Putnam's Sons, 1960.

Hamarneh, Sami. "The Rise of Professional Pharmacy in Islam." *Medical History* 6 (1962): 59–66.

Hamarneh, Sami, and Glenn Sonnedeker. *A Pharmaceutical View of Abulcasis al-Zahrawi in Moorish Spain*. Leiden: Brill, 1963.

Handler, Jerome. "Slave Medicine and Obeah in Barbados, circa 1650 to 1834." *Nieuwe West-Indische Gids–New West Indian Guide* 74 (2000): 57–90.

Havard, Gilles. *Empire et métissages. Indiens et Français dans le Pays d'en Haut, 1660–1715.* Sillery: Septentrion, 2003.

Havard, Gilles. "Les forcer à devenir Cytoyens." *Annales. Histoire, Sciences Sociales* 64, no. 5 (2009): 985–1018.

Havard, Gilles, and Cécile Vidal. *Histoire de l'Amérique française.* Paris: Flammarion, 2004.

Henare, Amiria, Martin Holbraad, and Sari Wastell, eds. *Thinking through Things: Theorising Artefacts Ethnographically.* London: Routledge, 2007.

Hett, F. P., ed. *The Memoirs of Sir Robert Sibbald 1641–1722.* London: Oxford University Press, 1932.

Higby, Gregory J. *In Service to American Pharmacy: The Professional Life of William Procter Jr.* Huntsville: University of Alabama Press, 1991.

Hoeniger, F., and J. Hoeniger. *The Development of Natural History in Tudor England.* Charlottesville: University of Virginia Press, 1973.

Howie, W. B. "Sir Archibald Stevenson, His Ancestry and the Riot in the College of Physicians at Edinburgh." *Medical History* 11, no. 3 (1967): 269–84.

Huguet-Termes, Teresa. "Islamic Pharmacology and Pharmacy in the Latin West: An Approach to Early Pharmacopoeias." *European Review* 16 (2008): 229–39.

Huguet-Termes, Teresa. "Standardising Drug Therapy in Renaissance Europe? The Florence (1499) and Nuremberg Pharmacopoeia (1546)." *Medicina & Storia—Saggi* VIII 15 (2008): 77–101.

Hunt, Nancy. *Suturing New Medical Histories of Africa.* Munster: LIT Verlag, 2013.

Hunter, Michael, Alison Walker, and Arthur MacGregor, eds. *From Books to Bezoars: Sir Hans Sloane and His Collections.* London: British Library, 2013.

Hunting, P. *A History of the Society of Apothecaries.* London: Society of Apothecaries, 1998.

Hutchison, Keith. "What Happened to Occult Qualities in the Scientific Revolution?" In *The Scientific Enterprise in Early Modern Europe: Readings from Isis,* edited by Peter Dear, 86–106. Chicago: University of Chicago Press, 1997.

Iannini, Christopher. *Fatal Revolutions: Natural History, West Indian Slavery, and the Routes of American Literature.* Chapel Hill: University of North Carolina Press, 2012.

Impey, Olver, and Arthur MacGregor, eds. *The Origins of Museums: The Cabinet of Curiosities in Sixteenth- and Seventeenth-Century Europe.* Oxford: Clarendon, 1985.

Jacob, Christian, and Edward H. Dahl. *The Sovereign Map: Theoretical Approaches in Cartography throughout History.* Chicago: University of Chicago Press, 2006.

Jacquart, Danielle. "Islamic Pharmacology in the Middle Ages: Theories and Substances," *European Review* 16 (2008): 219–27.

Jacquart, Danielle. *La medicine médiévale dans le cadre parisien: XIV–XV siècle.* Paris: Fayard, 1998.

Jacques, David. *Essential to the Pracktick Part of Physic: The London Apothecaries 1540–1617.* London: Society of Apothecaries, 1992.

Janni, Kevin D., and Joseph W. Bastien. "Exotic Botanicals in the Kallawaya Pharmacopoeia." *Economic Botany* 58 (2004): S274–S279.

Jarcho, Saul. *Quinine's Predecessor: Francesco Torti and the Early History of Cinchona.* Baltimore: Johns Hopkins University Press, 1993.

Jardine, N., J. A. Secord, and E. C. Spary, eds. *Cultures of Natural History.* Cambridge: Cambridge University Press, 1996.

Johns, Adrian. *Piracy: The Intellectual Property Wars from Gutenberg to Gates.* Chicago: University of Chicago Press, 2009.

Jones, Colin. "The Great Chain of Buying: Medical Advertisement, the Bourgeois Public Sphere, and the Origins of the French Revolution." *American Historical Review* 101, no. 1 (2009): 13–40.

Kafka, Ben. *The Demon of Writing: Powers and Failures of Paperwork.* New York: Zone Books, 2012.

Kananoja, Kalle. "Bioprospecting and European Uses of African Natural Medicine in Early Modern Angola." *Portuguese Studies Review* 23, no. 2 (2015): 45–69.

Kelton, Paul. *Cherokee Medicine, Colonial Germs: An Indigenous Nation's Fight against Smallpox, 1518–1824.* Norman: University of Oklahoma Press, 2015.

Kenny, K. *Ireland and the British Empire.* Oxford: Oxford University Press, 2004.

Kirkpatrick, T.P. *The Dublin Pharmacopoeias.* Dublin: Royal College of Physicians of Ireland, 1921.

Kremers, Edward, and George Urdang. *History of Pharmacy.* 4th ed. Revised by Glenn Sonnedecker. Philadelphia: Lippincott, 1976.

Kriz, Kay Dian. "Curiosities, Commodities, and Transplanted Bodies in Hans Sloane's *Voyage to . . . Jamaica*." In *An Economy of Color: Visual Culture and the North Atlantic World, 1660–1830,* edited by Geoff Quilley and Kay Dian Kriz, 85–105. Manchester: Manchester University Press, 2003.

Kucich, John J. "Sons of the Forest: Environment and Transculturation in Jonathan Edwards, Samson Occom and William Apess." In *Assimilation and Subversion in Earlier American Literature,* edited by Robin de Rosa, 5–23. Newcastle: Cambridge Scholars Press, 2006.

Kuethe, Allan J., and Kenneth J. Andrien. *The Spanish Atlantic World in the Eighteenth Century: War and the Bourbon Reforms 1713–1796.* Cambridge: Cambridge University Press, 2014.

Kupperman, Karen Ordahl. *Indians & English: Facing Off in Early America*. Ithaca: Cornell University Press, 2000.

Kury, Lorelai. "Plantas sem fronteiras: jardins, livros e viagens, séculos XVIII–XIX." In *Usos e circulaçao de plantas no Brasil; Séculos XVI–XIX*, edited by Lorelai Kury. Rio de Janeiro: Andrea Jakobsson Estúdio Editorial, 2013.

Kusukawa, Sachiko. *Picturing the Book of Nature: Image, Text, and Argument in Sixteenth-Century Human Anatomy and Medical Botany*. Chicago: University of Chicago Press, 2012.

Lafont, Olivier. *Échevins et apothicaires sous Louis XIV: la vie de Matthieu-François Geoffroy, bourgeois de Paris*. Paris: Pharmathèmes, 2008.

Lamontagne, Roland. "L'influence de Maurepas sur les sciences: Le botaniste Jean Prat à la Nouvelle-Orléans, 1735–1746." *Revue d'histoire des sciences* 49, no. 1 (1996): 113–24.

Lanning, John Tate. *The Royal Protomedicato: The Regulation of the Medical Professions in the Spanish Empire*. Edited by John TePaske. Durham: Duke University Press, 1985.

Lathem, Edward Connery. *Ten Indian Remedies: From Manuscript Notes on Herbs and Roots*. N.p.: n.p., 1954.

Latour, Bruno. *On the Modern Cult of the Factish Gods*. Durham: Duke University Press, 2011.

Le Paulmier, Claude-Stéphen. *L'Orviétan. Histoire d'une famille de charlatans du Pont-Neuf aux XVIIe et XVIIIe siècles*. Paris: Librairie illustrée, 1893.

Leiser, Gary, and Michael Dols. "Evliya Chelebi's Description of Medicine in Seventeenth-Century Egypt," *Sudhoffs Archiv* 71 (1987): 197–216.

Leiser, Gary, and Michael Dols. "Evliya Chelebi's Description of Medicine in Seventeenth-Century Egypt," *Sudhoffs Archiv* 72 (1988): 49–68.

Leong, Elaine. "Collecting Knowledge for the Family: Recipes, Gender and Practical Knowledge in the Early Modern English Household." *Centaurus* 55 (2013): 81–103.

Leong, Elaine. "Making Medicines in the Early Modern Household." *Bulletin of the History of Medicine* 82, no. 1 (2008): 145–68.

Leong, Elaine, and Sara Pennell. "Recipe Collections and the Currency of Medical Knowledge in the Early Modern 'Medical Marketplace," In *Medicine and the Market in England and Its Colonies, c. 1450–1850*, edited by Mark S. R. Jenner and Patrick Wallis, 133–52. New York: Palgrave Macmillan, 2007.

Lessard, Rénald. "Aux XVIIe et XVIIIe siècles: L'exportation de plantes médicinales canadiennes en Europe." *Cap-aux-Diamants*, no. 46 (1996): 20–24.

Lessard, Rénald. "Pratique et praticiens en contexte colonial: le corps médical canadien aux XVIIe et XVIIIe siècles." PhD diss., Université Laval, 1994.

Levey, Martin. *Early Arabic Pharmacology*. Leiden: Brill, 1973.
Levey, Martin. "Ibn Māsawaih and His Treatise on Simple Aromatic Substances." *Journal of the History of Medicine and Allied Sciences* 16 (1961): 394–410.
Lewis, Andrew J. "Gathering for the Republic: Botany in Early Republic America." In *Colonial Botany: Science, Commerce, and Politics in the Early Modern World*, edited by Londa Schiebinger and Claudia Swan, 66–80. Philadelphia: University of Pennsylvania Press, 2005.
Lima, Nisia Trindade, ed. *A Ciência dos Viajantes: natureza, populaces, e saúde em 500 anos de interpretações do Brasil*. Rio de Janeiro: Fundação Oswaldo Cruz, 2000.
Litalien, Raymonde, Denis Vaugeois, and Jean-François Palomino. *La Mesure d'un continent: Atlas historique de l'Amérique du Nord, 1492–1814*. Sillery, QC: Septentrion, 2007.
Lopenzina, Drew. *Red Ink: Native Americans Picking Up the Pen in the Colonial Period*. Albany: State University of New York Press, 2012.
Lorène, Simon. "Intérêt pharmaceutique des lettres adressées à l'apothicaire dieppois Féret par les religieuses de l'Hôtel-Dieu de Québec." PhD diss., Université de Rouen, 2014.
Loux, Françoise. *Pierre-Martin de La Martinière : un médecin au XVIIe siècle*. Paris: Imago, 1988.
Lozier, Jean-François. "In Each Other's Arms: France and the St. Lawrence Mission Villages in War and Peace, 1630–1730." PhD diss., University of Toronto, 2012.
Lu, Jin-Mei, et al. "Biogeographic Disjunction between Eastern Asia and North America in the Adiantum Pedatum Complex (Pteridaceae)." *American Journal of Botany* 98, no. 10 (2011): 1680–93.
Maclean, Ian. *Logic, Signs and Nature in the Renaissance: The Case of Learned Medicine*. Cambridge: Cambridge University Press, 2002.
Maehle, Andreas-Holger. *Drugs on Trial: Experimental Pharmacology and Therapeutic Innovation in the Eighteenth Century*. Amsterdam: Rodopi, 1999.
Magliocco, Concetta. "Antonio di Guccio della Scarperia." *Dizionario Biografico degli Italiani*, vol. 3 (1961) Trecani, La Cultura Italiana. Accessed August 17, 2018. http://www.treccani.it/enciclopedia/antonio-di-guccio-della-scarperia_(Dizionario-Biografico)/.
Makin, Martti. "Between Herbals et alia: Intertextuality in Medieval English Herbals." PhD diss., University of Helsinki, 2006.
Mapp, Paul W. *The Elusive West and the Contest for Empire, 1713–1763*. Chapel Hill: University of North Carolina Press, 2011.
Marcon, Federico. *The Nature of Knowledge and the Knowledge of Nature in Early Modern Japan*. Chicago: University of Chicago Press, 2017.

Margócsy, Dániel. *Commercial Visions: Science, Trade and Visual Culture in the Dutch Golden Age*. Chicago: University of Chicago Press, 2014.

McCarl, Mary Rhinelander. "Publishing the Works of Nicholas Culpeper, Astrological Herbalist and Translator of Latin Medical Works in Seventeenth-Century London." *Canadian Bulletin of Medical History* 13 (1996): 225–76.

McCrae, M. *Physicians and Society: A Social History of the Royal College of Physicians of Edinburgh*. Edinburgh: John Donald, 2007.

McHarg, J. F. "Dr John Maklure and the 1630 Attempt to Establish the College." *Proceedings of the Royal College of Physicians of Edinburgh Tercentenary Congress* (1981): 49.

McVaugh, Michael. "Practical Pharmacy and Medical Theory at the End of the Thirteenth Century." Unpublished paper.

McVaugh, Michael. "Theriac at Montpellier 1285–1325." *Sudhoffs Archiv* 56 (1972): 113–44.

Michaud, C. "Un anti-jesuite au service de Pombal: l'abbé Platel." In *Pombal Revisitado*, edited by M. H. Carvalho dos Santos. Lisbon: Editorial Estampa, 1984.

Miller, Joseph Calder. *Way of Death: Merchant Capitalism and the Angolan Slave Trade, 1730–1830*. Madison: University of Wisconsin Press, 1997.

Milne, G. W. A., ed. *Drugs: Synonyms & Properties*. Burlington, VT: Ashgate, 1999.

Ministério da Agricultura, Industria e Commércio; Serviço de Informações. "Producção, Commércio e Consumo de Cacáo." Rio de Janeiro: Imprensa Nacional, 1924.

Moerman, Daniel E. *Native American Ethnobotany*. Portland, OR: Timber Press, 1998.

Monk, W. *Roll of the College of Physicians*. Vol. 1. 2nd ed. London: Royal College of Physicians, 1878.

Mooney, James. *The Swimmer Manuscript: Cherokee Sacred Formulas and Medicinal Prescriptions*. Washington, DC: US Government Printing Office, 1932.

Morgan, K. O. *The Oxford History of Britain. Volume 3: The Tudors and Stuarts*. Oxford: Oxford University Press, 1989.

Morgan, Philip D. *Slave Counterpoint: Black Culture in the Eighteenth-Century Chesapeake and Lowcountry*. Chapel Hill: University of North Carolina Press, 1998.

Morgenstern, Sali. "The Ricettario Fiorentino. 1567. The Origin of the Art of the Apothecary and the Florentine Pharmacopoeias." *Academy Bookman* 30 (1977): 3–12.

Morrison, Kenneth M. "The Cosmos as Intersubjective: Native American Other-Than-Human Persons." In *Indigenous Religions: A Companion*, ed. Graham Harvey, 23–36. London: Cassell, 2000.

Motsch, Andreas. "Le ginseng d'Amérique: Un lien entre les deux Indes, entre curiosité et science." *Etudes Epistémè: Revue de littérature et de civilisation (XVIe–XVIIIe siècles)* 26 (2014). Accessed August 17, 2018. http://dx.doi.org/10.4000/episteme.331.

Müller-Jahncke, Wolf-Dieter. "Platon im Arzneibuch und der Heller am Tresen: Pharmazie im 16. Jahrhundert zwischen Humanismus, Stadtgesellschaft und Ökonomie." *Berichte zur Wissenschaftsgeschichte* 23 (2000): 1–15.

Mukerji, Chandra. *Impossible Engineering: Technology and Territoriality on the Canal du Midi.* Princeton: Princeton University Press, 2009.

Nagel, Thomas. *The View from Nowhere.* New York: Oxford University Press, 1986.

Nandy, Ashis. "History's Forgotten Doubles." *History and Theory* 34 (1995): 44–66.

Newson, Linda. *Making Medicines in Early Colonial Lima, Peru: Apothecaries, Science and Society.* Leiden: Brill, 2017.

Noiriel, Gérard. *État, nation et immigration. Vers une histoire du pouvoir.* Paris: Belin, 2001.

Noiriel, Gérard. "L'identification des Citoyens. Naissance de l'état civil républicain." *Genèses. Sciences Sociales et Histoire* 13 (1993): 3–28.

Noiriel, Gérard. "Surveiller les déplacements ou identifier les personnes? Contribution à l'histoire du passeport en France de la Ire à la IIIe République." *Genèses. Sciences Sociales et Histoire* 30 (1998): 77–100.

Norton, Marcy. *Sacred Gifts, Divine Pleasures: A History of Tobacco and Chocolate in the Atlantic World.* Ithaca: Cornell University Press, 2004.

Norton, Marcy. "Subaltern Technologies and Early Modernity in the Atlantic World." *Colonial Latin American Review* 26 (2017): 18–38.

Nutton, Vivian. "Ancient Mediterranean Pharmacology and Cultural Transfer." *European Review* 16 (2008): 211–17.

Nutton, Vivian. "Matthioli and the Art of the Commentary." In *La complessa Scienza dei Semplici. Atti delle Celebrazioni per il V Centenario della Nascita di Pietro Andrea Mattioli, Atti dell' Accademia delle Scienze di Siena detta De' Fisiocritici,* Series 15, Tome 20, Supplement 2001, edited by D. Fausti, 133–48. Siena, Accademia dei Fisiocritici Onlus, 2004.

Nutton, Vivian. "Scribonius Largus, The Unknown Pharmacologist." *Pharmaceutical Historian* 25 (1995): 5–8.

Nutton, Vivian, and John Scarborough, "The Preface of Dioscorides' Materia Medica: Introduction, Translation and Commentary," *Transactions and Studies of the College of Physicians of Philadelphia* IV (1982): 187–227.

Ong, Walter. *Orality and Literacy: The Technologizing of the Word.* New York: Methuen, 1988.

Opsomer, Carmélia. "La Pharmacopée Medieval: Images et Manuscrits." *Journal de Pharmacie de Belgique* 57 (2002): 3–10.

Osseo-Asare, Abena Dove. *Bitter Roots: The Search for Healing Plants in Africa*. Chicago: University of Chicago Press, 2014.

Palomino, Jean-François. "Pratiques cartographiques en Nouvelle-France: La prise en charge de l'État dans la description de son espace colonial à l'orée du XVIIIe siècle." *Lumen* 31 (2012): 21–39.

Paquette, Gabriel. *Enlightenment, Governance and Reform in Spain and Its Empire, 1759–1808*. New York: Palgrave Macmillan, 2008.

Park, Katharine. *Doctors and Medicine in Early Renaissance Florence*. Princeton: Princeton University Press, 1985.

Parrish, Susan Scott. *American Curiosity: Cultures of Natural History in the Colonial British Atlantic World*. Chapel Hill: University of North Carolina Press, 2006.

Parsons, Christopher M. "The Natural History of Colonial Science: Joseph-François Lafitau's Discovery of Ginseng and Its Afterlives." *William and Mary Quarterly* 73, no. 1 (2016): 37–72.

Pauly, Philip. *Biologists and the Promise of American Life: From Meriwether Lewis to Alfred Kinsey*. Princeton: Princeton University Press, 2002.

Peterson, Kristin. *Speculative Markets: Drug Circuits and Derivative Life in Nigeria*. Durham: Duke University Press, 2014.

Pickstone, J. V. *Ways of Knowing: A New History of Science, Technology and Medicine*. Chicago: Chicago University Press, 2000.

Pinto, António José de Sousa. *Materia Medica*. Ouro Preto, Brazil: Typografia de Silva, 1837.

Planchon, Gustave. "Notes sur l'histoire de l'Orviétan." *Journal de Pharmacie et de Chimie* 26, nos. 3–7 (1892): 97–103, 145–52, 193–98, 241–50, and 289–98.

Pollio, A., A. De Natale, E. Appetiti, G. Aliotta, and A. Touwaide. "Continuity and Change in the Mediterranean Medical Tradition: *Ruta* spp. (rutaceae) in Hippocratic Medicine and Present Practices." *Journal of Ethnopharmacology* 116 (2008): 469–82.

Pomata, Gianna. *Contracting a Cure: Patients, Healers and the Law in Early Modern Bologna*. Baltimore: Johns Hopkins University Press, 1998.

Pomata, Gianna. "Epistemic Genres or Styles of Thinking? Tools for the Cultural Histories of Knowledge," Lecture, University of Minnesota, May 10, 2013.

Pomata, Gianna. "The Medical Case Narrative: Distant Reading of an Epistemic Genre." *Literature and Medicine* 32, no. 1 (2014): 1–23.

Pomata, Gianna. "Observation Rising: Birth of an Epistemic Genre, 1500–1650." In *Histories of Scientific Observation*, edited by Lorraine Daston and Elizabeth Lunbeck, 45–80. Chicago: University of Chicago Press, 2011.

Pomata, Gianna. "Sharing Cases: The *Observationes* in Early Modern Medicine." *Early Science and Medicine* 15 (2010): 193–236.

Pringle, James S. "How 'Canadian' is Cornut's *Canadensium Plantarum Historia*?

A Phytogeographic and Historic Analysis." *Canadian Horticultural History: An Interdisciplinary Journal* 1, no. 4 (1988): 190–209.

Pritchard, James S. "Early French Hydrographic Surveys in the Saint Lawrence River." *International Hydrographic Review* 56, no. 1 (1979): 125–42.

Rabito-Wyppensenwah, Philip, and Robert Abiuso. "The Montaukett Use of Herbs: A Review of the Recorded Material." In *The History and Archaeology of the Montauk*, vol. 3 of *Readings in Long Island Archeology and Ethnohistory*, 2nd ed., edited by Gaynell Stone, 585–88. Stony Brook: Suffolk County Archeological Association, 1993.

Ramsey, Matthew. "Academic Medicine and Medical Industrialism: The Regulation of Secret Remedies in Nineteenth-Century France." In *French Medical Culture in the Nineteenth Century*, edited by Ann La Berge and Mordechai Feingold, 25–78. Amsterdam: Rodopi, 1994.

Ramsey, Matthew. *Professional and Popular Medicine in France, 1770–1830: The Social World of Medical Practice.* Cambridge: Cambridge University Press, 1988.

Ramsey, Matthew. "Property Rights and the Right to Health: the Regulation of Secret Remedies in France, 1789–1815." In *Medical Fringe and Medical Orthodoxy, 1750–1850*, edited by William F. Bynum and Roy Porter, 79–105. London: Croom Helm, 1987.

Ramsey, Matthew. "Traditional Medicine and Medical Enlightenment: The Regulation of Secret Remedies in the Ancien Régime." *Historical Reflections/Réflexions historiques* 9, nos. 1–2 (1982): 215–32.

Rankin, Alisha. "On Anecdote and Antidotes: Poison Trials in Sixteenth-Century Europe." *Bulletin of the History of Medicine* 91 (2017): 274–342.

Rankin, Alisha. *Panaceia's Daughters: Noblewomen as Healers in Early Modern Germany.* Chicago: University of Chicago Press, 2013.

Raubenheimer, Otto. "History of Substitutes and Substitution." *Journal of the American Pharmaceutical Association* 6 (1917): 50–55.

Ribeiro, Jorge Manuel Martins. "Comércio e Diplomacia nas Relações Luso-Americanas (1776–1822)." 2 vols. PhD diss., Universidade do Porto, Portugal, 1997.

Riddle, John. "The Introduction and Use of Eastern Drug." *Sudhoffs Archiv* 49 (1965): 185–98.

Riva, Ernesto. "Confronto tra le prime due edizioni del Ricettario Fiorentino e l'Antidotario mantovano del 1558." *Atti e Memorie* 8 (1991): 99–104.

Rivest, Justin. "Secret Remedies and the Rise of Pharmaceutical Monopolies in France during the First Global Age." PhD diss., Johns Hopkins University, 2016.

Rivest, Justin. "Testing Drugs and Attesting Cures: Pharmaceutical Monopolies and Military Contracts in Eighteenth-Century France." *Bulletin of the History of Medicine* 91, no. 2 (2017): 362–90.

Robinson, Martha. "New Worlds, New Medicines: Indian Remedies and English Medicine in Early America." *Early American Studies* 3 (Spring 2005): 94–110.

Round, Phillip H. *Removable Type: Histories of the Book in Indian Country, 1663–1880*. Chapel Hill: University of North Carolina Press, 2010.

Rousseau, François. *L'Oeuvre de chère en Nouvelle-France: Le régime des malades à l'Hôtel-Dieu de Québec*. Sainte-Foy, QC: Presses de l'Université Laval, 1983.

Safier, Neil. "Global Knowledge on the Move: Itineraries, Amerindian Narratives, and Deep Histories of Science." *Isis* 101 (2010): 133–45.

Safier, Neil. *Measuring the New World: Enlightenment Science and South America*. Chicago: University of Chicago Press, 2008.

Saller, Jacques. *La pharmacopée française, dans l'évolution scientifique, technique et professionnelle*. Metz: Maisonnueve, 1969.

Santos Filho, Licurgo de Castro. *História da Medicina no Brazil, do Século XVI ao Século XIX*. 2 vols. São Paulo: Editora Brasiliense, 1947.

Schalick, Walton O. "Add One Part Pharmacy to One Part Surgery and One Part Medicine: Jean De St. Amand and the Development of Medical Pharmacology in Thirteenth Century Paris." PhD diss, Johns Hopkins University, 1997.

Schepelern, H. D. "The Museum Wormianum Reconstructed: A Note on the Illustration of 1655." *Journal of the History of Collections* 2 (1990): 81–85.

Schiebinger, Londa, ed. "Focus: Colonial Science." *Isis* 96 (2005): 52–63.

Schiebinger, Londa. *Plants and Empire: Colonial Bioprospecting in the Atlantic World*. Cambridge, MA: Harvard University Press, 2004.

Schiebinger, Londa. *Secret Cures of Slaves: People, Plants, and Medicine in the Eighteenth-Century Atlantic World*. Palo Alto: Stanford University Press, 2017.

Schiebinger, Londa, and Claudia Swan. *Colonial Botany: Science, Commerce, and Politics in the Early Modern World*. Philadelphia: University of Pennsylvania Press, 2005.

Schwartz, Stuart B. *Sugar Plantations in the Formation of Brazilian Society*. Cambridge: Cambridge University Press, 1985.

Scott, James C. *Seeing Like a State: How Certain Schemes to Improve the Human Condition Have Failed*. New Haven, CT: Yale University Press, 1998.

Sebastian, Anton. "American Medical Journals." In *A Dictionary of the History of Medicine*, 39. Pearl River, NY: Pantheon, 1999.

Serrão, Joaquim Verríssimo. *História de Portugal*. 10 vols. Lisbon: Editora Verbo, 1996.

Shackelford, Jole. "Documenting the Factual and the Artifactual: Ole Worm and Public Knowledge." *Endeavour* 23 (1999): 65–71.

Shaw, James, and Evelyn Welch. *Making and Marketing Medicine in Renaissance Florence*. Amsterdam: Brill, 2011.

Silva, Cristobal. *Miraculous Plagues: An Epidemiology of Early New England Narrative*. New York: Oxford University Press, 2011.

Silva Dias, José Sebastião. "Portugal e a Cultura Europeia: Séculos XVI a XVIII." In *Biblos* XXVIII. Coimbra: Universidade de Coimbra, 1952.

Simon, Jonathan. *Chemistry, Pharmacy and Revolution in France, 1777–1809*. Aldershot: Ashgate, 2005.

Simon, William J. *Scientific Expeditions in the Portuguese Overseas Territories (1783–1808)*. Lisbon: Instituto de Investigação Científica Tropical, 1983.

Simmons, Donald C. "Efik Divination, Ordeals, and Omens." *Southwestern Journal of Anthropology* 12, no. 2 (July 1, 1956): 223–28.

Simmons, William S. *Spirit of the New England Tribes: Indian History and Folklore, 1620–1984*. Hanover: University Press of New England, 1986.

Singer, Charles. "The Herbal in Antiquity and Its Transmission to Later Ages." *Journal of Hellenic Studies* 47 (1927): 1–52.

Siraisi, Nancy. *Medieval and Early Renaissance Medicine: An Introduction to Knowledge and Practice*. Chicago: University of Chicago Press, 1990.

Sivasundaram, Sujit, ed. "Focus: Global Histories of Science." *Isis* 101 (2010): 95–158.

Skidmore, Thomas E. *Brazil: Five Centuries of Change*. Oxford: Oxford University Press, 1999.

Slattery, Brian. "French Claims in North America, 1500–59." *Canadian Historical Review* 59, no. 2 (1978): 139–69.

Slotkin, Richard. *Regeneration through Violence: The Mythology of the American Frontier, 1600–1860*. Norman: University of Oklahoma Press, 2000.

Smith, Pamela. *The Body of the Artisan: Art and Experience in the Scientific Revolution*. Chicago: University of Chicago Press, 2004.

Smith, Pamela. "Science on the Move: Recent Trends in the History of Early Modern Science." *Renaissance Quarterly* 62 (2009): 345–75.

Soll, Jacob. *The Information Master: Jean-Baptiste Colbert's Secret State Intelligence System*. Ann Arbor: University of Michigan Press, 2009.

Sonnedecker, Glenn. *Kremers and Urdang's History of Pharmacy*. 4th ed. Madison: American Institute of the History of Pharmacy, 1976.

Sonnedecker, Glenn. "The Founding Period of the US Pharmacopeia: I. European Antecedents." *Pharmacy in History* 35, no. 4 (1993): 151–62.

Sonnedecker, Glenn. "The Founding Period of the U.S. Pharmacopeia II. A National Movement Emerges." *Pharmacy in History* 36, no. 1 (1994): 3–25.

Sonnedecker, Glenn. "The Founding Period of the U.S. Pharmacopeia III. The First Edition." *Pharmacy in History* 36 no. 3 (1994): 103–21.

BIBLIOGRAPHY

Soto Laveaga, Gabriela. *Jungle Laboratories: Mexican Peasants, National Projects and the Making of the Pill*. Durham: Duke University Press, 2009.

Sousa Dias, José Pedro. "Documentos sobre duas boticas da Companhia de Jesus em Lisboa: Colégio de Santo Antão e Casa Professa de S. Roque." In *Economia e Sociologia* 88–89 (2009): 295–312.

Spary, Emma C. *Translations of Potency: Taking Drugs in the Sun King's Reign*. Forthcoming.

Speck, Frank G. "Medicine Practices of the Northeastern Algonquians." *Proceedings of the 19th International Congress of Americanists*, Washington, DC, 1917, 303–21.

Staley, Edgecumbe. *The Guilds of Florence*. New York: B. Blom, 1906.

Stephenson, M. "From Marvelous Antidote to the Poison of Idolatry: The Transatlantic Role of Andean Bezoar Stones during the Late Sixteenth and Early Seventeenth Centuries." *Hispanic American Historical Review* 90, no. 1 (2010): 3–39.

Stein, Stanley J., and Barbara H. Stein. *Apogee of Empire: Spain and New Spain in the Age of Charles III, 1759–1789*. Baltimore: Johns Hopkins University Press, 2005.

Stevens, Laura M. *The Poor Indians: British Missionaries, Native Americans, and Colonial Sensibility*. Philadelphia: University of Pennsylvania Press, 2004.

Styles, John. "Product Innovation in Early Modern London." *Past & Present* 168, no. 1 (2000): 124–69.

Sweet, James H. *Domingos Álvares, African Healing, and the Intellectual History of the Atlantic World*. Chapel Hill: University of North Carolina Press, 2011.

Taavitsainen, Irma, and Paivi Pahta. "Corpus of Early English Medical Writing, 1350-1750." *ICAME Journal: Computers in English Linguistics* 21 (1997): 71–78.

Taavitsainen, Irma, and Paivi Pahta. "Vernacularization of Scientific and Medical Writing in Sociohistorical Context." In *Medical and Scientific Writing in Late Medieval English*, edited by Irma Taavitsainen and Paivi Pahta, 1–18. Cambridge: Cambridge University Press, 2004.

Tantaquidgeon, Gladys. *Folk Medicine of the Delaware and Related Algonkian Indians*. Harrisburg: Pennsylvania Historical and Museum Commission, 2001.

Tantaquidgeon, Gladys. "Mohegan Medicinal Practices, Weather-Lore and Superstition." In *Annual Report of the Bureau of American Ethnology to the Secretary of the Smithsonian Institution, 1925–1926*, 246–76. Washington DC: 1926.

Tantaquidgeon, Gladys. "Notes on the Gay Head Indians of Massachusetts." *Indian Notes* 7, no. 1 (January 1930): 1–26.

Tantaquidgeon, Gladys, and Jayne G. Fawcett. "Symbolic Motifs on Painted Baskets on the Mohegan-Pequot." In *A Key into the Language of Woodsplint Baskets*, edited by Ann McMullen and Russell G. Handsman, 94–102. Washington, CT: American Indian Archaeological Institute, 1987.

Taussig, Michael. *Defacement: Public Secrecy and the Labor of the Negative*. Stanford: Stanford University Press, 1999.

Temkin, Oswei. "Byzantine Medicine: Tradition and Empiricism." *Dumbarton Oaks Papers* 16 (1962): 97–115.

Temkin, Oswei. *The Double Face of Janus and Other Essays in The History of Medicine.* Baltimore: Johns Hopkins University Press, 2006.

Temkin, Oswei. *Galenism: Rise and Decline of a Medical Philosophy.* Ithaca: Cornell University Press, 1973.

Temkin, Oswei. *'On Second Thought' and Other Essays in the History of Medicine and Science.* Baltimore: Johns Hopkins University Press, 2002.

Tésio, Stéphanie. "De La Croix-Avranchin à Québec, Jean-François Gaultier, médecin du roi, de 1742 à 1756." *Annales de Normandie* 55, no. 5 (2005): 403–26.

Tésio, Stéphanie. *Histoire de la pharmacie en France et en Nouvelle-France.* Sainte-Foy, QC: Les Presses de l'Université Laval, 2009.

Thomas, Nicholas. *Entangled Objects: Exchange, Material Culture and Colonialism in the Pacific.* Cambridge, MA: Harvard University Press, 1991.

Tobyn, Graeme, Alison Denham, and Margaret Whitelegg. *The Western Herbal Tradition: 2000 Years of Medicinal Plant Knowledge.* London: Elsevier Health Sciences, 2010.

Totelin, Laurence. "And to end on a poetic note: Galen's Authorial Strategies in the Pharmacological Books." *Studies in History and Philosophy of Science* 43 (2012): 307–15.

Trouillot, Michel-Rolph. *Silencing the Past: Power and the Production of History.* Boston: Beacon Press, 2015.

Turnbull, David. "Cartography and Science in Early Modern Europe: Mapping the Construction of Knowledge Spaces." *Imago Mundi* 48, no. 1 (1996): 5–24.

Ullman, Manfred. *Islamic Medicine.* Edinburgh: Edinburgh University Press, 1978.

Urdang, George. "The Development of Pharmacopoeias: A Review with Special Reference to the Pharmacopoea Internationalis." *Bulletin of the World Health Organization* 4 (1951): 577–603.

Urdang, George. *Pharmacopoeia Londinensis of 1618, with a Historical Introduction by G. Urdang.* Madison: State Historical Society of Wisconsin, 1944.

Urdang, George. "Pharmacopoeias as Witnesses of World History." *Journal of the History of Medicine* 1 (1946): 46–70.

Van Arsdall, Anne, and Timothy Graham, eds. *Herbs and Healers from the Ancient Mediterranean through the Medieval West: Essays in Honor of John M. Riddle.* Burlington: Ashgate, 2012.

Van der Eijk, Philip. "Principles and Practices of Compilation and Abbreviation in the Medical 'Encyclopedias' of Late Antiquity." In *Condensing Texts—Condensed Texts,* edited by M. Horster and C. Reitz, 519–54. Stuttgart: de Gruyter, 2010.

Varey, Simon, Rafael Chabrán, and Dora B. Weiner, eds. *Searching for the Secrets of Nature: The Life and Works of Dr. Francisco Hernández*. Stanford: Stanford University Press, 2002.

Viveiros de Castro, Eduardo. "Cosmological Deixis and Amerindian Perspectivism." *Journal of the Royal Anthropological Institute* 4 (1998): 469–88.

Voeks, Robert A. "Disturbance Pharmacopoeias: Medicine and Myth from the Humid Tropics." *Annals of the Association of American Geographers* 94 (2004): 868–88.

Voeks, Robert A. *Sacred Leaves of Candomblé: African Magic, Medicine, and Religion in Brazil*. Austin: University of Texas Press, 2010.

Vogel, Virgil J. *American Indian Medicine*. Norman: University of Oklahoma Press, 1970.

Vogt, Sabine. "Drugs and Pharmacology." In *The Cambridge Companion to Galen*, edited by R. J. Hankinson, 304–22. Cambridge: Cambridge University Press, 2008.

Voigts, Linda Ehrsam. "Scientific and Medical Books." In *Book Production and Publishing in Britain 1375–1475*, edited by Jeremy Griffiths and Derek Pearsall, 345–402. Cambridge: Cambridge University Press, 2007.

Walker, Timothy D. "The Medicines Trade in the Portuguese Atlantic World: Acquisition and Dissemination of Healing Knowledge from Brazil (c. 1580–1800)." *Social History of Medicine* 26, no. 3 (2013): 403–31.

Wall, Wendy. *Recipes for Thought: Knowledge and Taste in the Early Modern English Kitchen*. Philadelphia: University of Pennsylvania Press, 2016.

Wallis, Patrick. "Exotic Drugs and English Medicine: England's Drug Trade, c. 1550–c. 1800." *Social History of Medicine* 25 (2012): 25–46.

Warkentin, Germaine. "Aristotle in New France: Louis Nicolas and the Making of the Codex Canadensis." *French Colonial History* 11, no. 1 (2010): 71–107.

Warolin, Christian. "Le remède secret en France jusqu'à son abolition en 1926." *Revue d'histoire de la pharmacie* 90, no. 334 (2002): 229–38.

Warrior, Robert Allen. *The People and the Word: Reading Native Nonfiction*. Minneapolis: University of Minnesota Press, 2005.

Warrior, Robert Allen. *Tribal Secrets: Recovering American Indian Intellectual Traditions*. Minneapolis: University of Minnesota Press, 1995.

Watson, Gilbert. *Theriac and Mithridatium: A Study in Therapeutics*. London: Wellcome, 1966.

Wear, Andrew. "Medicine in Early Modern Europe, 1500–1700." In *The Western Medical Tradition 800 B.C. to A.D. 1800*, edited by Lawrence I. Conrad, Michael Neve, Vivian Nutton, Roy Porter, and Andrew Wear, 215–62. Cambridge: Cambridge University Press, 1995.

BIBLIOGRAPHY

White, Sophie. *Wild Frenchmen and Frenchified Indians: Material Culture and Race in Colonial Louisiana.* Philadelphia: University of Pennsylvania Press, 2012.

Widdess, J.D.H. *A History of the Royal College of Physicians of Ireland 1654–1963.* Edinburgh: E. & S. Livingstone, 1963.

Wisecup, Kelly. *Medical Encounters: Knowledge and Identity in Early American Literature.* Amherst: University of Massachusetts Press, 2013.

Whooley, Owen. *Knowledge in the Time of Cholera: The Struggle over American Medicine in the Nineteenth Century.* Chicago: University of Chicago Press, 2013.

Wootton, A. C. *Chronicles of Pharmacy.* Vols. 1 and 2. London: Macmillan, 1910.

Wyss, Hilary E. *English Letters and Indian Literacies: Reading, Writing, and New England Missionary Schools, 1750–1830.* Philadelphia: University of Pennsylvania Press, 2012.

Wyss, Hilary E. *Writing Indians: Literacy, Christianity, and Native Community in Early America.* Amherst: University of Massachusetts Press, 2003.

Young, James Harvey. *The Toadstool Millionaires: A Social History of Patent Medicines in America before Federal Regulation.* Princeton: Princeton University Press, 1961.

LIST OF CONTRIBUTORS

STUART ANDERSON is professor of the history of pharmacy at the Centre for History in Public Health at the London School of Hygiene and Tropical Medicine. He has been researching and writing on the history of pharmacy for over thirty years and is the editor of *Making Medicines: A Brief History of Pharmacy and Pharmaceuticals* (Pharmaceutical Press, 2005). He has contributed over twenty chapters to edited volumes and published upwards of fifty papers and articles. He is a former chair of the Society for the Social History of Medicine and is currently editor of the international journal *Pharmaceutical Historian*.

EMILY BECK is the assistant curator of the Wangensteen Historical Library of Biology and Medicine at the University of Minnesota. She earned her PhD in the history of medicine from the University of Minnesota in 2018. Her historical research focuses broadly on food, medicine, and material culture through the lens of manuscript recipe culture in early modern Europe. Her curatorial expertise is in the history of medical illustrations and the history of the book.

BENJAMIN BREEN is an assistant professor of history at the University of California, Santa Cruz. He received his PhD in history from the University of Texas at Austin in 2015 and was a member of Columbia University's Society of Fellows in 2015–2016. He is currently working on a book about the history of drugs in the early modern world called *The Age of Intoxication*.

MATTHEW JAMES CRAWFORD received his PhD in history and science studies from the University of California, San Diego and is currently an associate professor in the Department of History at Kent State University. He is the author of *The Andean Wonder Drug: Cinchona Bark and Imperial Science in the Spanish Atlantic, 1630–1820* (University of Pittsburgh Press, 2016). His research has been supported by the John Carter Brown Library, the Science History Institute, and the Lloyd Library and Museum.

CONTRIBUTORS

PAULA DE VOS teaches Latin American history and history of science and medicine at San Diego State University. Her research, which has received support from the ACLS, NIH, and NEH, focuses on the development of Galenic pharmacy in late medieval and early modern Europe and its transmission to the Americas, particularly Mexico, under the Spanish Empire. She is coeditor of *Science in the Spanish and Portuguese Empires* (Stanford University Press, 2009) and her most recent articles have appeared in *History of Science, Isis, Journal of Interdisciplinary History,* and *Journal of Ethnopharmacology.*

JOSEPH M. GABRIEL is an associate professor at Florida State University, where he holds joint appointments in the Department of History and the Department of Behavioral Sciences and Social Medicine. He is the author of *Medical Monopoly: Intellectual Property Rights and the Origins of the Modern Pharmaceutical Industry* (University of Chicago Press, 2014). He has also published essays in journals such as the *Raritan: A Quarterly Review, Rhetoric Society Quarterly, British Journal for the History of Science,* and the *Journal of the History of Medicine and Allied Sciences.* He lives in Tallahassee.

PABLO GÓMEZ is an associate professor in the Department of Medical History and Bioethics, and the Department of History at the University of Wisconsin, Madison. His work examines the history of science, race, health, and corporeality in the Caribbean and the Atlantic world. He is the author of *The Experiential Caribbean: Creating Knowledge and Healing in the Early Modern Atlantic* (University of North Carolina Press, 2017), and is currently working on an early history of bodies, disease, capital, and quantification in the Atlantic. He is also actively involved in projects of digital archival preservation in Colombia, Cuba, and Brazil.

ANTOINE LENTACKER is an assistant professor of history at the University of California, Riverside. His research is broadly dedicated to investigating the effects of changing communication technologies on the governing of people and things. He teaches courses in the history of science and medicine, the history of media, and the history of modern Europe.

CHRISTOPHER M. PARSONS teaches in the Department of History at Northeastern University. His research focuses on the history of science and the environment in the French Atlantic world. He is the author of *A Not-So-New World: Empire and Environment in French Colonial North America,* published by the University of Pennsylvania Press in 2018. He has also published articles in *Early American Studies, Environmental History,* and the *William & Mary Quarterly.*

CONTRIBUTORS

JUSTIN RIVEST is a Leverhulme Trust Early Career Fellow in the Faculty of History at the University of Cambridge. From 2016 to 2018 he collaborated on Leverhulme Research Project RPG-2014-289, "Selling the exotic in Paris and Versailles, 1670-1730," directed by Emma Spary. His recent work explores the ways in which early pharmaceutical monopolies were intertwined with the medical needs of the French military under Louis XIV. He earned his Bachelor of Humanities (2008) and MA (2010) from Carleton University and his PhD (2016) from Johns Hopkins University.

WILLIAM J. RYAN (PhD, Rutgers University, 2015) is assistant professor of English at Queensborough Community College of the City University of New York, where he teaches courses in early American literature, literature and medicine, and illness and narrative. He is completing a book manuscript on the medical case study in the eighteenth-century Atlantic World, portions of which have appeared or will appear in *Eighteenth Century: Theory and Interpretation* and *Literature and Medicine*.

TIMOTHY D. WALKER (PhD, Boston University, 2001) is professor of history at the University of Massachusetts Dartmouth, fellow at the Center for Portuguese Studies and Culture, and graduate faculty in Luso-Afro-Brazilian Studies and Theory. He is also an affiliated researcher at the Centro de História d'Aquém e d'Além-Mar (CHAM) at the Universidade Nova de Lisboa, Portugal and was a visiting professor at the Universidade Aberta in Lisbon (1994–2003) and at Brown University (2010). His publications include *Doctors, Folk Medicine and the Inquisition: The Repression of Magical Healing in Portugal during the Enlightenment* (Brill, 2005), and "The Medicines Trade in the Portuguese Atlantic World: Dissemination of Plant Remedies and Healing Knowledge from Brazil, c. 1580–1830," in *Social History of Medicine* 26, no. 3 (2013).

KELLY WISECUP is an associate professor of English at Northwestern University, where she is also affiliated faculty with the Center for Native American and Indigenous Research and the Science in Human Culture Program. She is the author of *Medical Encounters: Knowledge and Identity in Early American Literatures* (2013) and editor of a scholarly edition of Edward Winslow's 1624 *Good News from New England* (2014). Her articles have appeared in the *Journal of Native American and Indigenous Studies, Early American Literature, Early American Studies, Atlantic Studies, Studies in Travel Writing, Literature and Medicine*, and *Southern Literary Journal*.

INDEX

aAbriz, Dr. José, 107, 108
Abulcasis. *See* al-Zahrāwī, Abū al-Qāsim Khalaf ibn al-ʿAbbās
Académie Royale de Sciences, 87–88, 160
Africa, 104–8, 116, 119, 127, 151; cures from, 139, 144, 156, 158; healers in, 129, 152, 154, 155, 158–59, 295n20; materia medica from, 71, 117, 131–36, 153, 155, 295n20, 298n47; medicaments from, 134, 139, 146, 147, 149; slave trade, 9–10, 109, 157. *See also* Angola; West Africa
African knowledge, 105, 126, 131, 135, 136, 138, 186, 295n18; botanical, 144, 152
Aguilera, Antonio de, 40, 41
ailments: appetite loss, 135; "asthmatic fluxes," 151; bleeding, 241; bleeding lungs, 183; blood problems, 173; "Bloody Fluxes of the Belly [and] Uterus," 157; cankers, 182; chest problems, 173; colic, 7, 88; coughs, 175; deafness, 168; digestion problems, 88, 116; dysentery, 84; epilepsy, 168; eye problems, 33, 204; fevers, 84, 115, 130, 182, 183, 246; hair loss, 115; heartburn, 182; intestinal problems, 134; itch, 137; lice, 183; "lousy Disease," 157; pectoral disease, 175; phrenzy, 137; plague, 88; pleurisy, 244; pulmonary afflictions, 174; scurvy, 137–38, 301n39; skin disorders, 112, 115; smallpox, 88; snake bite, 244; sore throat, 246; stomach problems, 173, 182; toothache, 116; ulcers, 241, 248; urinary problems, 115; venereal diseases, 49, 112, 241; worms, 184
alchemy, 89, 272n8, 276n46
Algonquian: language, 177, 183, 187, 195; medical practice, 182, 190
allspice berry, 121, 122
Álvares, Domingos, 148, 149, 152
al-Zahrāwī, Abū al-Qāsim Khalaf ibn al-ʿAbbās, 28, 35, 36

"Ambettaway," 135, 137
American Indian physic (plant), 244, 259
American Indians: doctors, 93, 256, 257, 258; medical knowledge of, 7, 242–43, 247, 253–56, 260; in advertising, 244, 257, 259, 260. *See also* Native Americans
American medical science, 252, 254, 258
American medicaments, 75, 76, 145, 169, 255, 272n7, 299n13
American New Dispensatory. See Coxe, John Redman; dispensatories
Americas, the, 145, 203; materia medica from, 5, 32, 71, 72, 74, 76–77, 240
Amerindians, 109, 110, 125, 126, 131, 295n18
amulets, 152–53, 155, 156, 157, 159
Angola, 145, 151, 153, 154, 300n36, 302n55; Europeans and, 149–50, 106, 118; Luanda, 147–48; Mal de Loanda, 151, 300n39
animals (*animalia*), 4, 5, 107
anticolics, 110, 290n31
antidotaries: medieval, 25; *Antidotario Nicolao*, 28, 43; *Antidotario* of Mantova, 279n20
antispasmodics, 110, 290n31
apothecaries, 67, 91, 102, 150, 158, 176, 205; American, 250, 259; and the buen boticario, 38–39; the Contugis and, 90, 100; in Dieppe, 170; early modern, 168, 265; ecclesiastical, 111; English, 124, 136, 143, 203, 205–6; European, 145, 149, 157; Faculty of Surgeon-Apothecaries, Edinburgh, 210; guilds of, 225, 227; Irish, 214–15, 216; in Lisbon, 143, 144, 147, 149, 302n62; and materia medica, 20, 84, 138–39, 155, 163; and the Muslim world, 224; and orviétan, 92, 93; Royal College of Apothecaries, Madrid, 19; of royal households, 82, 95; Scottish, 208, 209; spicer-, 202; surgeons, and physicians, 130, 219, 220; training of, 21, 227, 228, 231; urban, 94, 147

INDEX

apothecary shops, 5, 46, 48, 146, 202, 204, 215; inventories of, 157, 169

Arabic authorities, 28, 32, 45, 52, 53–54, 55, 61; rejected, 56–57; works by, 25, 35, 39, 273n20. *See also* formularies, Arabic

Asia, 32, 71, 299n13

asthma, 115, 116, 151

astringents, 33–34, 110, 290n31

Augsburg, Germany, 19, 92, 284n44

Avicenna (Ibn Sina), 25, 28, 36, 45, 48

Bache, Franklin, 242, 251–54, 259, 260

Balfour, Andrew, 210, 211

balsams, 49, 71, 193

barbers, 147, 150, 158, 208, 209, 215

Barton, Benjamin Smith, 240, 242–43, 246, 247

Bartram, John, 179, 180

Berlu, John Jacob, 143, 144, 147, 298n3

best practices, 42, 277n76

Biblioteca Riccardiana, Florence, 47, 50, 280n24, 280n25; manuscripts in, 54, 60, 61

Bigelow, Jacob, 245, 247–48

"Bitter Root," 182, 183

"bitters," 153, 261

Black Atlantic, 146, 148, 149

bolsas de mandinga, 148, 153

Bosman, Willem, 144, 148, 155

botanicals, 128, 144, 148, 156–57, 169, 240, 243, 253; classification of, 32, 192, 225, 291n49, 298n47; names of, 124, 138, 139, 297n44

botany, 102, 109–10, 125–26, 132, 136–37, 139, 172, 178, 248; medical, 101, 110

Bouillon-Lagrange, Edmé-Jean-Baptiste, 229, 231–32

Boulogne, Antoine, 82, 91–95, 99, 100

Bourbon government, 64, 68, 69, 104

Brazil, 106, 115, 116, 118, 148, 150, 154; Bahia, 109, 110, 114, 151; Biblioteca Nacional of Brazil, Rio de Janeiro, 105, 109, 115; colonial, 119, 152; flora and fauna of, 108, 109, 117, 134, 172, 291n49; independence movements in, 117, 120; Indigenous peoples in, 104–5; remedies, 112, 113; slaves in, 109, 144, 155

British Isles, 127, 199, 201. *See also* Ireland; Scotland

British, the, 116, 123, 147, 179, 180, 299n15

Buytrago, Francisco de, 153, 154–55, 302n55

Byzantine authors, 24, 28

cacao, 71, 116, 117, 131–32, 145

Cadillac, Antoine Laumet de Lamothe, sieur de, 169, 174

Caesar (slave), 151–53, 158, 301n40, 301n43

Canada, 5, 161, 163–65, 168, 171, 174–76

capillaire, 161, 162, 167, 168, 173–74, 175; du Canada, 162–63, 170, 172, 306n73. *See also* maidenhair

Caribbean, the, 13, 144, 147, 153, 165, 240, 243

Carolina pink, 241, 251, 254

casca da vida, 153, 154

Castilian language, 39–40, 41, 276n61

cathartics, 33–34, 246, 253

Catholics, 154, 191, 204, 211, 212, 214

Charas, Moyse, 66, 92, 93

charlatans, 21, 52, 82, 83, 86, 89, 90, 92, 93, 285n11

chemical and pharmaceutical industry, 239, 243, 262

chemical medicine, 43, 66, 68, 155, 156, 158, 200, 205, 206, 219, 220, 302n62; chemotherapy drugs, 299n10. *See also* Paracelsus

Cherokee Indians, 254, 258, 259

China, 104–5, 126, 127

Chomel, Pierre-Jean-Baptiste, 161, 172, 174

Christianity: Occom's, 179, 187, 188, 189, 193, 194, 308n6; Occom's letter to unnamed Christian, 178, 190, 191–92

cinchona: bark, 77, 112, 113, 118, 139, 272n7, 296n22; Peruvian, 147–48, 153; *quinquina*, 75; Hans Sloane and, 130, 131–32; tree, 116, 117, 129. *See also* quina; quinine

cinnamon, 111, 121, 122, 134, 169, 240

Cockburn, William, 129, 130

Codex Canadensis (Louis Nicolas), 166, 167, 169

Codex Medicamentarius seu Pharmacopoea Parisiensis, 65–66, 226, 229, 230, 232

Codex Medicamentarius, sive Pharmacopoea Gallica, 12–13, 114, 202, 222–24, 228, 231, 237; commission, 231, 232, 236; in French, 232; in Latin, 238–39

coffee, 111, 117, 162, 170

Colbert, Jean-Baptiste, 163–64, 168

College of Physicians of Florence, 45–47, 48, 50, 54–55, 56, 61, 62, 64, 278n11, 279n21. See also *Nuovo Receptario Fiorentino*; *Ricettario Fiorentino*

colleges of physicians, 66, 77, 81, 200; in Dublin, 212, 213–14, 215, 217; in Edinburgh, 208, 209, 210, 211, 213; in London, 6, 203, 205, 209–10, 213; Royal College of Physicians and Surgeons of Glasgow, 208; Royal College of Physicians of Ireland, 218. *See also* College of Physicians of Florence; Royal College of Physicians of London

colonial space, 14; French, 113, 161–65, 66; in-

366

INDEX

stability of, 137, 138, 265–66; Portuguese, 118, 293n5; racial intimacies of, 130, 131
colonialism, 125–26, 135, 136, 164, 247; British, 124, 127, 129, 133, 186; French, 119, 164–65; knowledge and, 178, 180, 186, 187, 188, 243; materia medica and, 10, 114, 181; Portuguese, 107, 109, 118, 300n36
compound medicines, 33, 35, 36, 38, 51, 66–67, 69, 74, 78, 138, 217, 275n40; confections, 34, 204, 275n44; recipes for, 42–43, 277n76
Conselho Ultramarino, Portugal, 102, 106, 107, 108, 112, 113, 115; reports sent to, 156
Contugi family, 12–13, 82–83, 88–100, 246; advertisement by, 95–97; privileges of, 287n54. *See also* Richard, Roberte
Contugi, Christophe, 82, 88, 89, 91, 96
Contugi, Louis-Anne, 90–91, 99
Cowen, David L., 199, 200, 201, 211, 220
Coxe, John Redman, 240–42, 246, 248, 252, 253
Creek Indians, 249, 254–55
creoles, 121–22, 126, 131, 165, 268, 294n11
Cromwell, Oliver, 209, 210
Culpeper, Nicholas, 6–7, 179
Cumming, Dr. Duncan, 214, 215

Dampier, George, 133, 134, 296n30
decoctions, 37, 42, 68, 174, 204, 259, 275n44
"Decree of the Royal Protomedicato," 63–64, 68
Delisle, Guillaume, 163, 165
della Scarperia, Antonio, 54, 58, 60
diaphoretics, 112, 290n31
Dioscorides, 25, 28, 35, 36, 44, 49, 54; *De materia medica*, 27, 30–32, 272n17
diseases, 88, 105–7, 112, 123–25, 151, 152, 179, 181–85, 241, 258. *See also* ailments
dispensatories, 33, 65, 200, 204, 214, 225–27, 250, 259; *American Dispensatory*, 240, 241, 246, 248; *American New Dispensatory*, 248, 252; *Dublin Dispensatory*, 215, 216; *Edinburgh New Dispensatory*, 240, 241, 246; *The Indian Doctor's Dispensatory*, by Peter Smith, 257; *London Dispensatory*, 215; Nuremberg's *Dispensatorium*, 225, 316n6. *See also Dispensatory of the United States of America, The*
Dispensatory of the United States of America, The, 200, 242, 251–54, 259, 260
diuretics, 246, 253
dog-mercury, 132–33, 296n27
dogwood, 241, 251
drug merchants, 81, 83–88, 99, 143, 147, 157

drug substitutions, 32, 38, 43, 75–76, 274n33
drug trade, 72–73, 82, 95, 97, 105, 143, 144
drugs, 144, 202, 224–25, 227, 238, 239; magistral, 84, 229, 231, 239; naming of, 225, 234, 235, 274n37; officinal, 84, 85, 229, 231, 237; "specifics," 83–87, 119, 131, 133, 134, 146, 180, 257; vendors, 83, 84, 85, 86, 87. *See also* remedies, "secret"
Dublin, 132; College of Physicians, 212, 214; *Dublin Dispensatory*, 215; *Dublin Pharmacopoeia*, 14, 201, 212, 216, 217–18, 221
Dubois, Jacques, 225, 226
Duncan, Arnold, 240, 246
Duplessis sisters Marie-Andrée de Sainte Hélène and Geneviève de l'Enfant Jésus, 170, 174
Dutch commerce, 116, 119, 144; Dutch West India Company, 155; slave traders, 147, 299n15

East Indies, 126, 143, 146, 297n32
Edinburgh Pharmacopoeia, The, 201, 208, 211, 219, 220, 221, 246; first edition, 212, 213
Edinburgh pharmacopoeias, 14, 201, 251
electuaries, 34, 57, 68, 88, 204, 275n44
elite physicians, 154, 201; in America, 241, 243, 246, 248, 251, 256, 258
emetics, 33–34; bone-set, 241; dog-mercury, 296n27; ipecacuanha, 112, 133–34; Seneca snakeroot, 246
empiricks, 129, 130
empirics, 83, 85, 86, 87, 205; apothecaries and, 129, 130; and charlatans, 21, 60, 83
Enlightenment, the, 104–9, 114, 120, 123, 159, 195, 228, 253; ideals, 103, 223; rationality, 181, 221
extracts, 68, 69, 217, 232

Ferreira de Castro, Manuel, 143, 144
Ferreira, Alexandre Rodrigues, 114, 115
Ferreira, Luís Gomes, 150, 151
fetish objects, 156, 264
Florence, 19; Florentine College of Doctors and Apothecaries, 48. *See also* College of Physicians of Florence; *Nuovo Receptario Fiorentino*; *Ricettario Fiorentino*
formularies, 21, 25, 30, 31, 33, 44, 226, 227, 229, 230, 232; Arabic, 23, 33–34, 35, 277n76; military, 246, 248
Fourcroy, Antoine François de, 228, 229, 235
Fowler, Mary, 177, 193, 311n65
Fragoso, Juan, 32, 33

367

INDEX

France, 28, 81, 95, 127; Old Regime, 228–30, 233, 234, 235. *See also* French Revolution; King Louis XIV; Paris Faculty of Medicine
French Codex. See *Codex Medicamentarius, sive Pharmacopoea Gallica*
French Revolution, 86, 114, 202, 222–24, 228, 234, 235
Fuchs, Leonhart, 53–54

Galen, 20, 31, 48, 53, 66, 204–5, 221; referenced, 28, 35, 37, 54, 118–19; works by, 25, 33, 212, 275n40
Galenic medicine, 76, 94, 200, 206, 212, 220
Galenic pharmacy, 20, 21, 24, 26–27, 31, 34, 44, 66–68, 107, 272n7
glossaries, 31, 32, 42–44
grains of paradise, 139, 143, 147, 158, 302n66
Great Lakes, 161, 163, 165, 166, 170
Greco-Roman texts, 20, 24, 27–28, 32–33, 226
Grégoire, the Abbé, 223, 224
Grew, Nehemiah, 146, 147
Grocers' Company, London, 203–4, 205
guaiacum, 9, 71, 75, 145, 272n7
guilds, 50, 92, 99–100, 215, 234; abolished in France, 228, 233; Guild of St. Luke, apothecaries of Dublin, 215; Parisian apothecaries, 99, 230; surgeons and apothecaries, France, 81; surgeons and barbers, Scotland, 208
Guinea, 135, 144, 150, 298n3
Guybert, Philbert, 174, 226

Haudenosaunee, the. *See* Iroquois
healing knowledge, 120, 128, 134, 136, 144, 163, 243, 246, 293n5; development of, 64–65, 78, 123, 130, 220, 221; and healers, 46, 77, 150; in Occom, 182, 188, 192–93
Hellenistic world, 24, 33. *See also* Galenic pharmacy
Henry, Samuel, 249, 250, 251
herbals, 30–32, 44, 182, 187, 190, 202, 248, 267
herbs, 35–36, 144, 152, 180, 181, 185–88, 192, 193
Hernandez, Francisco, 6, 32, 154
Herrera y Campos, Don Jacinto de, 19, 20
Hôtel-Dieu de Québec, 169, 170, 307n82

iatrochemistry, 155, 220
iatromechanics, 212, 220, 221
Iberian Peninsula, 28, 75
India, 32, 49, 104–5, 111, 117, 119, 120; Goa, 106, 107–8, 112, 145
Indian tobacco. *See* lobelia
Indigenous medical substances, 101, 103, 240–41; names of, 111; plants, 115, 168, 242, 243, 247–49, 252
Indigenous peoples, 10–11, 104–5, 165, 174–75, 244, 254, 262; in Brazil, 115, 118; healing culture of, 106, 108, 118, 119, 180, 251, 256; in India, 107, 118; in North America, 7, 145, 170, 241, 247
infusions, 37, 42, 68, 275n44
Inquisition, the, 112, 148, 158
ipecac / ipecacuanha, 9, 71, 112, 113, 133–34, 137
Ireland, 200, 201, 217, 219–20; Irish Parliament, 212, 214, 215, 216
Iroquois, 163, 168, 170
Islamic world, 20, 21, 24 25, 32
Italy, 20, 28, 52, 61, 127, 132; Tuscan dialect, 48, 51, 280n24

jalap / jalapa, 75, 112, 113, 139, 169
Jamaica, 5, 148, 157; African healers in, 159, 295n18; Hans Sloane and, 13, 110, 121, 124, 125, 126, 131, 135
Jamaica pepper, 124, 137, 138, 139, 240; tree, 121, 122, 136
Jesuits, 119, 168, 246, 266; André João Antonil, 116; Father Davaugour, 170; Juan Eusebio Nieremberg, 166; Louis Nicolas, 166; and medical information, 102, 103, 104, 106, 107; Pierre-François-Xavier de Charlevoix, 171; "pouder," 129, 130. *See also* Society of Jesus
Journal des Sçavans, Le, 93, 94, 127
Jubera, Alonso de, 38, 40, 41–42
juices. See *succi* (juices)
Jurin, James, 128, 130

King Charles I, 209, 203; and Queen Henrietta, 206
King Charles II, 210, 211, 214
King Henry VII, 202, 208, 212
King Henry VIII, 203, 213
King James I of England (King James VI of Scotland), 201, 204, 206, 208, 209, 210, 213
King James II, 210, 211, 214
King Louis XIV, 97, 99, 163–64, 168, 233, 287n54
King Philip II of Spain, 68, 69, 271n3
Kwasi (Quashe), 152, 153, 158

La Martinière, Pierre-Martin de, 88–90, 92–94
Lahontan, baron de, 170, 174
lambatives, 34, 275n44
landed gentry, 201, 210
languages: African, 138; Algonquian, 187, 195; Amerindian, 265; and paleography,

368

280n24; arcane, 38; Bantu, 265; Castilian, 40, 265, 276n61; different, 32, 33; Dutch, 265; English, 265; European, 107; Ewe/Gbe/Fon, 265; French, 224, 265, 276; in lists, 267; Indian, 107; Latin, 39–41, 43, 195, 200, 206, 225, 230, 276n61; medicines and, 226; Ojibwa, 187; pharmaceutical, 225; Portuguese, 265; purity of, 228; reforming, 202, 223, 224; similar across recipes, 58; vernacular, 39–40, 48, 222, 223, 230, 280n24

Lavoisier, Antoine, 229, 232
lay practitioners, 50–54, 60–61, 62
Lémery, Nicolas, 74–75, 161, 172, 173, 227
letters patent, 83, 90, 91, 99
Lichen Cinereus terrestris, 138, 296n30
Lima, 73, 284n35
Linacre, Thomas, 202, 219
liniments, 34, 204
Linnaeus, Carl, 152, 195
Lisbon apothecaries, 147; Caetano de Santo Antonio, 147, 149, 302n62; João Curvo Semedo, 149, 158; Manuel Ferreira de Castro, 143, 144
Lisbon, 102, 105, 111, 115–16; Academia Real das Ciências de Lisboa, 110, 111, 114–15; National Archive (Torre do Tombo), 112; Royal Academy of Sciences, 106–7; royal botanical garden, Ajuda Palace, 113, 114. *See also* Lisbon apothecaries
lobelia, 241, 243, 244
London Pharmacopoeia, 203–4, 206, 221, 296n22; editions, 202, 207, 214, 215, 216, 219. *See also Pharmacopoeia Londinensis*
Love medicine, 185–87
lozenges, 94, 204, 257, 275n44
Lusophone world, 102, 111–14, 118
Lyon, 92, 93, 225, 226, 237

magic, 148, 157, 186, 188, 268, 272n8
magistral drugs. *See* drugs, magistral
maidenhair: Canadian, 13–14, 171; northern, 113, 160, 161–62, 167–71, 173–75. *See also* capillaire
maladies, 180, 181, 182, 184, 185. *See also* ailments
malaria, 112, 117, 129
Malaysia, 104–5, 146
manuscript recipe books, 46–47, 50, 51, 54, 55, 57, 61, 62, 151, 280n25, 312n73
marina (marine objects), 4, 5, 71
Mateo, Pedro Benedicto, 28, 41
materia medica, 4–5, 7, 31, 65, 67, 72–77, 113, 122, 125, 133, 159, 162, 167, 226, 264; African, 71, 117, 120, 134, 131–36, 145–53, 156–59; American, 68, 71, 76, 241, 248, 249, 252; Atlantic lists of, 265, 266; authority over, 138–39; colonial, 126, 130, 136; European, 69, 73, 273n17; heterogeneous, 119–20; indigenous, 108–9, 111, 112, 115–16, 118–20, 126, 130, 170, 241–42, 258, 262; in Lusophone world, 102–4; in medieval Arabic formularies, 32–33, 277n76; misuse of, 130; New World, 137, 266, 296n26; sharing of, 129; Hans Sloane and, 121–22, 124, 127, 132, 137; in written texts, 4–5, 20–21, 42–43, 75, 136–37, 226, 252
mayapple, 241, 251
Mayerne, Sir Theodore Turquet de, 206, 219
medical authority, 138–39, 242; Occom's, 189, 190, 192, 196
medical corporations, France, 81, 85
medical education, 108, 110. *See also* universities
medical elites, 153–56, 158, 183–84, 201, 241, 242–43, 246, 248, 251, 256, 257, 258, 259
medical knowledge, 129, 133, 178, 188, 237, 248, 258, 262; "codification of," 12–14; different kinds of, 125, 178, 187; from multiple origins, 311n43; networks, 126–27, 135–36, 137, 298n47; Occom's, 189–90, 193; "operative," 276n46; tradition, 296n22; transmitting, 118, 123, 124, 148, 181, 190, 293n1. *See also* African knowledge; American Indians
medical politics, 202, 209, 220
medical practice, 85, 139–40, 144, 201, 208, 221; European, 128–29, 229, 296n20
medical practitioners, 54, 160, 184, 193, 194–95, 265; control of, 48, 199; literate, 13, 58, 61, 129; Native, 180, 181, 182, 185, 187, 190; of New France, 169, 170; and plants, 179, 189, 181, 185; and profession, 103, 107, 108, 237, 241; relations among, 200, 201, 219; Hans Sloane and, 132
medical publishing, 124, 125, 132, 255. *See also* Sloane, Sir Hans; *Transactions*
medical schools, 28, 221, 315n3, 273n20; in America, 243, 253, 255; in Bologna, 25, 205; in Montpellier, 25, 161; in Padua, 25; in Paris, 232; in Salerno, 25, 28. *See also* Paris Faculty of Medicine
medical science, 241, 242, 243, 244, 250
medical secrecy, 88–89, 90, 95, 99. *See also* secrecy
medical societies, 243, 255, 256
medical texts, 104, 128, 131, 156, 162, 179, 244;

medical texts (*cont.*): field guidebook, 109. *See also* medical writing
medical writing, 30–31, 66, 78, 89, 123, 180, 293n5. *See also* medical texts
medicaments, 85, 139, 145, 203, 204. *See also* American mediccaments; materia medica; medicinal plants; oinments; remedies; syrups
medicinal plants, 103, 106, 107, 115, 117, 180, 192; and herbs, 132, 180–81, 185
medicinal substances, 36, 67, 117, 147, 143, 177, 265; grave dirt, 148, 157, 159
medicinal virtues, 182, 187, 188
medicine, 109, 126, 220, 244, 248, 273n20; legislation and, 69, 153, 229; practice of, in America, 250, 253; in *Transactions*, 136–37, 139
medicines, 83, 149, 169, 210, 219, 237; preparation of, 138, 153, 224, 251; solimão, 156, 302n62. *See also* proprietary drugs; proprietary medicine; remedies, "secret"
Mediterranean, the, 21, 24–25, 33
Mesue, John [Yuhanna ibn Masawaih], 25, 27, 28, 45, 58, 226, 272n16, 278n2; *Canons*, 36, 37, 38; *Grabadin*, 34; referenced, 44, 52–53, 54, 57
metallica (metallic objects), 4, 5
Mexican pharmacies, 38, 272n13
Mexico City, 19, 73, 145
Middle East, 49, 71
mineralia (minerals), 4, 107
mithridatium, 89, 93
Mohegan Indian, 185, 187, 193, 195, 266; medical practitioners, 180, 182, 190
monarchies, 67, 201, 208; absolutist, 227, 228; Catholic, 211; European, 83, 199, 200; French, 227, 234, 251; Portuguese, 101, 102, 105
Monardes, Nicolas, 9–10, 32
Montauk, Long Island, 177, 183, 190, 195, 266
Mozambique, 106, 112, 118, 156
Muslim world, 224, 299n13

naming, 49, 54, 69, 161, 225, 257; as government tool, 224, 232–34, 235; Occom, 182, 183. *See also* nomenclature
nation-state, 14, 20, 103, 265, 268; rise of, 9, 64
Native Americans, 183, 186, 188–91, 194, 246, 267; medical knowledge of, 103, 178, 179, 182, 242, 262, 308n6; medical practitioners, 180, 181, 185–86, 187. *See also* American Indians; Amerindians

Native peoples, 108, 110, 117–20. *See also* Native Americans
natural history, 126, 132, 294n9, 294n11, 309n13; and *Transactions*, 136–37, 139
natural philosophers, 102, 127, 146, 149, 154, 295n15
natural philosophy, 109, 122–23, 125, 128, 137
New England, 126, 183, 189–90
New France, 113, 160–65, 167–71, 174, 175, 266
New Georgia, 189, 190
New World, the, 137, 139, 145, 244, 272n7, 298n47
New York Medical Society, 250, 255
Niccholao, 45, 48–49, 54, 57, 58
Nicolas, Louis, 166, 167, 169
nomenclature, 217, 222, 236, 241, 253, 263, 298n47
North America, 7, 106, 116–17, 127; French, 166–68, 169; plants in, 112–13, 161–62
northern maidenhair fern. *See* capillaire; maidenhair
Nuovo Receptario Fiorentino [renamed *Ricettario Fiorentino*], 45, 48, 64, 202, 219, 225
nutmeg, 134, 169

obeah magic, 148, 153, 157, 159
Occom, Mohegan Samson, 14, 177, 183, 191–92, 194–95; "Account of the Montauk Indians, on Long Island," 187, 188; and herbs, 182–83, 185, 186, 308n6; "Herbs & Roots," 177–81, 184, 188, 193; works of, 191, 193, 308n1, 308n9
Ocus, 181, 184, 189–90, 192, 308n4
officinal remedies. *See* drugs, officinal
oils, 34, 69, 204, 275n44
ointments, 34, 204, 275n44
Oldenburg, Henry, 127, 128, 130
opium, 143, 232
Orta, Garcia da, 32, 106, 145, 154
orviétan, 82, 88, 92, 94, 97, 99, 287n47; broadsheet, 96; recipes, 89, 93
Overseas Council. *See* Conselho Ultramarino, Portugal
Oviedo, Luis de, 30, 37–38, 40

Palacios, Félix: *Palestra Pharmaceutica Chymico-Galenica* by, 19, 42, 43, 66, 68, 74–75
Paracelsus, 66, 204–5, 221
Paris Faculty of Medicine, 99, 206, 225, 226, 230, 231. *See also Codex Medicamentarius seu Pharmacopoea Parisiensis*

INDEX

Parrish, Susan Scott, 186, 294n11
passionflower, 166, 168
patent medicines, 255, 258, 259, 260
patronage, 66–67, 84, 87, 111; royal, 114, 201, 203, 219
pedagogical texts, 21, 31, 38
Persian world, 32, 278n2
Peru, 5, 134
Petiver, James, 134, 135, 136, 137, 138, 181, 297n32
pharmacists, 67, 69, 199, 227, 265; and the Codex, 231, 232, 236, 237; French, 228, 229, 230
Pharmacopeia of the United States of America, 14, 202, 241, 251–52, 253, 259, 260. See also *Dispensatory of the United States of America, The*
Pharmacopoeia Augustana, 19, 65, 66, 68, 75, 283n14, 284n44
Pharmacopoeia Londinensis, 4, 6, 65–67, 122–24, 131, 138. See also *London Pharmacopoeia*
Pharmacopoeia Matritensis, 12, 19–20, 63, 67–70, 74, 75–76, 78, 276n61, 284n44
"pharmacopoeia" (the term), 42, 200, 225
pharmacopoeias, 4–5, 8, 9, 20, 87, 99, 200. See also pharmacopoeias, herbals, and reference works cited
pharmacopoeias, herbals, and reference works cited: *Abregé d'histoire des plantes usuelles*, by Pierre-Jean-Baptiste Chomel, 172; *Almansore*, by Rhazes, 48–49, 54; *Antidotarium Nicolai*, by Niccholao, 45, 48–49, 54, 57, 58; *Bald's Leechbook, or Medicinale Anglicum*, 202; "The Book of Drugs, or, The French Pharmacopoeia," 230; "Botanical Guide to Some Plants from the Interior of Piauí," by Vicente Jorge Dias Cabral, 115; *British Pharmacopoeia*, 221; *Canadensium Plantarum*, by Jacques-Philippe Cornut, 171; *Cartilla Pharmaceutica*, by Pedro Vinaburu (in Castilian), 40; "Catalogue of Some Guinea-Plants, with their Native Names and Virtues," by James Petiver, 134, 181; *Cherokee Physician, or Indian Guide to Health, The*, by Jas. W. Mahoney, 259; *Codex Canadensis*, by Louis Nicolas, 166, 167, 169; *Coloquios dos simples e drogas da India*, by Garcia da Orta, 145; *Compendium aromatarium*, 30, 39, 41, 43; *Concordia Pharmacopolarum Barchinonensium*, 68; *De Historia Stirpium*, by Leonhart Fuchs, 54; *Description de plusieurs plantes du Canada*, by Jean-François Gaultier, 160–61; *De succedaneis medicamentis*, by Juan Fragoso, 32, 33; *Dictionnaire universel de commerce*, by Jacques Savary des Bruslons, 173; *Discurso pharmaceutico sobre los canones de Mesue*, by Miguel Martínez de Leache, 21; *Examen Pharmaceutico Galeno-Chimico Teorico-Practico*, by Francisco de Brihuega, 272n14; *Farmacopea Lisbonense*, by Manuel Joaquim Henriques de Paiva, 113, 114; *Histoire veritable et naturelle*, by Pierre Boucher, 168; *Historae naturae*, by Juan Eusebio Nieremberg, 166–67; *Historia dos Reinos Vegital, Animal e Mineral*, by Francisco António de Sampaio, 105, 106, 109; *Indian Doctor's Dispensatory, The*, by Peter Smith, 257; *Jurisprudence de la medecine*, by Jean Verdier, 85–86; *Liber in Examen Apothecariorum*, by Pedro Benedicto Mateo, 28, 41; *Libro de Sinonimas*, by Simon of Genoa, 28; *Medicina Britannica*, by Thomas Short, 179; *Metodo de la coleccion y reposicion de las medicinas simples y de su correccion y preparacion*, by Luis de Oviedo, 30, 37–38; "Nine Herbs Charm," 152; *Nouveau plan de constitution pour la medecine*, by Félix Vicq d'Azyr, 235; *Officina Medicamentorum*, published by the College of Apothecaries in Valencia, 68; *On Compounds (De Compositiones)*, by Scribonius Largus, 33; *Opera Medicinalia*, by Francisco Bravo, 145; *Pharmacopea Valentina*, 19; *Pharmacopoée royale galenique et chemique*, by Moyse Charas, 66; *Pharmacopée universelle*, by Nicolas Lémery, 173, 227; *Pharmacopoeia Bateana*, by George Bate, 66; pharmacopoeia, by Herford (1667), 93; pharmacopoeia, by Valerius Cordus, 226; *Pharmacopoeia Hispana*, 69, 276n61; *Pharmacopoeia medico-chymica*, by Johannes Schröder, 88, 92, 93; pharmacopoeia of Prévost, 93; *Pharmacopoeia of the Massachusetts Medical Society*, 248; *Pharmacopoeia of the New-York Hospital, The*, by Valentine Seaman, 250; *Pharmacopoeia regia*, by Johann Zwelfer, 66; *Polyanthea Medicinal*, by João Curvo Semedo, 149, 158; *Poor Richard's Almanack*, by Richard Saunders, 244, 246; *Present Uncertainty in the Knowledge of Med'cines, The*, by William Cockburn, 129; *Singularitez de la France antarctique*, by André Thevet, 165–66;

INDEX

pharmacopoeias, herbals, and reference works cited (*cont.*): *Ten Indian Remedies: From Manuscript Notes on Herbs and Roots*, by Edward Connery Lathem, 178; *Traité de pharmacie*, by Julien-Joseph Virey, 232; *Tyrocinio pharmacopeo methodo medico y chimico*, by Fuente Pierola, 42; *Tyrocinium pharmaceuticum*, by Juan de Loeches, 42. See also *Codex Medicamentarius, sive Pharmacopoea Gallica*; Dublin, *Dublin Pharmacopoeia*; *Edinburgh Pharmacopoeia, The*; Edinburgh pharmacopoeias; *Pharmacopoeia Augustana*; *Pharmacopoeia Londinensis*; *Pharmacopoeia Matritensis*

pharmacy, 20, 21–23, 73, 223, 224, 225; in America, 250, 252, 253; in France, 229, 230, 235, 236, 239; and surgery, 69, 209. See also apothecaries; apothecary shops

Philosophical Transactions, 122–28, 130, 131; contributions to, 132–34, 295n12, 296n26, 296n30; Hans Sloane and, 13, 121, 125–27, 132, 135, 136

physicians, 93, 132, 200, 203, 220, 227; academic, 237, 238; in England, 130, 202; European, 76, 145, 157, 228, 229, 231; in Ireland, 214, 215, 216, 219; Negro, 129; in Scotland, 204, 208, 209, 210, 219; Lusophone colonial, 102; and materia medica, 138–39; Persian, 278n2; royal, 85, 86, 87, 160, 206. See also American Indians; elite physicians

pills, 68, 69, 204, 275n44

plants, 179, 180, 181, 246; classification of, 195, 249, 306n73; naming of, 49, 54, 181–82; in Occom, 181, 184; occult virtues of, 86, 166, 182, 186, 187, 188–89, 264. See also materia medica

plasters, 34, 275n40, 275n44

Platerius, Mattheus, 25, 28

Pliny, 25, 28, 31, 36, 54, 158

poisons, 132–33, 145, 146, 150, 186, 235, 296n26; and antidotes, 110, 145, 148, 149, 156, 158–59, 290n31; Caesar's cures for, 151, 152

political turmoil, 202, 204, 214

Portugal, 111, 143, 146, 149–50, 154; and slave trade, 147, 150, 155, 299n15. See also Conselho Ultramarino, Portugal; Lisbon

Portuguese colonies, 103, 112, 113, 116, 119, 150

Portuguese Crown, the, 13, 101–6, 111, 114, 116

Portuguese *feiticeiro/a*, 148, 150

poultices, 34, 181

powaws, 187–88

powders, 69, 204, 275n44

prescriptions, 241, 225

privileges. See royal privileges

procedural texts, 21, 31, 34–38

proprietary drugs, 84, 237, 238, 239

proprietary medicine, 83, 86, 95, 97, 129, 132, 266

Protestants, 204, 208, 211, 214, 216, 219, 233

publishing, 28, 127, 202, 232; medical, 124, 125, 243, 255

purgatives, 33–34, 110, 122, 290n31

quackery, 60, 67, 129, 130, 132, 242, 255–62

Quashe. See Kwasi (Quashe)

Quatro libros. De la naturaleza y virtudes de las plantas y animales que estan recivides en el uso de Medicina en la Nueva Espana, by Francisco Ximénez, 6, 73

Queen Elizabeth I, 201, 203, 204

Queen Maria I, 104, 106–7, 114

quina, 76, 117. See also cinchona; quinia

quinine, 159, 237. See also cinchona; quinine

rattlesnake bite, 151, 184, 244, 246, 253, 296n26

Real Tribunal del Protomedicato. See Royal Protomedicato

recipes, 58, 110, 231, 277n76; authors of, 52, 280n32, 281n38; provenance of, 51–52, 62. See also manuscript recipe books

religious troubles, 202, 204, 208, 211, 214. See also Catholics; Protestants

remedies, 7, 86, 199, 226, 236, 299n13; "secret," 81–85, 87–88, 99, 234–38, 317n32

Renaissance, the, 24, 28, 40, 225

res publica medica, 121, 123, 124, 130, 292n1, 296n20

Rhazes, 45, 48–49, 54, 278n2

Ricettario Fiorentino, 12, 19, 45–61, 66, 264, 278n7, 279n20; analysis of, 281n49; first edition (1498), 50, 56, 57, 281n36; Latinized, 279n12; second edition (1550), 57, 279n20; third edition (1567), 49, 50. See also *Nuovo Receptario Fiorentino*

Richard, Roberte, 90–91, 92, 99

Rio de Janeiro, 111, 113, 143

Roman authors, 20, 24, 32, 93

roots, 71, 181, 187

Rouvière, Henry, 92, 93–94, 95

royal charters, 200, 202, 205–6, 208, 219; in Dublin, 214, 215; in Edinburgh, 210, 211

Royal College of Physicians of London, 131, 202–3, 205, 207–8; presidents of, 121, 123, 138, 206; "Twelve Reformers," 278n4. See also colleges of physicians

372

INDEX

royal commissions, 106, 107, 109, 112–14
royal patronage, 114, 201, 203, 219
royal privileges, 82, 84, 85, 86, 89–90, 233
Royal Protomedicato, 63, 69, 72, 75
Royal Society, London, 136, 137, 139, 146–47, 210; Hans Sloane and, 121, 123, 126

Saint Lawrence River, 161, 163, 164, 169
Saladino da Ascoli, 28, 38–39, 40, 44
salts: *sales*, 5; *salia*, 4
Sampaio, Francisco António de, 105, 106, 109, 110, 111, 114
Santo Antonio, Caetano de, 149, 302n62
sarsaparilla, 49, 232
sassafras, 7, 247
Saunders, Richard, 244, 246
Schiebinger, Londa, 124, 195
Schröder, Johannes, 88, 92, 93
scientific publishing, 127, 133; *Journal de Pharmacie*, 231–32. See also *Journal des Scavans, Le*; *Philosophical Transactions*
Scotland, 200, 208, 219; Glasgow, 208, 209; Scottish Parliament, 209, 210, 211; universities in, 209, 210–11. See also *Edinburgh Pharmacopoeia, The*; Edinburgh pharmacopoeias
secrecy, 12, 83, 91, 93, 129, 257, 264, 266; Indian, 243, 246, 247, 260, 262; medical, 84, 99, 132, 255–56; Portuguese and, 13, 102–3, 112; prohibition of, 235, 242, 256
secret remedies. See remedies, "secret"
seeds, 71; *semina* (seeds), 4; *semina sive grana* (seeds or grains), 5
Semedo, João Curvo, 149, 150, 158
Seneca snakeroot, 244, 246, 251, 253, 259
Senegal, 147, 298n3
Serapion (Ibn Wafid), 25, 28, 54
Shaw, Nathaniel, 181, 193, 311n65
Short, Thomas, 157, 179, 180
Sibbald, George, 209, 210
Sibbald, Robert, 210, 211, 212, 219
simple remedies ("simples"), 21, 36, 68, 73, 134, 155, 167; African, 138, 139; books of, 25, 30; catalogued, 32, 37, 78, 124, 133, 138; "exotic," 71–72; lists of, 38, 43, 69; names of, 139, 253; "official," 71, 72
slaves, 144, 148, 151, 152, 153, 244; in Jamaica, 122; trade in, 146, 147, 150, 155, 157, 299n15. See also Caesar (slave); Kwasi (Quashe)
Sloane, Sir Hans, 110, 121, 130, 131, 136, 138, 154, 295n18; *Catalogus Planatarum Quae In Insula Jamaica*, 123, 128, 135, 298n44; as editor, 13, 137, 293n6, 294n10; *A Voyage to the Islands of Madera, Barbadoes, Nieves, S. Christophers, and Jamaica* . . . 123, 128–29, 294n8; works authored or translated, by, 121, 123, 295n12. See also *Philosophical Transactions*, Hans Sloane and
Smyth, John, 134, 135, 136, 138
Society of Jesus, 70, 102, 103, 104, 166. See also Jesuits
South America, 105, 108, 117, 120, 127, 130; medicinal plants from, 109, 112, 114, 118, 148, 169
South Asia, 118, 145, 146, 147
Spain, 12, 20, 39, 67–68, 104, 202–3; King Philip II of, 68, 69, 271n3; and pharmacy texts, 28, 34, 38, 276n61. See also Spanish America; Spanish Crown
Spanish America, 73–74, 76, 143, 147, 166, 298n3
Spanish Crown, 6, 63–64, 67, 69, 104, 203; and *Pharmacopoeia Matritensis*, 65, 72–73
Speck, Frank, 182, 183
Stainpeis, Michael, 225, 315n3
sub-Saharan Africa, 143, 144, 145, 147, 154, 157–58
succi (juices), 4, 5
surgeons, 102, 150, 219, 220; in Scotland, 208, 209, 210; in Ireland, 215, 216
surgery, 69, 209
syphilis, 49, 112, 242, 243, 246
syrups, 34, 69, 204, 275n43, 275n44

Tantaquidgeon, Gladys, 180, 181, 182, 187, 190, 310n32, 311n55
Tennent, John, 244, 246, 248, 253
theorems, 25, 36
therapeutic innovation, 81, 87, 99
therapeutics, 67, 110, 133, 243
theriac, 34, 57, 82, 89, 93, 94, 100
tobacco, 75, 116, 129, 145, 299n15
Tocqueville, Alexis de, 114, 222
Toledo, 25, 273n20
Tournefort, Joseph Pitton de, 161, 172, 173
trademarks, 95, 97, 100
Trail of Tears, the, 254, 258
Transactions. See *Philosophical Transactions*
tropical diseases, 101, 106, 118
Tufts, Henry, 256, 257

universities: University of Coimbra, 104, 106, 107, 114, 115; University of Copenhagen, 5; University of Pennsylvania, 240, 242
Urdang, George, 64–65, 77, 201, 208

venereal disease, 84, 110, 241, 290n31
vernacular. *See* languages, vernacular
Villa, Esteban de, 21, 30
violence, 106, 136, 157, 191, 200, 254, 258
Villanova, Arnald de, 25, 28, 36
Voeks, Robert, 144, 152

Warrior, Robert, 189, 195
washing process, 35, 42
water dock, 248, 251
waters, 69, 204
weights and measures, 43, 217, 277n76
West Africa, 5, 126, 135, 148, 154, 157; coast of, 117, 147; and slave trade, 144, 299n15

West Indies, 49, 126, 127, 131–32, 135, 143, 146; British, 148
Wheelock, Eleazar, 178, 187, 191, 193, 194, 311n65
wild cherry, 241, 251
wild cinnamon, 121, 240
witchcraft, 148, 150, 188, 255
Wood, George B. 200, 242, 251–54, 259, 260
Worm, Ole, 3, 4, 5–6, 7

Ximénez, Francisco, 6, 73

Yuhanna ibn Masawaih. *See* Mesue, John